ANTENNAS AND RADAR FOR ENVIRONMENTAL SCIENTISTS AND ENGINEERS

This book gives a complete overview of the scientific and engineering aspects of radio and radar pertaining to studies of the Earth environment. The book opens with an analysis of wire antennas, antenna arrays, and aperture antennas suitable for radar applications. Following a treatment of sources of noise, the book moves on to give a detailed presentation of the most important scattering mechanisms exploited by radar. It then provides an overview of basic signal processing strategies, including coherent and incoherent strategies. Pulse compression, especially binary phase coding and frequency chirping, are then analyzed, and the radar range-Doppler ambiguity function is introduced. This is followed by a comprehensive treatment of radio wave propagation in the atmosphere and ionosphere. The remainder of the book deals with radar applications. The book will be valuable for graduate students and researchers interested in antenna and radar applications across the Earth and environmental sciences and engineering.

PROFESSOR DAVID HYSELL is a professor in the Department of Earth and Atmospheric Sciences at Cornell University. His research focuses on ionospheric plasma physics, plasma instabilities, ionospheric irregularities, and their effects on radio wave propagation. An important research tool for studying the equatorial ionosphere is the Jicamarca Radio Observatory near Lima, Peru, which is the world's largest radar. Hysell is currently the Principal Investigator for the National Science Foundation (NSF) award that helps support Jicamarca.

ANTENNAS AND RADAR FOR ENVIRONMENTAL SCIENTISTS AND ENGINEERS

DAVID HYSELL
Cornell University

CAMBRIDGE UNIVERSITY PRESS

CAMBRIDGE
UNIVERSITY PRESS

University Printing House, Cambridge CB2 8BS, United Kingdom

One Liberty Plaza, 20th Floor, New York, NY 10006, USA

477 Williamstown Road, Port Melbourne, VIC 3207, Australia

314–321, 3rd Floor, Plot 3, Splendor Forum, Jasola District Centre, New Delhi – 110025, India

79 Anson Road, #06–04/06, Singapore 079906

Cambridge University Press is part of the University of Cambridge.

It furthers the University's mission by disseminating knowledge in the pursuit of education, learning, and research at the highest international levels of excellence.

www.cambridge.org
Information on this title: www.cambridge.org/9781107195431
DOI: 10.1017/9781108164122

© David Hysell 2018

This publication is in copyright. Subject to statutory exception and to the provisions of relevant collective licensing agreements, no reproduction of any part may take place without the written permission of Cambridge University Press.

First published 2018

Printed in the United States of America by Sheridan Books, Inc.

A catalogue record for this publication is available from the British Library.

Library of Congress Cataloging-in-Publication Data
Names: Hysell, David.
Title: Antennas and radar for environmental scientists and engineers / David Hysell.
Description: Cambridge, United Kingdom ; New York, NY : Cambridge University Press, 2018. | Includes bibliographical references and index.
Identifiers: LCCN 2017028181 | ISBN 9781107195431
Subjects: LCSH: Environmental monitoring–Remote sensing. | Antennas (Electronics) | Synthetic aperture radar.
Classification: LCC GE45.R44 H97 2018 | DDC 621.382–dc23
LC record available at https://lccn.loc.gov/2017028181

ISBN 978-1-107-19543-1 Hardback

Cambridge University Press has no responsibility for the persistence or accuracy of URLs for external or third-party internet websites referred to in this publication and does not guarantee that any content on such websites is, or will remain, accurate or appropriate.

Contents

Preface		xi
1	**Introduction**	1
	1.1 History	1
	1.2 Motivation and Outline	6
	1.3 Mathematical Preliminaries	9
	1.3.1 Phasor Notation	9
	1.3.2 Fourier Analysis	11
	1.3.3 Probability Theory	18
	1.4 Notes and Further Reading	26
	Exercises	27
2	**Introduction to Antenna Theory**	29
	2.1 Hertzian Dipole	29
	2.1.1 Vector Potential: Phasor Form	29
	2.1.2 Electromagnetic Fields	32
	2.1.3 Radiation Pattern	33
	2.1.4 Total Radiated Power	35
	2.1.5 Radiation Resistance	36
	2.1.6 Directivity	37
	2.1.7 Gain	39
	2.1.8 Beam Solid Angle	40
	2.1.9 Antenna Effective Area	40
	2.2 Reciprocity Theorem	41
	2.3 Practical Wire Antennas	44
	2.3.1 Half-Wave Dipole	46
	2.3.2 Driving-Point Impedance	48
	2.3.3 Antenna Effective Length	50
	2.3.4 Full-Wave Dipole	52
	2.3.5 Quarter-Wave Monopole	52

		2.3.6	Folded Dipole Antenna	54
		2.3.7	Short Dipole	55
		2.3.8	Small Current Loop	55
		2.3.9	Duality	57
	2.4	Polarization		58
		2.4.1	Linear Polarization	59
		2.4.2	Circular Polarization	59
		2.4.3	Elliptical Polarization	60
		2.4.4	Stokes Parameters	63
		2.4.5	Helical Antennas	63
	2.5	Balanced and Unbalanced Transmission Lines and Balun Transformers		65
	2.6	Notes and Further Reading		67
	Exercises			68
3	**Antenna Arrays**			71
	3.1	Uniform Linear Arrays		73
		3.1.1	Broadside Array	75
		3.1.2	Endfire Array	77
		3.1.3	Hansen–Woodyard Array	79
	3.2	Binomial Arrays		80
	3.3	Inter-Element Coupling and Parasitic Arrays		81
	3.4	Uniform Two-Dimensional Arrays		83
	3.5	Three-Dimensional Arrays		85
	3.6	Very Long Baseline Interferometry (VLBI)		86
	3.7	Preliminary Remarks on SAR		86
	3.8	Beam Synthesis		88
	3.9	Adaptive Arrays		89
	3.10	Acoustic Arrays and Non-Invasive Medicine		91
	3.11	Notes and Further Reading		93
	Exercises			94
4	**Aperture Antennas**			98
	4.1	Simple Apertures		99
		4.1.1	Example – Rectangular Aperture	103
		4.1.2	Fourier Transform Shortcuts	104
		4.1.3	Example – Circular Aperture	105
		4.1.4	Aperture Efficiency	107
	4.2	Reflector Antennas		109
		4.2.1	Spillover, Obstruction, and Cross-Polarization Losses	112
		4.2.2	Surface Roughness	114
		4.2.3	Cassegrain, Gregorian, and Spherical Reflectors	114
	4.3	Notes and Further Reading		116
	Exercises			117

Contents

5 Noise — 121
- 5.1 Blackbody Radiation — 121
- 5.2 Nyquist Noise Theorem — 123
  - 5.2.1 Noise in Passive Linear Networks — 124
  - 5.2.2 Reciprocity Theorem, Revisited — 125
- 5.3 Antenna and System Noise Temperature — 127
  - 5.3.1 Atmospheric Absorption Noise and Radiative Transport — 129
  - 5.3.2 Feedline Absorption — 131
  - 5.3.3 Cascades — 133
  - 5.3.4 Receiver Noise Temperature, Noise Figure — 134
  - 5.3.5 Receiver Bandwidth — 136
  - 5.3.6 Bandwidth and Channel Capacity — 136
- 5.4 The Matched Filter — 137
- 5.5 Two Simple Radar Experiments — 140
- 5.6 Quantization Noise — 143
- 5.7 Notes and Further Reading — 144
- Exercises — 144

6 Scattering — 147
- 6.1 Rayleigh Scatter — 149
  - 6.1.1 Polarization and Scattering Matrices — 150
  - 6.1.2 Scattering Matrices with Stokes Parameters — 151
- 6.2 Mie Scatter — 152
  - 6.2.1 Scattering Matrix — 155
  - 6.2.2 Scattering, Extinction, and Absorption Efficiency — 156
  - 6.2.3 Mie-Scatter Phase Functions — 157
- 6.3 Specular Reflection — 158
- 6.4 Corner Reflectors — 160
- 6.5 Thomson Scatter — 161
- 6.6 Bragg Scatter — 162
- 6.7 Volume Scatter — 163
- 6.8 Notes and Further Reading — 165
- Exercises — 166

7 Signal Processing — 167
- 7.1 Fundamentals — 167
- 7.2 Power Estimation and Incoherent Processing — 170
- 7.3 Spectral Estimation and Coherent Processing — 171
- 7.4 Summary of Findings — 174
- 7.5 CW Radars — 175
- 7.6 FMCW Radars — 177
- 7.7 Pulsed Radars — 179
  - 7.7.1 Range Aliasing — 179

viii   *Contents*

		7.7.2	Frequency Aliasing	183
	7.8	\multicolumn{2}{l	}{Example – Planetary Radar}	185
	7.9	\multicolumn{2}{l	}{Range and Frequency Resolution}	186
	7.10	\multicolumn{2}{l	}{MTI Radars}	188
	7.11	\multicolumn{2}{l	}{Notes and Further Reading}	191
	\multicolumn{3}{l	}{Exercises}	192
8	\multicolumn{3}{l	}{**Pulse Compression**}	**195**
	8.1	\multicolumn{2}{l	}{Phase Coding}	195
		8.1.1	Complementary (Golay) Codes	198
		8.1.2	Cyclic Codes	199
		8.1.3	Gold Codes	200
		8.1.4	Polyphase Codes	201
	8.2	\multicolumn{2}{l	}{Frequency Chirping}	201
		8.2.1	Frank Codes	202
		8.2.2	Costas Codes	203
	8.3	\multicolumn{2}{l	}{Range-Doppler Ambiguity Functions}	204
		8.3.1	Costas Codes Revisited	209
		8.3.2	Volume Scatter Revisited	210
	8.4	\multicolumn{2}{l	}{Passive Radar}	211
	8.5	\multicolumn{2}{l	}{Notes and Further Reading}	214
	\multicolumn{3}{l	}{Exercises}	214
9	\multicolumn{3}{l	}{**Propagation**}	**218**
	9.1	\multicolumn{2}{l	}{Tropospheric Ducting}	220
	9.2	\multicolumn{2}{l	}{Ionospheric Reflection and Refraction}	222
		9.2.1	Vertical-Incidence Radio Sounding	224
		9.2.2	Oblique Sounding and the Theorem of Breit and Tuve	226
		9.2.3	Maximum Usable Frequency (MUF)	227
	9.3	\multicolumn{2}{l	}{Generalized Wave Propagation in Inhomogeneous Media}	229
		9.3.1	Stratified Media	231
		9.3.2	Radius of Curvature	232
	9.4	\multicolumn{2}{l	}{Birefringence (Double Refraction)}	233
		9.4.1	Fresnel Equation of Wave Normals	235
		9.4.2	Astrom's Equation	236
		9.4.3	Booker Quartic	237
		9.4.4	Appleton–Hartree Equation	237
		9.4.5	Phase and Group Velocity	240
		9.4.6	Index of Refraction Surfaces	243
		9.4.7	Ionospheric Sounding	244
		9.4.8	Attenuation and Absorption	248
		9.4.9	Faraday Rotation	249
	9.5	\multicolumn{2}{l	}{Propagation Near Obstacles – Knife-Edge Diffraction}	251

	9.5.1	Geometric Interpretation	253
	9.5.2	Other Diffraction Models	255
9.6	Notes and Further Reading		260
	Exercises		261

10 Overspread Targets — 264

10.1	Conventional Long-Pulse Analysis	265
10.2	Amplitude Modulation and the Multipulse	268
10.3	Pulse Compression and Binary Phase Coding	270
	10.3.1 Coded Long Pulse	273
	10.3.2 Alternating Code	274
10.4	Range-Lag Ambiguity Functions	276
10.5	Error Analysis	280
	10.5.1 Noise and Self Clutter	283
10.6	Notes and Further Reading	285
	Exercises	286

11 Weather Radar — 288

11.1	Radar Cross-Section and Radar Equation	289
	11.1.1 Reflectivity	291
	11.1.2 Relative Permittivity of Hydrometeors	294
	11.1.3 Mixed-Phase Hydrometeors and Effective Permittivity	296
	11.1.4 Attenuation	298
11.2	Z-R Relationships	302
11.3	Doppler Radar	304
	11.3.1 Spectral Moments and Errors	305
	11.3.2 Spectral Noise Estimates	307
	11.3.3 Wind Field Estimation	308
	11.3.4 Least-Squares Estimation	310
11.4	Polarimetry	311
	11.4.1 Z_{DR}	312
	11.4.2 ρ_{HV}	313
	11.4.3 LDR	313
	11.4.4 ϕ_{DP}	314
11.5	Clear-Air Echoes	314
	11.5.1 Wind Profiling and Momentum Flux	315
	11.5.2 Scattering from Irregular Media	316
11.6	Notes and Further Reading	320
	Exercises	322

12 Radar Imaging — 323

12.1	Radar Interferometry	323
	12.1.1 Ambiguity	324

	12.1.2 Volume Scatter	325
12.2	Aperture Synthesis Imaging	327
	12.2.1 CLEAN	329
	12.2.2 Maximum Entropy Method	330
12.3	Synthetic Aperture Radar (SAR)	335
	12.3.1 Wavenumber-Domain (WD) Processing	339
	12.3.2 Practical Implementation Issues	341
	12.3.3 InSAR	343
	12.3.4 ISAR	347
12.4	Notes and Further Reading	349
Exercises		350
Appendix A	*Radio Frequency Designations*	352
Appendix B	*Review of Electromagnetics*	353
References		377
Index		383

Preface

This textbook is a compendium of material covered in a course offered by D. L. Hysell to advanced undergraduates and graduate students in the Department of Earth and Atmospheric Sciences in the College of Engineering at Cornell University. Hysell's course evolved from another one offered in the Department of Electrical and Computer Engineering for many years by Professor Emeritus D. T. Farley. In developing his course, Farley combined material from three other courses that were offered at one time in the Electrical Engineering curriculum: one on radar, another on antennas, and a third on radio wave propagation. The fundamentals of signal processing were also introduced on an as-needed basis. The goal of combining material in this way was to provide students with an integrated understanding of all of the factors involved in the design of radar experiments and the interpretation of the measurements they produce.

The present-day course at Cornell serves students with interests ranging from radar meteorology to planetary radar to satellite-based remote sensing to ground-penetrating radar. As the students come from different colleges armed with diverse backgrounds, the course and this text endeavor to build a radar curriculum from first principles, from the ground up, imposing minimal requirements on familiarity with signal processing, electromagnetics, radio hardware, or specific radar applications. A solid grounding in undergraduate mathematics is taken for granted, however.

A typical, semester-long course might cover the material in Chapters 1–8, inclusive. This material is rather fundamental and common to most any radar application. Subsequent chapters address more specialized content that might be covered occasionally to varying levels of depth, based on student interest. Chapter 9, for example, presents a fairly detailed treatment of magneto-ionic theory, which would normally only be of interest to students studying the Earth's upper atmosphere. Chapter 10 addresses overspread targets, another highly specialized topic that is seldom treated in introductory radar texts but is crucial for certain problems. Chapters 11 and 12 deal, respectively, with weather radar and radar imaging, two very popular applications. While neither chapter is meant to offer a complete treatment of the subject in question, either should be sufficient to propel interested students in the direction of more advanced and specialized classes. A review of electromagnetics sufficient for most radar-related intents and purposes can be found in the Appendix, and should be useful for students without background in this area.

The problems (Exercises) at the end of every chapter are meant to help students test their understanding of the most important concepts. Many of them are best solved with the help of numerical computations. Over the years, different computer languages and packages have fallen in and out of favor with our students. Present-day students seem to have little trouble tackling the problem sets using Python.

1
Introduction

This text investigates how radars can be used to explore the environment and examine natural and manufactured objects in it remotely. Everyone is familiar with popular portrayals of radar engineers straining to see blips on old-style radar cathode ray tubes. However, the radar art has evolved beyond this cliché. Radars are now used in geology and archaeology, ornithology and entomology, oceanography, meteorology, and studies of the Earth's upper atmosphere, ionosphere, and magnetosphere. Radars are also used routinely to construct fine-grained images of the surface of this planet and others. Astonishingly detailed information, often from otherwise inaccessible places, is being made available to wide audiences for research, exploration, commerce, transportation, security, and commercial purposes. New analysis methods are emerging all the time, and the radar art is enjoying a period of rapidly increasing sophistication.

The invention of radar cannot be attributed to any single individual or event but instead involved scientific and engineering breakthrough occurring across the earth throughout the early part of the twentieth century. By the start of World War II, operational radars could be found in the US, Britain and other Commonwealth countries, France, Germany, Italy, the Soviet Union, Switzerland, The Netherlands, and Japan. Like many technologies, radar was created for niche civilian and scientific applications, developed extensively during wartime, and then exploited for all manner of purposes, from the profound to the mundane. The story of radar is rich and colorful and gives insights into the serendipitous routes by which new technologies emerge.

1.1 History

In 1904, German inventor Christian Hülsmeyer obtained patents in multiple countries for an "obstacle detector and ship navigation device" that can be considered a basic kind of radar. The device operated at UHF frequencies and incorporated directional, resonant antennas for transmission and reception along with a rudimentary detection system. The device could detect ships at sea to distances of several km when visibility was poor and give an indication of their distance (indirectly) and bearing. What makes the invention particularly remarkable is that it predated Lee de Forest's vacuum tube radio detector by two years. However, even though the device was demonstrated successfully to the German

Navy, its potential was not recognized. The device had no compelling application in its day and was not adopted. It would be another two decades before a branch of the military would pursue radar.

At about the same time that Hülsmeyer was detecting ships in foggy channels, Italian physicist Guglielmo Marconi was attempting transatlantic radio communications using radio technology developed by Heinrich Hertz and Nicola Tesla. It was far from clear at the turn of the twentieth century whether radio waves could propagate long distances "over the horizon" without being impeded by the curvature of the earth. Successful demonstration of transatlantic propagation would amount to the detection of a conducting layer in the upper atmosphere, the "ionosphere," and it is in that regard that Marconi's work pertains to radar.

On December 12, 1901, Marconi reported hearing the letter "S" in Morse code at a receiving station on Signal Hill, St. John's, Newfoundland, sent by his station in Poldhu, Cornwall, England, approximately 3,500 km away. The experiment in question followed earlier, successful experiments involving shorter propagation paths, as well as numerous mishaps and unsuccessful transatlantic attempts involving different North American receiving sites. (The Cornwall–St. John's route is about the shortest transatlantic path possible.) For the 1901 experiments, Marconi employed a high-power (for its day–about 13 kW) spark-gap transmitter feeding a resonant, fan-beam antenna at Poldhu and a kite-borne, longwire antenna and broadband receiver on Signal Hill.

Marconi's 1901 results have been met with serious skepticism, both then and now. The experiments were performed at solar minimum under geomagnetically quiet conditions during the daytime at a wavelength that would fall today within the MF band (the AM radio band, more precisely). We now know that the signal would have experience strong ionospheric absorption under the experimental conditions described. It seems unlikely that the signal could have been detected using the receiving technology available at the time. (How many transatlantic signals can you hear in the AM broadcast band during the day?) Furthermore, it would have been very difficult to distinguish the Morse letter "S" from atmospheric noise in any case.

Subsequently and in response to his critics, Marconi would go on to perform more systematic and better-documented propagation experiments, including experiments with shipboard receivers, demonstrating that over-the-horizon radio communication in the MF band was possible and practical, particularly at night. Within a few years, regular, long-distance commercial broadcasting would be underway. Marconi would receive the 1909 Nobel Prize with Karl Braun for his contributions to radio.

The idea of a conducting layer in the upper atmosphere was not new in the early twentieth century, having been proposed in 1882 by Balfour Stewart as a means of explaining rapid fluctuations in the terrestrial magnetic field. In 1902, Oliver Heaviside and Arthur E. Kennelly proposed that the existence of such a layer could explain Marconi's results. Experimental evidence for the hypothesis would ultimately come from Lee de Forest and Leonard Fuller, working for the Federal Telegraph Co. in San Francisco, and later, in a more widely publicized series of experiments, from Cambridge physicist Edward Appleton.

Based on experiments performed in 1924, Appleton attributed fading in the reception of shortwave radio stations (BBC broadcasts) to interference between waves from the transmitter traveling along the earth (ground waves) and waves reflecting from the upper atmosphere (sky waves). Careful studies of the wavelength dependence of the interference confirmed the existence of the Kennelly–Heaviside layer (now called the E layer) at about 100 km altitude as well as another, denser layer at about 300–400 km altitude (originally named for Appleton but now called the F layer). Appleton would go on to attribute radio reflection to ionization and free electrons in the atmosphere, to quantify how the free electrons modify the index of refraction of the medium, and to understand how absorption due to collisions between the electrons and the neutral gas makes communications in the MF band difficult during the day. For this work, Appleton would receive the 1947 Nobel Prize in Physics.

If Appleton showed how the ionosphere affects radio communications, Gregory Breit and Merle A. Tuve showed how pulsed radio signals could be used to measure ionospheric structure. Their pivotal 1925 and 1926 papers described and demonstrated an instrument capable of measuring the time it takes for radio signals transmitted from the ground to be reflected by the ionosphere and then detected back on the ground, all as a function of frequency. The ionospheric sounder or "ionosonde" they described was essentially a radar. (The authors even noted how air traffic near a local Navy airfield interrupted their soundings.) By the late 1930s, scientists like Henry Booker were using ionosondes to study ionospheric phenomena at different sites around the world and extending Appleton's theory to account for what they were seeing.

At the same time, Hülsmeyer's ideas for object detection and tracking were being rediscovered. At the US Naval Research Laboratory, Alfred Taylor and Leo Young noted in 1922 how a radio signal originating on one side of the Potomac River and received on the other was interfered with by a passing ship. (The phenomenon was actually discovered by Russian radio pioneer Alexander Popov, noting the interference caused to naval communications by an intervening ship in 1897, although nothing more was made of the finding at the time.) In 1930, similar results for intervening aircraft were reported by Lawrence A. Hyland, working with Taylor and Young. In either case, what would today be called a "bistatic" system with the transmitter and receiver separated was at work. It was also a continuous wave or "CW" system, lacking any sort of modulation. Absent sophisticated signal processing, such "interference" radars could provide little information about the target position or velocity, and so were of limited practical utility to the Navy.

Young and Hyland persevered and considered adding amplitude modulation in the form of pulsing to the system in order to provide information about target range. It was their pulsed transmitter that Breit and Tuve used for their ionospheric research. By 1934, Robert Page had implemented Young's pulsing idea and also developed a duplexer, allowing the radar transmitter and receiver to be collocated and to share a common antenna. The resulting system, which operated in the VHF band, was perhaps the first "true" radar, capable of making direct measurements of the range and bearing of a target. Practical radar sets were in production by 1938, and a portable version was installed on the battleship USS New York for testing in 1939.

Parallel developments in radar were taking place at about this time all over the world as signs of impending war were becoming increasingly clear. In 1935, Robert Watson Watt, director of the Radio Research Station (home of an ionosonde) in Slough, was contacted by the British Air Ministry to investigate the practicality of building a "death ray" of the kind Nicola Tesla had claimed to have invented. Watson Watt argued that using radio for detection would be a more promising endeavor. In bistatic experiments involving BBC transmissions reminiscent of Appleton's pioneering work, Watson Watt demonstrated that radio signals reflected by aircraft could readily be detected on the ground. Interferometry was used to null the direct-path signal so that only reflected signals could be received. The simple but elegant work was done in collaboration with his assistant, Arnold Wilkins.

The results of the "Daventry Experiment" set in motion the development of a coastal chain of radar stations intended for the air defense of the British Isles. The "Chain Home Low" radars grew from the pulsed-radar technology found in ionosondes and operated initially at the upper end of the HF band in the main. The detection of a flying boat by a Chain Home prototype in 1938 is perhaps the first true demonstration of radar in the fullest sense. The peak power of the production Chain Home transmitters was at first 350 kW and later 750 kW, easily outclassing all forerunners. The radars were bistatic, employing dipole arrays draped between pairs of tall towers for transmission and separate towers with crossed dipoles for reception. Bearing information could be deduced with the knowledge of the radiation patterns of dipoles with different orientations and heights above ground. At the start of WWII, 21 Chain Home Low stations were in operation. The operational success of the Chain Home system is widely credited for the outcome of the Battle of Britain, see Appendix A for a list of radio frequency designations.

During the course of WWII, a premium was placed on radar operations at ever-higher frequencies. Higher radar frequencies allow higher antenna gain, given a fixed antenna size, or, equivalently, allow smaller antennas, given a fixed gain requirement. Small antennas in turn are more steerable and portable and may be mounted in ships and even aircraft. Higher frequencies also permit higher bandwidths, shorter pulses, and better range resolution. Finally, cosmic noise decreases with increasing frequency, while the scattering cross-section of large, flat-faced metal targets increases. This makes high-frequency radars potentially more sensitive in many (although not all) military applications.

The Chain Home Low stations were ultimately joined by Chain Home High and Chain Home Extra Low stations operating at 200 MHz and 3 GHz, respectively. The main German counterpart radar system, Freya, was introduced in 1938, and operated at 250 MHz. Its cousin, the Wurzburg radar, operated at 560 MHz and was used for targeting anti-aircraft guns. The SCR-268 radar deployed by the US Signal Corps for coastal defense operated at 205 MHz. (The successor to the SCR-268, the SCR-270, was the radar that spotted waves of aircraft on their way to attack Pearl Harbor.) The CXAM radars developed by NRL for surface ships used 400 MHz.

At frequencies above UHF, ordinary vacuum tubes become ineffective for generating high-power RF. One of the first devices capable of producing high power at microwave frequencies was the magnetron. A magnetron is a vacuum tube with a strong magnetic field aligned along its axis. The two-pole version of this device was invented in 1920 in Schenectady, New York, by Albert Hull at General Electric, who was attempting to bypass patents held by Lee de Forest. August Zacek in Prague and Erich Habann in Jena saw potential in the device for generating powerful microwave signals, and many others soon followed. The first high-power centimeter (cm) waves were generated using a split-anode magnetron developed by Kinjiro Okabe in Japan in 1929. The first multi-cavity resonant magnetron was constructed in Berlin by Hans Erich Hollmann in 1935. His device exhibited frequency instability, however, and was not pursued for radar applications in Germany, where the klystron was perceived to have even greater potential. (A klystron is a traveling wave vacuum tube that operates without an axial magnetic field. It was invented by Russel and Sigurd Varian in 1937 at Stanford University and saw development during and after WWII.)

By 1940, working at the University of Birmingham, John Turton Randall and Harry Boot had replicated Hollmann's results and gone on to produce a much more powerful device. Continued development efforts ultimately overcame the frequency instability problem. The technology was transferred the same year to the US, where the resources to fully exploit it were available. (The actual transfer occurred in the so-called "Tizard Mission," when a magnetron prototype was borne across the Atlantic by the cruise ship Duchess of Richmond.) The Radiation Laboratory on the campus of MIT was set up to oversee the utilization of the new technology, which found its way into, among other things, airborne radar capable of locating submarines and tracing coastlines and highly accurate fire-control radars for surface ships.

Magnetrons continued to see service during the cold war in the lines of radars spanning North America and intended as a deterrent against bomber raids from the Soviet Union. Radars along the northern-most line, the "distant early warning" or DEW Line radars, operated at L band frequencies and had a peak power of 160 kW. After the launch of Sputnik, the relevance of the DEW Line radars was diminished. The replacement "ballistic missile early warning system" or BMEWS radars relied upon newly perfected klystrons capable of generating 2.5 MW of peak power at UHF frequencies. These radars, too, have become obsolete and have been replaced by phased-array UHF radars like the PAVE PAWS system. Phased-array radars utilize solid-state technology and have the distinct advantage of electronic steerability. A phased-array antenna is typically made of a number of independent modules, each one a transistor amplifier mated to an antenna element. The peak power of the antenna is determined by the number of modules, since each module can generate no more than about 1 kW of peak power in practice. Multiple antenna sections can be combined to increase the overall steering range.

Most of the development in radar technology and applications since the end of WWII has come outside of the realm of defense. Below, some of the more visible and important applications are itemized.

6   Introduction

- Air traffic control
- Automotive radar
- Weather and boundary-layer radar
- Mesosphere-stratosphere-troposphere (MST) radar
- Meteor-scatter radar
- Ionospheric sounders
- Coherent- and incoherent-scatter radar for ionospheric and space-weather research
- Over-the-horizon radar
- Radar imaging, including SAR, InSAR, and ISAR
- Planetary radar
- Radars for ornithology, entomology, and other wildlife
- Ground-penetrating radar for archaeology, paleontology, and Earth science
- Hand-held radar
- Passive and stealth radar

To better appreciate the elements on the list, we must read on.

1.2 Motivation and Outline

Our objective is to understand how radars can be used to investigate the environment. It is useful to begin the study of radar, however, with an elementary question: how much of the power transmitted by a radar will be returned in the echoes from the target and received? The answer to this question, which culminates in the derivation of the so-called "radar equation," illustrates some important aspects of radar design and application. The derivation therefore serves as a vehicle for introducing basic ideas and terminology. It also illustrates the central importance of antenna performance. Much of the early material in this text will therefore emphasize antenna theory, analysis, and design.

We can sketch out the basic ideas behind radar systems with some common-sense reasoning. The analysis begins by considering a simple communications link involving distant transmitting and receiving stations and then considers what happens when a radar target is introduced into the problem. The relationships given below are defined rather than derived, and serve as the basis for more rigorous material to follow. More detailed treatments will come later, when we will also relax the condition that the stations be distant.

The power budget for the simple communication link illustrated in Figure 1.1 is governed by

$$P_{rx} = \underbrace{P_{tx} \frac{D_{tx}}{4\pi R^2}}_{P_{inc}} A_{\text{eff}},$$

with the received power P_{rx} being proportional to the transmitted power P_{tx}, assuming that the media between the stations is linear. The incident power density falling on the

1.2 Motivation and Outline

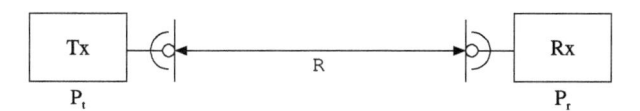

Figure 1.1 RF communication link.

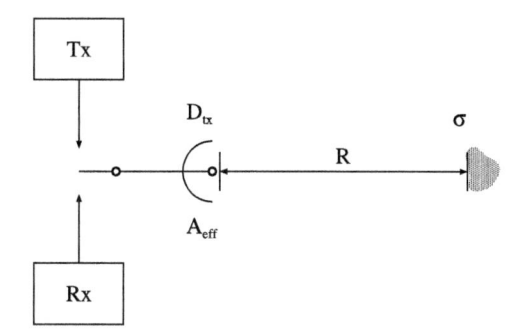

Figure 1.2 Radar link.

receiving antenna P_{inc} would just be $P_{tx}/4\pi R^2$ if the illumination were isotropic. However, isotropic radiation cannot be achieved by any antenna and is seldom desirable in any case, and the transmitting antenna is usually optimized to concentrate radiation in the direction of the receiver. The transmitting antenna directivity D_{tx} is a dimensionless quantity that represents the degree of concentration achieved. Rewriting the equation above provides a definition of directivity:

$$D_{tx} = \frac{P_{inc}}{P_{tx}/4\pi R^2} = \frac{P_{inc}}{P_{iso}} = \frac{P_{inc}(W/m^2)}{P_{avg}(W/m^2)},$$

where the denominator can be interpreted as either the power density that would fall on the receiving station if transmission were isotropic or the average power density across the sphere of radius R on which the receiver lies. So the directivity is the ratio of the power density incident on the receiving station at a distance or range R to the equivalent isotropic power density or the average power density radiated in all directions at that range.

The receiving antenna intercepts some of the power incident upon it, which is delivered to the receiver. The factor that expresses how much power is its effective area, $A_{eff} = P_{rx}(W)/P_{inc}(W/m^2)$. The effective area has units of area and is sometimes (but not always) related to the physical area of the antenna.

This picture is modified somewhat for a radar link that also involves some kind of distant target. The transmitter and receiver are also frequently collocated in a radar, as shown in Figure 1.2. Here, the transmitter and receiver share a common antenna connected through a transmit-receive (T/R) switch. The power budget for such a radar is governed by

$$P_{inc}(W/m^2) = P_{tx}\frac{D_{tx}}{4\pi R^2}$$
$$P_{scat}(W) = P_{inc}\underbrace{\sigma_{radar}}_{(m^2)}.$$

The first line above is the power density incident on the radar target. We next define a radar-scattering cross-section as the ratio of the total power scattered by the target to the incident power density illuminating it. The cross-section has units of area and may or may not be related to the physical area of the target itself. The definition is such that we take the scattering to be isotropic, simplifying the calculation of the power received back at the radar station:

$$P_{rx}(W) = P_{scat}\frac{A_{\text{eff}}}{4\pi R^2}.$$

Altogether, these expressions for the received power can be combined to form the *radar equation*:

$$P_{rx}(W) = \underbrace{\underbrace{P_{tx}\frac{D_{tx}}{4\pi R^2}}_{P_{inc}(W/m^2)}\sigma_{radar}\frac{A_{\text{eff}}}{4\pi R^2}}_{P_{scat}(W)} \propto R^{-4},$$

where we have neglected any losses in the system and the propagation channel. The radar equation is useful for evaluating different contributions to the power budget in the radar link. Notice in particular the inverse dependence on the fourth power of the range, which may make long-range detection rather challenging. While the equation is essentially complete, it has a number of components that require further investigation. These will make up the body of this content of this text. They are

1. Antenna directivity D_{tx}: What is it, and how is it calculated?
2. Antenna effective area A_{eff}: What is it, how is it calculated, and how is it related to directivity?
3. Scattering cross-section σ_{radar}: How is it defined, and how is it calculated for different kinds of scatterers?
4. Noise, statistical errors: From where does it come, how is it estimated, and how is it mitigated?
5. Propagation effects, polarization, etc.: How do radar signals propagate from place to place?
6. Signal processing: How are radar data acquired and handled?

The first two of these items concern antenna performance. More than any other characteristic of a radar, its antenna determines the applications for which it will be well suited. Chapters 2–4 of the text consequently analyze antennas in detail. Estimating the cross-section of a radar target requires some understanding of the physics of scattering. This topic is addressed in Chapter 6. Whether the incident power density at the target actually falls off with range as R^{-2} and the backscatter power density as R^{-4} depends on the range, the characteristics of the radar antenna, and the scattering medium, and needs further examination as well. Noise in the environment and the radar equipment itself limit the ability to detect a scattered signal. Environmental and instrumental noise will be discussed in Chapter 5. For a target to be detected, the radar signal has to propagate from the radar to the target and back, and the manner in which it does so is affected by the intervening medium. Radio wave propagation in the Earth's atmosphere and ionosphere is therefore explored in Chapter 9. Perhaps most importantly, a considerable amount of data processing is usually required in order to extract meaningful information about the radar target from the received signal. Signal processing is therefore described in detail in Chapters 7 and 8. The text concludes with the exploration of a few specialized radar applications pertaining to overspread targets (Chapter 10), weather radar (Chapter 11), and radar imaging (Chapter 12).

Fourier analysis and probability and statistics appear as recurring themes throughout the text. Both concepts are reviewed immediately below. The review culminates with a definition of the power spectral density of a signal sampled from a random process. The Doppler spectrum is one of the most important outcomes of a radar experiment, and its meaning (and the meaning of the corresponding autocorrelation function) warrants special examination.

1.3 Mathematical Preliminaries

The remainder of this chapter concerns powerful mathematical concepts that will be helpful for understanding material in successive chapters of the text. The first is a review of phasor notation. A brief treatment of Fourier analysis is then presented. After that, basic ideas in probability theory are reviewed. Fourier analysis and probability theory culminate in the definition of the power spectral density of a signal, something that plays a central role in radar signal processing.

1.3.1 Phasor Notation

Phasor notation is a convenient way to analyze linear dynamic systems. If we force any component of a system described by a second-order linear differential equation with constant coefficients (e.g., the wave equation) at a single frequency, all components of the system will respond at that frequency once the transient response has decayed. We need then only be concerned with the amplitude and phase of the responses of the different components of the system. The approach can be extended trivially to forcing at multiple frequencies or to broadband forcing using the principle of superposition.

Let us represent a signal of the form $q(t) = a\cos(\omega t + \phi)$ as the real part of the complex quantity, i.e.,

$$\begin{aligned} q(t) &= a\Re(e^{j(\omega t + \phi)}) \\ &= \Re(ae^{j\phi}e^{j\omega t}) \\ &= \Re(Ae^{j\omega t}), \end{aligned}$$

where a is the real signal amplitude, ϕ is the phase with respect to a defined reference, ω is the frequency, and $A \equiv ae^{j\phi}$ is the complex phasor amplitude that contains both the amplitude and phase information. The phasor representation for $q(t)$ is then just A. The introduction of complex notation may seem like an unnecessary complication, but it actually simplifies things considerably. Writing "A" is simpler than writing "$a\cos(\omega t + \phi)$." Furthermore, differentiation and integration become algebraic operations for exponential functions. We can most often regard the real part of the phasor quantity as being physically significant, the imaginary part being a mathematical contrivance.

In phasor notation, A can be viewed as a vector in the complex plane. When dealing with waves, we expect both the amplitude and phase (the magnitude and angle of A) to vary in space. The task is then usually to solve for A everywhere. There is no explicit time variation in phasor notation; time is contained entirely in the implicit $\exp(j\omega t)$ term, which we often neglect to write but understand to be present, just as we understand the $\Re$ to be present.

The rules of arithmetic for phasors are somewhat different for other number systems. Addition and subtraction behave as intuition would suggest. The phasor representation for $q(t) = a\cos(\omega t + \phi_a) + r(t) = b\cos(\omega t + \phi_b)$ is just $A + B$, with $A \equiv ae^{j\phi_a}$ and $B \equiv be^{j\phi_b}$. This implies that vectors can be constructed from phasors, since $\mathbf{q}(t) \equiv q_x(t)\hat{\mathbf{x}} + q_y(t)\hat{\mathbf{y}} + q_z(t)\hat{\mathbf{z}}$ can be represented in phasor notation as $\mathbf{A} \equiv A_x\hat{\mathbf{x}} + A_y\hat{\mathbf{y}} + A_z\hat{\mathbf{z}}$, so long as all the components of $\mathbf{q}(t)$ oscillate at the same frequency. As above, the real part of a phasor vector is what is physically significant.

Phasor products require special care. This can be seen by considering the product:

$$q(t)r(t) = a\cos(\omega t + \phi_a)b\cos(\omega t + \phi_b)$$
$$= \frac{1}{2}ab\left\{\cos(2\omega t + \phi_a + \phi_b) + \cos(\phi_a - \phi_b)\right\}.$$

The product of two sinusoidally varying signals yields a constant term plus another term varying sinusoidally in time at twice the original frequency. Since one of the purposes of phasors is to express time dependence implicitly, their utility would seem to falter when evaluating products and quotients. Consider, however, what happens when we evaluate the time average of the product of p and q, denoted below by angle brackets:

$$\langle q(t)r(t)\rangle = \frac{1}{2}ab\cos(\phi_a - \phi_b)$$
$$= \frac{1}{2}\Re(ae^{j\phi_a}be^{-j\phi_b})$$
$$= \frac{1}{2}\Re AB^*.$$

The last expression therefore supplies the recipe for calculating the average of the product of two sinusoidally varying quantities. Average power and energy calculations are of this form, and so the utility of phasor notation remains high. Vector inner and outer products follow intuitive rules of computation. We will see that the average Poynting flux of an electromagnetic wave, defined as $\mathbf{P(t)} = \mathcal{E}(t) \times \mathcal{H}(t)$, has an average value expressible in phasor notation as $\langle \mathbf{P} \rangle = \frac{1}{2}\Re(\mathbf{E} \times \mathbf{H}^*)$, for example.

Some other, subtle issues arise with the use of phasor notation. For example, the norm of a vector in phasor notation is $|\mathbf{E}|^2 = \mathbf{E} \cdot \mathbf{E}^*$. Consider the norm of the electric field associated with a circularly polarized wave traveling in the $\hat{\mathbf{z}}$ direction such that $\mathbf{E} = (\hat{\mathbf{x}} \pm j\hat{\mathbf{y}}) \exp(j(\omega t - kz))$. The length of the electric field vector is clearly unity, and evaluating the Euclidean length of the real part of the electric field indeed produces a value of 1. The phasor norm of the electric field, meanwhile, gives $\mathbf{E} \cdot \mathbf{E}^* = 2$. We can see that this is because the real part of inner and outer conjugate products in phasor notation give twice the average value of the given product.

1.3.2 Fourier Analysis

Phasor notation can be viewed as a kind of poor man's Fourier analysis, a fundamental mathematical tool used in a great many fields of science and engineering, including signal processing. The basis of Fourier analysis is the fact that a periodic function in time, for example, can be expressed as an infinite sum of sinusoidal basis functions:

$$f(t) = \frac{1}{2}a_\circ + a_1 \cos \omega_\circ t + a_2 \cos 2\omega_\circ t + \cdots + b_1 \sin \omega_\circ t + b_2 \sin 2\omega_\circ t + \cdots,$$

where the period of $f(t)$ is τ and $\omega_\circ \equiv 2\pi/\tau$. The series converges to $f(t)$, where the function is continuous and converges to the midpoint, where it has discontinuous jumps. The Fourier coefficients can be found by utilizing the orthogonality of sine and cosine functions over the period τ:

$$a_n = \frac{\omega_\circ}{\pi} \int_{-\tau/2}^{\tau/2} f(t) \cos n\omega_\circ t \, dt$$

$$b_n = \frac{\omega_\circ}{\pi} \int_{-\tau/2}^{\tau/2} f(t) \sin n\omega_\circ t \, dt,$$

where the integrals can be evaluated over any complete period of $f(t)$. Fourier decomposition applies equally well to real and complex functions. With the aid of the Euler theorem, the decomposition above can be shown to be equivalent to

$$f(t) = \sum_{n=-\infty}^{\infty} c_n e^{jn\omega_\circ t} \quad (1.1)$$

$$c_n = \frac{\omega_\circ}{2\pi} \int_{-\tau/2}^{\tau/2} f(t) e^{-jn\omega_\circ t} dt. \quad (1.2)$$

Written this way, it is clear that the Fourier coefficients c_n can be interpreted as the amplitudes of the various frequency components or harmonics that compose the periodic function $f(t)$. The periodicity of the function ultimately permits its representation in terms of discrete frequencies.

It can be verified easily that the sum of the squares of the harmonic amplitudes gives the mean squared average of the function,

$$\langle |f(t)|^2 \rangle = \sum_{n=-\infty}^{\infty} |c_n|^2, \tag{1.3}$$

which is a simplified form of Parseval's theorem. This shows that the power in the function $f(t)$ is the sum of the powers in all the individual harmonics, which are the normal modes of the function. Parseval's theorem is sometimes called the completeness relation because it shows how all the harmonic terms in the Fourier expansion are required in general to completely represent the function. Other forms of Parseval's theorem are discussed below.

Fourier Transforms

The question immediately arises: is there a transformation analogous to (1.1) and (1.2) that applies to functions that are not periodic? Let us regard a non-periodic function as the limit of a periodic one where the period is taken to infinity. We can write

$$c_n = \frac{\omega_\circ}{2\pi} \int_{-\tau/2}^{\tau/2} f(u) e^{-jn\omega_\circ u} du$$

and substitute into (1.1), yielding

$$f(t) = \sum_{n=-\infty}^{\infty} \frac{\omega_\circ}{2\pi} \int_{-\tau/2}^{\tau/2} f(u) e^{jn\omega_\circ(t-u)} du$$

$$= \frac{1}{2\pi} \sum_{n=-\infty}^{\infty} g(\omega_n) \Delta\omega$$

$$g(\omega_n) \equiv \int_{-\tau/2}^{\tau/2} f(u) e^{j\omega_n(t-u)} du,$$

where $\omega_n = n\omega_\circ$ and $\Delta\omega = \omega_{n+1} - \omega_n = \omega_\circ$ have been defined. Now, in the limit that $\tau \to \infty$, the sum $\sum_{n=-\infty}^{\infty} g(\omega_n)\Delta\omega$ formally goes over to the integral $\int_{-\infty}^{\infty} g(\omega)d\omega$, ω representing a continuous independent variable. The corresponding expression for $g(\omega)$ just above becomes

$$g(\omega) = \int_{-\infty}^{\infty} f(u) e^{j\omega(t-u)} du.$$

Substituting into the equation above for $f(t)$ yields

$$f(t) = \frac{1}{2\pi} \int_{-\infty}^{\infty} e^{j\omega t} d\omega \int_{-\infty}^{\infty} f(u) e^{-j\omega u} du.$$

Finally, if we change variables and define $F(\omega)$ as (analogous to (1.2))

$$F(\omega) = \int_{-\infty}^{\infty} f(t) e^{-j\omega t} dt, \tag{1.4}$$

1.3 Mathematical Preliminaries

we are left with (analogous to (1.1))

$$f(t) = \frac{1}{2\pi} \int_{-\infty}^{\infty} F(\omega) e^{j\omega t} d\omega, \tag{1.5}$$

where the continuous function $F(\omega)$ replaces the discrete set of amplitudes c_n in a Fourier series. Together, these two formulas define the Fourier transform and its inverse. Note that the placement of the factor of 2π is a matter of convention. The factor appears in the forward transform in some texts, in the inverse in others. In some texts, both integral transforms receive a factor of the square root of 2π in their denominators. What is important is consistency. Similar remarks hold for the sign convention in the exponents for forward and inverse transforms.

Now $f(t)$ and $F(\omega)$ are said to constitute a continuous Fourier transform pair which relate the time- and frequency-domain representations of the function. Note that, by using the Euler theorem, the forward and inverse transformations can be sinusoidal and cosinusoidal parts that can be used to represent anti-symmetric and symmetric functions, respectively. Fourier transforms are often easy to calculate, and tables of some of the more common Fourier transform pairs are available. Fourier transforms are very useful for solving ordinary and partial differential equations, since calculus operations (differentiation and integration) transform to algebraic operations. The Fourier transform also exhibits numerous properties that simplify evaluation of the integrals and manipulation of the results, linearity being among them. Some differentiation, scaling, and shifting properties are listed below, where each line contains a Fourier transform pair.

$$\frac{d}{dt} f(t) \leftrightarrow j\omega F(\omega) \tag{1.6}$$

$$-jt f(t) \leftrightarrow \frac{d}{d\omega} F(\omega) \tag{1.7}$$

$$f(t - t_\circ) \leftrightarrow e^{-j\omega t_\circ} F(\omega) \tag{1.8}$$

$$e^{j\omega t} f(t) \leftrightarrow F(\omega - \omega_\circ) \tag{1.9}$$

$$f(at) \leftrightarrow \frac{1}{|a|} F(\omega/a) \tag{1.10}$$

$$f^*(-t) \leftrightarrow F^*(\omega). \tag{1.11}$$

Convolution Theorem

One of the most important properties of the Fourier transform is the way the convolution operator transforms. The convolution of two functions f and g is defined by

$$f \otimes g = \int_0^t f(t - \tau) g(\tau) d\tau.$$

Convolution appears in a number of signal processing and radar contexts, particularly where filters are concerned. While the integral can sometimes be carried out simply, it is just as often difficult to calculate analytically and computationally. The most expedient means of calculation usually involves Fourier analysis.

Consider the product of the frequency-domain representations of f and g:

$$F(\omega)G(\omega) = \int_{-\infty}^{\infty} f(\tau')e^{-j\omega\tau'}d\tau' \int_{-\infty}^{\infty} g(\tau)e^{-j\omega\tau}d\tau$$

$$= \int_{-\infty}^{\infty}\int_{-\infty}^{\infty} e^{-j\omega(\tau'+\tau)} f(\tau')g(\tau)d\tau'd\tau.$$

With the change of variables $t = \tau' + \tau$, $dt = d\tau'$, this becomes

$$F(\omega)G(\omega) = \int_{-\infty}^{\infty}\int_{-\infty}^{\infty} e^{-j\omega t} f(t-\tau)g(\tau)dt d\tau$$

$$= \int_{-\infty}^{\infty} e^{-j\omega t} \left[\int_{-\infty}^{\infty} f(t-\tau)g(\tau)d\tau \right] dt.$$

The quantity in the square brackets is clearly the convolution of $f(t)$ and $g(t)$. Evidently, then, the Fourier transform of the convolution of two functions is the product of the Fourier transforms of the functions. This suggests an easy and efficient means of evaluating a convolution integral. It also suggests another way of interpreting the convolution operation, which translates into a product in the frequency domain.

Parseval's Theorem

The convolution theorem proven above can be used to reformulate Parseval's theorem for continuous functions. First, it is necessary to establish how the complex conjugate operation behaves under Fourier transformation. Let

$$f(t) = \frac{1}{2\pi} \int_{-\infty}^{\infty} F(\omega)e^{j\omega t} d\omega.$$

Taking the complex conjugate of both sides gives

$$f^*(t) = \frac{1}{2\pi} \int_{-\infty}^{\infty} F^*(\omega)e^{-j\omega t} d\omega.$$

Next, replace the variable t with $-t$:

$$f^*(-t) = \frac{1}{2\pi} \int_{-\infty}^{\infty} F^*(\omega)e^{j\omega t} d\omega,$$

which is then the inverse Fourier transform of $F^*(\omega)$.

Let us use this information to express the product $F^*(\omega)G(\omega)$, borrowing from results obtained earlier:

$$F^*(\omega)G(\omega) = \int_{-\infty}^{\infty} e^{-j\omega t} \left[\int_{-\infty}^{\infty} f^*[-(t-\tau)]g(\tau)d\tau \right] dt.$$

Taking the inverse Fourier transform gives

$$\frac{1}{2\pi}\int_{-\infty}^{\infty} F^*(\omega)G(\omega)e^{j\omega t}d\omega = \int_{-\infty}^{\infty} f^*(\tau-t)g(\tau)d\tau.$$

This expression represents a general form of Parseval's theorem. A more narrow form is given considering $t = 0$ and taking $f = g$, $F = G$:

$$\frac{1}{2\pi}\int_{-\infty}^{\infty} |F(\omega)|^2 d\omega = \int_{-\infty}^{\infty} |f(t)|^2 dt, \tag{1.12}$$

where t has replaced τ as the dummy variable on the right side of the last equation.

Parseval's theorem is an intuitive statement equating the energy in a signal in its time- and-frequency-domain formulations. It is a continuous generalization of (1.3). As $|f(t)|^2$ is the energy density per unit time or power, so $|F(\omega)|^2$ is the energy density per unit frequency, or the energy spectral density (ESD). The ESD is a convenient quantity to consider in situations involving filters, since the output ESD is equal to the input ESD times the frequency response function of the filter (modulus squared).

However, (1.12) as written here is problematic in the sense that the integrals as written are not guaranteed to exist. One normally differentiates at this point between so-called "energy signals," where the right side of (1.12) is finite and there are no existence issues, and more practical "power signals," where the average power over some time interval of interest T is instead finite:

$$P = \frac{1}{T}\int_{-T/2}^{T/2} |f(t)|^2 dt < \infty.$$

Parseval's theorem can be extended to apply to power signals with the definition of the power spectral density (PSD) $S(\omega)$, which is the power density per unit frequency. Let us write an expression for the power in the limiting case of $T \to \infty$ incorporating Parseval's theorem above:

$$\lim_{T\to\infty}\frac{1}{T}\int_{-T/2}^{T/2} |f(t)|^2 dt = \lim_{T\to\infty}\frac{1}{2\pi T}\int_{-\infty}^{\infty} |F(\omega)|^2 d\omega$$

$$= \frac{1}{2\pi}\int_{-\infty}^{\infty} S(\omega)d\omega$$

$$S(\omega) \equiv \lim_{T\to\infty}\frac{1}{T}|F(\omega)|^2. \tag{1.13}$$

This is the conventional definition of the PSD, the spectrum of a signal. In practice, an approximation is made where T is finite but long enough to incorporate all of the behavior of interest in the signal. If the signal is stationary, the choice of the particular time interval of analysis is arbitrary.

Sampling Theorem

Using a Fourier series, we saw in (1.1) how a continuous, periodic function can be expressed in terms of a discrete set of coefficients. If the function is also band limited, then those coefficients form a finite set of nonzero coefficients. Only these coefficients need be known to completely specify the function. The question then arises, can the function be expressed directly with a finite set of samples? This is tantamount to asking if the integral in (1.2) can be evaluated exactly using discrete mathematics. That the answer to both questions is "yes" is the foundation for the way data are acquired and processed.

Discrete sampling of a signal can be viewed as multiplication with a train of Dirac delta functions separated by an interval T. The sampling function in this case is

$$s(t) = T \sum_{n=-\infty}^{\infty} \delta(t - nT)$$

$$= \sum_{n=-\infty}^{\infty} e^{j2\pi nt/T},$$

where the second form is the equivalent Fourier series representation. Multiplying this by the sampled function $x(t)$ yields the samples $y(t)$,

$$y(t) = \sum_{n=-\infty}^{\infty} x(t) e^{j2\pi nt/T}$$

$$Y(\omega) = \sum_{n=-\infty}^{\infty} X\left(\omega - \frac{2\pi n}{T}\right),$$

where the second line this time is the frequency-domain representation (see Figure 1.3). Evidently, the spectrum of the sampled signal $|Y(\omega)|^2$ is an endless succession of replicas of the spectrum of the original signal, spaced by intervals of the sampling frequency $\omega_\circ = 2\pi/T$. The replicas can be removed easily by low-pass filtering. None of the original information is lost, so long as the replicas do not overlap or alias onto one another. Overlap can be avoided so long as the sampling frequency is as great or greater than the total bandwidth of the signal, a number bounded by twice the maximum frequency component. This requirement defines the Nyquist sampling frequency $\omega_s = 2|\omega|_{max}$. In practice, ideal low-pass filtering is impossible to implement, and oversampling at a rate above the Nyquist frequency is usually performed.

Discrete Fourier Transform

The sampling theorem implies a transformation between a discrete set of samples and a discrete set of Fourier amplitudes. The discrete Fourier transform (DFT) is defined by

$$F_m = \sum_{n=0}^{N-1} f_n e^{-j2\pi nm/N}, \quad m = 0, \cdots, N-1$$

1.3 Mathematical Preliminaries

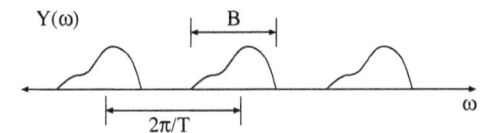

Figure 1.3 Illustration of the sampling theorem. The spectrum of the sampled signal is an endless succession of replicas of the original spectrum. To avoid frequency aliasing, the sampling theorem requires that $B \leq 2\pi/T$.

$$f_n = \frac{1}{N} \sum_{m=0}^{N-1} F_m e^{j2\pi nm/N}, \quad n = 0, \cdots, N-1,$$

where the sequences F_m and f_n can be viewed as frequency- and time-domain representations of a periodic, band-limited sequence with period N. The two expressions follow immediately from the Fourier series formalism outlined earlier in this section, given discrete time sampling. Expedient computational means of evaluating the DFT, notably the fast Fourier transform or FFT, have existed for decades and are at the heart of many signal processing and display numerical codes. Most often, N is made to be a power of 2, although this is not strictly necessary.

DFTs and FFTs in particular are in widespread use wherever spectral analysis is being performed. Offsetting their simplicity and efficiency is a serious pathology arising from the fact that time series in a great many real-world situations are not periodic. Power spectra computed using periodograms (algorithms for performing DFTs that assume the periodicity of the signal) suffer from artifacts as a result, and these may be severe and misleading. No choice of sample rate or sample length N can completely remedy the problem.

The jump between the first and last samples in the sequential time series will function as a discontinuity upon analysis with a DFT, and artificial ringing will appear in the spectrum as a result. A non-periodic signal can be viewed as being excised from longer, periodic signal by means of multiplication by a gate function. The Fourier transform of a gate function is a sinc function. Consequently, the DFT spectrum reflects something like the convolution of the true $F(\omega)$ with a sinc function, producing both ringing and broadening. Artificial sidelobes down about -13.5 dB from the main lobe will accompany features in the spectrum, and these may be mistaken for actual frequency components of the signal.

The most prevalent remedy for this so-called "spectral leakage" is windowing, whereby the sampled signal is weighted prior to Fourier analysis, an operation that de-emphasizes samples taken near the start and end of the sample interval. This has the effect of attenuating the apparent discontinuity in the signal. It can also be viewed as replacing the gate function mentioned above with another function with lower sidelobes. Reducing the sidelobe level comes at the expense of broadening the main lobe and degrading the frequency resolution. Various tradeoffs have to be weighed when selecting the spectral window. Popular choices include the Hann window, which has a sinusoidal shape, and the Hamming window, which is like the Hann window except with a pedestal.

1.3.3 Probability Theory

Probability is a concept that governs the outcomes of experiments that are random rather than deterministic. Noise in radar systems is an example of a random process and can only be described probabilistically. The unpredictability of noise is a consequence of the complexity of the myriad processes that contribute to it. Radar signals may reflect both random and deterministic processes, depending on the nature of the target, but the sum of signal plus noise is always a random process, and we need probability theory to understand it.

We should consider the operation of a radar as the performance of an experiment with outcomes that cannot be predicted because it is simply impossible to determine all of the conditions under which the experiment is performed. Such an experiment is called a random experiment. Although the outcomes cannot be predicted, they nevertheless exhibit statistical regularity that can be appreciated after the fact. If a large number of experiments are performed under similar conditions, a frequency distribution of all the outcomes can be constructed. Because of the statistical regularity of the experiments, the distribution will converge on a limit as the number of trials increases. We can assign a probability $P(A)$ to an outcome A as the relative frequency of occurrence of that outcome in the limit of an infinite number of trials. For any A, $0 \leq P(A) \leq 1$. This is the frequentist interpretation of statistics. Other interpretations of probability related to the plausibility of events as determined using information theory also exist, but are beyond the scope of this text.

Random Variables

Since the outcome of a radar experiment is finally a set of numbers, we are interested in random variables that label all the possible outcomes of a random experiment numerically. A random variable is a function that maps the outcomes of a random experiment to numbers. Particular numbers emerging from an experiment are called realizations of the random variables. Random variables come in discrete and continuous forms.

Probabilities are often stated in terms of cumulative distribution functions (CDFs). The CDF $F_x(a)$ of a random variable x is the probability that the random variable x takes on a value less than or equal to a, viz.

$$F_x(a) = P(x \leq a).$$

$F_x(a)$ is a non-decreasing function of its argument that takes on values between 0 and 1. Another important function is the probability density function (PDF) $f_x(x)$ of the random variable x, which is related to the CDF through integration/differentiation:

$$\begin{aligned} F_x(a) &= P(x \leq a) = \int_{-\infty}^{a} f_x(x)dx \\ \frac{dF_x(x)}{dx} &= f_x(x). \end{aligned}$$

The PDF is the probability of observing the variable x in the interval $(x, x+dx)$. Since a random variable always has some value, we must have

$$\int_{-\infty}^{\infty} f_x(x)dx = 1.$$

Note that, for continuous random variables, a PDF of 0 (1) for some outcome does not mean that that outcome is prohibited (or guaranteed). Either the PDF or the CDF completely determines the behavior of the random variable, and they are therefore equivalent.

Random variables that are normal or Gaussian occur frequently in many contexts and have the PDF

$$f_N(x) = \frac{1}{\sigma\sqrt{2\pi}} e^{-\frac{1}{2}(x-\mu)^2/\sigma^2},$$

which is parametrized by μ and σ. Normal distributions are often denoted by the shorthand $N(\mu,\sigma)$. Another common distribution is the Rayleigh distribution, given by

$$f_r(r) = \begin{cases} \frac{r}{\sigma^2} e^{-r^2/2\sigma^2}, & r \geq 0 \\ 0, & r < 0 \end{cases},$$

which is parametrized just by σ. It can be shown that a random variable defined as the length of a 2-D Cartesian vector with component lengths given by Gaussian random variables has this distribution. In terms of radar signals, the envelope of quadrature Gaussian noise is Rayleigh distributed. Still another important random variable called χ^2 has the PDF

$$f_{\chi^2} = \frac{1}{2^{\nu/2}\Gamma(\nu/2)} x^{\frac{1}{2}\nu-1} e^{-x/2},$$

where Γ is the Gamma function, $\Gamma(x) = \int_0^\infty \xi^{x-1} e^{-x i} d\xi$, and where the parameter ν is referred to as the number of degrees of freedom. It can be shown that if the random variable x_i is $N(0,1)$, then the random variable

$$z = \sum_{i=1}^n x_i^2$$

is χ^2 with n degrees of freedom. If x_i refers to voltage samples, for example, then z refers to power estimates based on n voltage samples or realizations.

Mean and Variance

It is often not the PDF that is of interest but rather just its moments. The expectation or mean of a random variable x, denoted $E(x)$ or μ_x, is given by

$$E(x) = \int_{-\infty}^{\infty} x f_x(x) dx.$$

For example, it can readily be shown that the expected value for a normal distribution $N(\mu,\sigma)$ is $E(x) = \mu$. The variance of a random variable, denoted $Var(x)$ or σ_x^2, is given by

$$\begin{aligned}Var(x) &= E[(x-\mu_x)^2] \\ &= E(x^2) - \mu_x^2 \\ &= \int_{-\infty}^{\infty} (x-\mu_x)^2 f_x(x)dx.\end{aligned}$$

The standard deviation of x, meanwhile, which is denoted σ_x, is the square root of the variance. In the case of the same normal distribution $N(\mu,\sigma)$, the variance is given by σ^2, and the standard deviation by σ.

It is important to distinguish between the mean and variance of a random variable and the sample mean and sample variance of the realizations of a random variable. The latter are experimental quantities constructed from a finite number of realizations of the random variable x. The latter will depart from the former due to the finite sample size and perhaps because of estimator biases. The deviations constitute experimental errors, and understanding them is a substantial part of the radar art.

Joint Probability Distributions

Two random variables x and y may be described by a joint probability density function (JPD) $f(x,y)$ such that

$$P(x \leq a \text{ and } y \leq b) = \int_{-\infty}^{a} \int_{-\infty}^{b} f(x,y) dy dy.$$

In the event the two random variables are independent, then $f(x,y) = f_x(x)f_y(y)$. More generally, we can define the covariance of x and y by

$$\begin{aligned}Cov(x,y) &= E[(x-\mu_x)(y-\mu_y)] \\ &= E(xy) - \mu_x\mu_y.\end{aligned}$$

If x and y are independent, their covariance will be 0. It is possible for x and y to be dependent and still have zero covariance. In either case, we say that the random variables are uncorrelated.

The correlation function is defined by

$$\rho(x,y) = \frac{Cov(x,y)}{\sqrt{Var(x)Var(y)}}.$$

The variance, covariance, and correlation function conform to the following properties (where a is a scalar and x and y are random variables):

1. $Var(x) \geq 0$
2. $Var(x+a) = Var(x)$
3. $Var(ax) = a^2 Var(x)$

4 $\text{Var}(x+y) = \text{Var}(x) + 2\text{Cov}(x,y) + \text{Var}(y)$
5 $\rho(x,y) \leq 1$

These ideas apply not only to different random variables but also to realizations of the same random variable at different times. In this case, we write of autocovariances and autocorrelation functions evaluated at some time lag (see below).

Conditional Probability

It is often necessary to determine the probability of an event, given that some other event has already occurred. The conditional probability of event A given event B having occurred is

$$P(A|B) = P(A \cap B)/P(B).$$

Since $P(A \cap B) = P(B \cap A)$, the equation could equally well be written with the order of A and B reversed. This leads immediately to Bayes' theorem, which gives a prescription for reversing the order of conditioning:

$$P(B|A) = \frac{P(A|B)P(B)}{P(A)}.$$

Given conditional probabilities, one can construct conditional PDFs $f_{x|y}(x)$, conditional CDFs $F_{x|y}(a)$, and also their moments.

Multivariate Normal Distribution

Consider now the joint probability distribution for a column vector $\mathbf{x}$ with n random variable components. If the components have a multivariate normal (MVN) distribution, their joint probability density function is

$$f(\mathbf{x}) = \frac{1}{(2\pi)^{n/2}} \frac{1}{\sqrt{\det(\mathbf{C})}} e^{-\frac{1}{2}(\mathbf{x}-\mathbf{u})^t \mathbf{C}^{-1}(\mathbf{x}-\mathbf{u})}, \tag{1.14}$$

where $\mathbf{u}$ is a column vector composed of the expectations of the random variables $\mathbf{x}$ and $\mathbf{C}$ is the covariance matrix for the random variables. This definition assumes that the covariance matrix is nonsingular.

Normally distributed (Gaussian) random variables possess a number of important properties that simplify their mathematical manipulation. One of these is that the mean vector and covariance matrix completely specify the joint probability distribution. This follows from the definition given above. In general, all of the moments (and not just the first two) are necessary to specify a distribution. Another property is that if $x_1, x_2, \cdots, x_n$ are uncorrelated, they are also independent. Independent random variables are always uncorrelated, but the reverse need not always be true. A third property of jointly Gaussian random variables is that the marginal densities $p_{x_i}(x_i)$ and the conditional densities $p_{x_i x_j}(x_i, x_j | x_k, x_l, \cdots, x_p)$ are also Gaussian.

A fourth property is that linear combinations of jointly Gaussian variables are also jointly Gaussian. As a consequence, the properties of the original variables specify the properties of any other random variable created by a linear operator acting on **x**. If $\mathbf{y} = \mathbf{A}\mathbf{x}$, then

$$E(\mathbf{y}) = \mathbf{A}\mu$$
$$\text{Cov}(\mathbf{y}) = \mathbf{A}\mathbf{C}\mathbf{A}^t.$$

Using these properties, it is straightforward to generate vectors of random variables with arbitrary means and covariances beginning with random variables that are MVN.

Central Limit Theorem

The central role of MVN random variables in a great many important processes in nature is borne out by the central limit theorem, which governs random variables that hinge on a large number of individual, independent events. Underlying noise, for example, is a random variable that is the sum of a vast number of other random variables, each associated with some minuscule, random event. Signals can reflect MVN random variables as well, for example, signals from radar pulses scattered from innumerable, random fluctuations in the index of refraction of a disturbed air volume.

Let x_i, $i = 1, n$ be independent, identically distributed random variables with common, finite μ and σ. Let

$$z_n = \frac{\sum_{i=1}^n x_i - n\mu}{\sqrt{n}\sigma}.$$

The central limit theorem states that, in the limit of n going to infinity, z_n approaches the MVN distribution. Notice that this result does not depend on the particular PDF of x_i. Regardless of the physical mechanism at work and the statistical distribution of the underlying random variables, the overall random variable will be normally distributed. This accounts for the widespread appearance of MVN statistics in different applications in science and engineering.

Rather than proving the central limit theorem, we will just argue its plausibility. Suppose a random variable is the sum of two other random variables, $z = x + y$. We can define a CDF for z in terms of a JDF for x and y as

$$\begin{aligned}
F_z(a) = P(z \leq a) &= P(x \leq \infty, y \leq a - x) \\
&= \int_{-\infty}^{\infty} \int_{-\infty}^{a-x} f_{xy}(x,y) dx dy \\
&= \int_{-\infty}^{\infty} dx \int_{-\infty}^{a-x} f_{xy}(x,y) dy.
\end{aligned}$$

The corresponding PDF for z is found through differentiation of the CDF:

$$f_z(z) = \frac{dF_z(z)}{dz} = \int_{-\infty}^{\infty} f_{xy}(x, z-x) dx.$$

1.3 Mathematical Preliminaries

Now, if x and y are independent, $f_{xy}(x, z-x) = f_x(x)f_y(z-x)$, and

$$f_z(z) = \int_{-\infty}^{\infty} f_x(x)f_y(x, z-x)dx,$$

which is the convolution of the individual PDF. This result can be extended for z being the sum of any number of PDFs, for which the total PDF is the convolution of all the individual PDFs.

Regardless of the shapes of the individual PDFs, the tendency upon multiple convolutions is to tend toward a net PDF with a Gaussian shape. This can easily be verified by successive convolution of uniform PDFs, for example, which have the form of gate functions. By the third of fourth convolution, it is already difficult to distinguish the result from a Bell curve.

Random Processes

A random or stochastic process is an extension of a random variable, the former being the latter but also being a function of an independent variable, usually time or space. The outcome of a random process is therefore not just a single value but rather a curve (a waveform). The randomness in a random process refers to the uncertainty regarding which curve will be realized. The collection of all possible waveforms is called the ensemble, and a member of the ensemble is a sample waveform.

Consider a random process that is a function of time. For each time, a PDF for the random variable can be determined from the frequency distribution. The PDF for a time t is denoted $f_x(x;t)$, emphasizing that the function can be different at different times. However, the PDF $f_x(x;t)$ does not completely specify the random process, lacking information about how the random variables representing different times are correlated. That information is contained in the autocorrelation function, itself derivable from the JDF. In the case of a random process with zero mean and unity variance, we have

$$\rho(t_1, t_2) = E[x(t_1)x(t_2)] = E(x_1, x_2)$$
$$= \int_{-\infty}^{\infty}\int_{-\infty}^{\infty} x_1 x_2 f_{x_1, x_2}(x_1, x_2) dx_1 dx_2.$$

(For a complex random process, the autocorrelation function is defined as $E(x_1^*, x_2)$.) Consequently, not only the PDF but also the second-order or JDF must be known to specify a random process. In fact, the nth order PDF $f_{x1,x2,\cdots,x_n}(x_1, x_2, \cdots, x_n; t_1, t_2, \cdots, t_n)$ must be known to completely specify the random process with n sample times. Fortunately, this near-impossibility is avoided when dealing with Gaussian random processes and linear systems, for which it is sufficient to deal with first- and second-order statistics only. Furthermore, lower-order PDFs are always derivable from a higher-order PDF, e.g.,

$$f_{x_1}(x_1) = \int_{-\infty}^{\infty} f_{x_1, x_2}(x_1, x_2) dx_2,$$

so one need only specify the second-order statistics to completely specify a Gaussian random process.

Among random processes is a class for which the statistics are invariant under a time shift. In other words, there is no preferred time for these processes, and the statistics remain the same regardless of where the origin of the time axis is located. Such processes are said to be stationary. True stationarity requires invariance of the nth order PDF. In practice, this can be very difficult to ascertain. Stationarity is often asserted not on the basis of the sample but instead in terms of the known properties of the generating mechanism.

A less restrictive requirement is satisfied by processes that are wide-sense or weakly stationary. Wide-sense stationarity requires only that the PDF be invariant under temporal translation through second order. For such processes, expectations are constant, and the autocorrelation function is a function of lag only, i.e., $\rho(t_1,t_2) = \rho(t_2 - t_1)$. Even this less restrictive characterization of a random variable is an idealization and cannot be met by a process that occurs over a finite span of times. In practice, the term is meant to apply to a process over some finite time span of interest.

Demonstrating stationarity or even wide-sense stationarity may be difficult in practice if the PDF is not known *a priori*, since one generally does not have access to the ensemble and so cannot perform the ensemble average necessary to compute, for example, the correlation function. However, stationary random processes of a special class, called ergodic processes, lend themselves to experimental interrogation. For these processes, all of the sample waveforms in the ensemble are realized over the course of time and an ensemble average can therefore be replicated by a time average (sample average):

$$E(x(t)) \equiv \lim_{T \to \infty} \frac{1}{T} \int_{-T/2}^{T/2} x(t)dt$$

$$\rho(\tau) \equiv \lim_{T \to \infty} \frac{1}{T} \int_{-T/2}^{T/2} x(t)x(t+\tau)dt.$$

An ergodic process obviously cannot depend on time and therefore must be stationary.

Gaussian Random Processes

The properties of random variables were stated in Section 1.5.5. A Gaussian random process is one for which the random variables $x(t_1)$, $x(t_2)$, $\cdots$, $x(t_n)$ are jointly Gaussian according to (1.14) for every n and for every set $(t_1, t_2, \cdots, t_n)$. All of the simplifying properties of jointly normal random variables therefore apply to Gaussian random processes. In particular, since the complete statistics of Gaussian processes are determined by their second-order statistics (autocorrelation functions), a Gaussian random process that is wide-sense stationary is automatically stationary in the strict sense. Moreover, the response of any linear system to a Gaussian random process will also be a Gaussian random process. This fact is crucial for evaluating the propagation of stochastic signals and noise through the components of communication and radar systems.

Power Spectral Density of a Random Process

Communication and radar engineers are often concerned with the frequency content of a signal. If the signal in question is a stationary random variable, at least in the wide sense, then it must exist over an infinite time domain and therefore would have to be a power signal. But what is the PSD for such a signal? The sample functions of the process that occupy the ensemble are potentially all different, with different power spectral densities, and no one of them is preferred over another.

The obvious definition of the PSD of a stationary random process is the ensemble average of the power spectral densities for all the members in the ensemble (compare with (1.13)),

$$S_x(\omega) = \lim_{T \to \infty} \frac{\overline{|X_T(\omega)|^2}}{T},$$

where the bar denotes an ensemble average and $X_T(\omega)$ is the Fourier transform of the truncated random process $x(t)\Pi(t/T)$:

$$X_T(\omega) = \int_{-T/2}^{T/2} x(t) e^{-j\omega t} dt.$$

For a complex process, we have

$$\begin{aligned}|X_T(\omega)|^2 &= X_T^*(\omega) X_T(\omega) \\ &= \int_{-T/2}^{T/2} x^*(t_1) e^{j\omega t_1} dt_1 \int_{-T/2}^{T/2} x(t_2) e^{-j\omega t_2} dt_2 \\ &= \int_{-T/2}^{T/2} \int_{-T/2}^{T/2} x^*(t_1) x(t_2) e^{-j\omega(t_2 - t_1)} dt_1 dt_2,\end{aligned}$$

so that

$$S_x(\omega) = \lim_{T \to \infty} \frac{1}{T} \overline{\int_{-T/2}^{T/2} \int_{-T/2}^{T/2} x^*(t_1) x(t_2) e^{-j\omega(t_2 - t_1)} dt_1 dt_2}.$$

The ensemble averaging and integrating operations may be interchanged, yielding

$$S_x(\omega) = \lim_{T \to \infty} \frac{1}{T} \int_{-T/2}^{T/2} \int_{-T/2}^{T/2} \rho_x(t_2 - t_1) e^{-j\omega(t_2 - t_1)} dt_1 dt_2,$$

where the general complex form of the autocorrelation function $\rho_x(\tau)$ has been used with the implicit assumption of wide-sense stationarity, so that the particular time interval span in question is irrelevant.

The integral can be evaluated using a change of variables, with $\tau \equiv t_2 - t_1$ and $\tau' = (t_2 + t_1)/2$. The corresponding region of integration will be diamond shaped, with only the τ coordinate appearing within the integrand:

$$S_x(\omega) = \lim_{T \to \infty} \frac{1}{T} \int_{-T}^{T} \int_{-|\tau|/2}^{|\tau|/2} \rho_x(\tau) e^{-j\omega\tau} d\tau' d\tau$$
$$= \lim_{T \to \infty} \int_{-T}^{T} \rho_x(\tau)(1 - |\tau|/T) e^{-j\omega\tau} d\tau$$
$$= \int_{-\infty}^{\infty} \rho_x(\tau) e^{-j\omega\tau} d\tau.$$

Consequently, we find that the PSD for a wide-sense stationary random process is the Fourier transform of its autocorrelation. This is known as the Wiener–Khinchine relation. It illustrates the fundamental equivalence of the PSD and the autocorrelation function for characterizing wide-sense stationary random processes.

Even-Moment Theorem for GRVs

There is a very useful theorem that applies to the even moments of Gaussian random variables. According to this theorem, any $2n$th moment of a set of GRVs can be expressed in terms of all the permutations of the nth moments. This is best illustrated by example. Consider the fourth moment $\langle x_1 x_2 x_3 x_4 \rangle$. According to the theorem, we may write

$$\langle x_1 x_2 x_3 x_4 \rangle \equiv \langle x_1 x_2 \rangle \langle x_3 x_4 \rangle + \langle x_1 x_3 \rangle \langle x_2 x_4 \rangle + \langle x_1 x_4 \rangle \langle x_2 x_3 \rangle.$$

The theorem holds for complex GRVs such as are represented by the outputs of quadrature radio receivers, and is especially useful for predicting the mean and variance of quantities derived from them, like the signal power and autocorrelation function.

In the next chapter, we begin the main material of the text, starting with the basic tenets of antenna theory. More complicated aspects of the theory emerge in subsequent chapters. Material from this introductory chapter will mainly be useful in the second half of the text, which covers noise, signals, and radar signal processing.

1.4 Notes and Further Reading

An enormous amount has been written about the history of radar as told from varying points of view. Notable are the histories by R. C. Watson and R. Buderi. First-person accounts from M. A. Tuve and A. Wilkins of some of the earliest research help establish the historical timeline. Contributions of the US Naval Research Laboratory to radar and other facets of ionospheric research have been described more recently in a review paper by H. Friedman. Activities at the Radiation Laboratory are documented in the review by D. K. van Keuren.

Basic signal processing and probability theory are two more well-trodden fields of applied mathematics and engineering. The textbook by B. P. Lathi offers an excellent introduction to the first of these, and the text by A. Papoulis a superb review of the second. For an overview of radar, readers have many choices, but should certainly consider the well-known and insightful introductory texts by M. I. Skolnik, N. L. Levanon, and B. R. Mahafza.

E. V. Appleton and M. A. F. Barnett. Local reflection of wireless waves from the upper atmosphere. *Nature*, 115:333–334, 1925.

G. Breit and M. A. Tuve. A radio method for estimating the height of the conducting layer. *Nature*, 116:357, 1925.

A test of the existence of the conducting layer. *Phys. Rev.*, 28:554, 1926.

R. Buderi. *The Invention that Changed the World*. Simon and Schuster, New York, 1996.

L. de Forest. Absorption (?) of undamped waves. *Electrician (letter to the editor)*, 69:369–370, 1912.

Recent developments in the work of the Federal Telegraph Company. *Proc. IRE*, 1:37–57, 1913.

H. Friedman. From ionosonde to rocket sonde. *J. Geophys. Res.*, 99(A10):19143–19153, 1994.

D. K. van Keuren. Science goes to war: The Radiation Laboratory, radar, and their technological consequences. *Rev. American Hist.*, 25:643–647, 1997.

C. Latham and A. Stobbs. *The Birth of British Radar: The memoirs of Arnold 'Skip' Wilkins*, 2nd ed. Radio Society of Great Britain, Reading, 2011.

B. P. Lathi. *Modern Digital and Analog Communication Systems*. Holt, Rinehart and Winston, New York, 1983.

N. L. Levanon. *Radar Principles*. Wiley, New York, 1988.

B. R. Mahafza. *Introduction to Radar Analysis*. CRC Press, New York, 1998.

A. Papoulis. *Probability, Random Variables, and Stochastic Processes*. McGraw-Hill, New York, 1984.

M. I. Skolnik. *Introduction to Radar Systems*, 3rd ed. McGraw-Hill, New York, 2001.

M. A. Tuve. Early days of pulse radio at the Carnegie Institution. *J. Atmos. Terr. Phys.*, 36:2079–2083, 1974.

R. C. Watson, Jr. *Radar Origins Worldwide: History of Its Evolution in 13 Nations Through World War II*. Trafford Publishing, Victoria, BC, Canada, 2009.

Exercises

1.1 For the electric fields specified below,

$$\mathbf{E}_1 = E_\circ(2j\hat{\mathbf{x}} + \hat{\mathbf{y}})e^{j(\omega t - kz)}$$

$$\mathbf{E}_2 = E_\circ(\hat{\mathbf{x}} + e^{j\pi/4}\hat{\mathbf{y}})e^{j(\omega t - kz)}$$

Find $|\mathbf{E}_1|^2$, $|\mathbf{E}_2|^2$, $\mathbf{E}_1 \cdot \mathbf{E}_2^*$, and the average power density associated with $\mathbf{E}_2$ if this is the electric field of an electromagnetic wave propagating in vacuum in the z direction. Also, sketch or plot $\Re[\mathbf{E}_2(t, z=0)]$.

1.2 Suppose a function $f(x)$ is defined as a trapezoid, increasing linearly from 0 to 1 in the interval $[0,a]$, keeping the value of 1 in the interval $[a,b]$, and decreasing back to 0 linearly in the interval $[b,c]$. Calculate $\langle f(x)\rangle^2$ and $\langle f(x)^2\rangle$ over the interval [0,c]. What is the ratio $\langle f(x)\rangle^2/\langle f(x)^2\rangle$, and for what values of a and b is it a maximum? Interpret the results.

1.3 Launched in 1977, Voyager I is the most distant manufactured object, at over 20 billion km distance. Consider designing a data link to communicate with the spacecraft. Assume that the spacecraft has a 20-watt transmitter operating at 10 GHz and driving a 3.7-meter diameter parabolic antenna with a 55 percent aperture efficiency. (This figure, which is the ratio of the effective area of an antenna to its physical area, is typical for practical parabolic antennas.) Assume also that the noise temperature of your receiving system is 30 K. As we will see later, this means that the noise power into the receiver is $N = KTB$ watts, where K is Boltzmann's constant (1.38×10^{-23} Joules/Kelvin), T is the noise temperature in Kelvins, and B is the receiver bandwidth in Hz. Design a parabolic receiving antenna (i.e., determine its size) that will allow you to receive data on Earth over a 46 Hz bandwidth link with a 10 dB signal-to-noise ratio (implying a 160 bit-per-second data rate, as we will also see later). As the figures above are accurate ones, the antenna you design should have a reasonable size.

2
Introduction to Antenna Theory

This chapter discusses simple wire antennas based on electric and magnetic dipoles. Starting from Maxwell's equations, the electromagnetic fields due to an ideal current element are calculated. The ideal current element then serves as a building block for more complicated antennas. A variety of practical wire antennas will be analyzed and evaluated. None of these will turn out by themselves to have very high gain, however. Methods of achieving the high gain required by radars will be taken up in Chapters 3 and 4. While much of the analysis concerns using antennas for transmission, the reciprocity theorem will provide a recipe for evaluating performance for reception.

2.1 Hertzian Dipole

All time-varying currents radiate as antennas, and the simplest radiating system is an ideal current element or a Hertzian electric dipole. A Hertzian dipole is a uniform current density **J** flowing in an infinitesimal volume with cross-section A and differential length dl. The current is assumed to have a sinusoidal time variation given by the implicit $\exp(j\omega t)$ factor, and so applying phasor notation is natural. The differential length of the element is regarded as being very small compared to a wavelength. We also regard the current density as being distributed uniformly in space in the infinitesimal volume. This is an idealization; one does not set out to construct an ideal dipole, although larger antennas can be regarded as being assembled from them. It is expedient to use a combination of spherical and rectangular coordinates and to alternate between them as needed. Be aware that the unit vectors associated with spherical coordinates vary with position.

Calculating the fields arising from a Hertzian dipole is a fundamental physics problem addressed by most textbooks on antennas and electricity and magnetism. An abbreviated, intuitive derivation is provided below. Readers interested in more details should consult the Appendix B or one of the titles under Notes and Additional Reading at the end of this chapter.

2.1.1 Vector Potential: Phasor Form

Maxwell's equations for electromagnetic fields in their native form are not amenable to solution by familiar mathematical methods, and it is expedient to reformulate them in

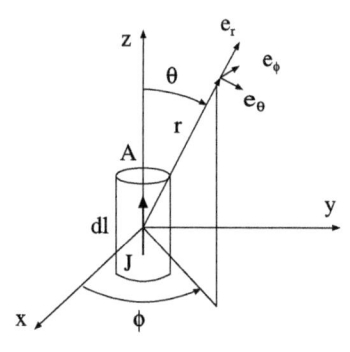

Figure 2.1 Diagram of a Hertzian dipole antenna.

terms of vector and scalar potentials along with a *gauge* or condition that completes their definitions. The magnetic vector potential **A**, scalar potential ϕ, and the Lorenz condition relating them are specified by (see Appendix B)

$$\mathbf{B} = \nabla \times \mathbf{A}$$

$$\mathbf{E} = -\nabla \phi - \frac{\partial \mathbf{A}}{\partial t}$$

$$\nabla \cdot \mathbf{A} = -\frac{1}{c^2} \frac{\partial \phi}{\partial t}.$$

Working with potentials rather than electric and magnetic fields allows Maxwell's equations in a vacuum to be recast into the following second-order partial differential equations (in phasor notation):

$$\nabla^2 \mathbf{A} + k^2 \mathbf{A} = -\mu_\circ \mathbf{J}$$

$$\nabla^2 \phi + k^2 \phi = -\rho/\varepsilon_\circ,$$

where the explicit time dependence is bound in k, the free-space wavenumber:

$$k = \omega/c$$
$$= \omega \sqrt{\mu_\circ \varepsilon_\circ}$$
$$= 2\pi/\lambda.$$

The potential equations are coupled by the fact that the current density **J** and charge density ρ obey the continuity equation $\nabla \cdot \mathbf{J} + j\omega\rho = 0$. They are examples of the inhomogeneous Helmholtz equation, which has well-known solutions.

Notice that the equation for the vector potential separates into three scalar equations for the three vector components of **A** and **J** in Cartesian coordinates. In the case of the Hertzian dipole where the current flows in one Cartesian direction, it is sufficient, therefore, to solve one scalar equation:

$$\nabla^2 A_z + k^2 A_z = -\mu_\circ J_z. \tag{2.1}$$

2.1 Hertzian Dipole

We view the right side of (2.1) as a source and the left side as a response (see Fig. 2.1). The general method of solution involves finding the response to a point source excitation, called a Green's function, and then constructing the solution to a particular problem from an appropriate superposition of point sources. In this case, the excitation is already a point source, and the Green's function is itself the solution we seek.

We first solve the homogeneous (source-free) problem. The overall symmetry of the problem suggests that the solution should depend only on the variable r. In spherical coordinates, this implies the following ordinary differential equation:

$$\frac{1}{r^2}\frac{d}{dr}r^2\frac{dA_z}{dr} + k^2 A_z = 0.$$

The solution is facilitated with the change of variables given by $A_z = \psi/r$, yielding

$$\frac{d^2\psi}{dr^2} + k^2\psi = 0,$$

with the general solution

$$\psi(r) = c_1 e^{-jkr} + c_2 e^{jkr}.$$

The two solution components represent spherically expanding and contracting wavefronts, and while both are mathematically valid, only the former is physically meaningful, given a source at the origin. The solution to the homogeneous problem, which holds everywhere except the origin where the Hertzian dipole is located, is then

$$A_z = c_1 \frac{e^{-jkr}}{r}.$$

All that remains is to determine the constant c_1. This is done by reconsidering the inhomogeneous equation (with the source restored). Integrating the equation over a small spherical volume surrounding the origin with a radius $r_\circ$ yields

$$\oint_S \nabla A_z \cdot \hat{r}\, ds + \int_v k^2 A_z\, dv = -\mu_\circ I dl,$$

where the divergence theorem has been used to convert one volume integral to a surface integral over the sphere and where the volume integral over the totally enclosed current element is Idl. By considering the limit where the radius $r_\circ$ is made arbitrarily small, it is possible to make the term immediately to the left of the equal sign vanish, since the product $A_z dV \sim r_\circ^2$. The surface integral term, meanwhile, evaluates to $-4\pi c_1$ in the limit that $r_\circ \ll \lambda$. The constant is thereby set, and the solution for the vector potential due to a Hertzian dipole at the origin is found to be

$$A_z = \mu_\circ I dl \frac{e^{-jkr}}{4\pi r}. \tag{2.2}$$

Recall finally that the $\exp(j\omega t)$ time dependence is implicit.

2.1.2 Electromagnetic Fields

The electric and magnetic fields can both be found from the vector potential, since $\mathbf{H} = \nabla \times \mathbf{A}/\mu_\circ$ and $\mathbf{E} = -j\omega\mathbf{A} + \nabla\nabla\cdot\mathbf{A}/j\omega\mu_\circ\varepsilon_\circ$ (invoking the potential definitions and the Lorenz condition in phasor form – see the Appendix B for details). A little more calculation gives

$$H_\phi = Idl\sin\theta\frac{1}{4\pi}\left[\frac{jk}{r} + \frac{1}{r^2}\right]e^{j(\omega t - kr)}$$

$$E_r = Idl\cos\theta\frac{Z_\circ}{2\pi k}\left[\frac{k}{r^2} + \frac{1}{jr^3}\right]e^{j(\omega t - kr)}$$

$$E_\theta = Idl\sin\theta\frac{jZ_\circ}{4\pi k}\left[\frac{k^2}{r} - \frac{jk}{r^2} - \frac{1}{r^3}\right]e^{j(\omega t - kr)},$$

where $Z_\circ$ is called the impedance of free space and has a numerical value of $\sqrt{\mu_\circ/\varepsilon_\circ}$ or about $120\pi \approx 377$ Ohms in MKS units.

The terms that vary as r^{-1} are known as the *radiative* terms. These terms alone carry power away from the source on average. The radiative fields dominate at distances much greater than a wavelength ($r \gg \lambda$). Here, the electric and magnetic fields are normal to one another and to the (radial) direction of wave propagation, E and H are in phase, and their ratio is $E_\theta/H_\phi = Z_\circ$.

Power flow is represented by the *Poynting vector*

$$\mathbf{P} \equiv \mathbf{E} \times \mathbf{H}.$$

This is the instantaneous definition, not yet written in phasor notation. The Poynting vector has units of power per unit area and represents the power density carried by electromagnetic radiation along with the direction. In phasor notation, the time-averaged Poynting vector is given by

$$\langle\mathbf{P}\rangle = \frac{1}{2}\Re(\mathbf{E} \times \mathbf{H}^*).$$

The average power density propagating away from an antenna centered at the origin is given by the dot product of the Poynting vector with the unit radial direction, $\hat{\mathbf{r}}$. For the Hertzian dipole, this is

$$\langle\mathbf{P}\rangle\cdot\hat{\mathbf{r}} = P_r = |I|^2(dl)^2 Z_\circ k^2 \sin^2\theta \frac{1}{32\pi^2 r^2} \; (W/m^2). \tag{2.3}$$

Only the radiative fields contribute to this result, although the derivation is completely general. Terms that decay faster that r^{-1} with radial distance are called the *reactive* terms and represent energy stored in the electric and magnetic fields in the vicinity of the dipole. The storage fields reconfigure themselves over the course of a wave period, and instantaneous power flow is associated with the reconfiguration. The reactive fields do not contribute to the time-averaged flow of power away from the antenna, however. This must be the case; power density must decay as r^{-2} (inverse square law) in the Hertzian dipole problem since the area of the spherically expanding wavefronts that carry it increases as r^2. Only the radiation field terms combine to give an expression that obeys the inverse square law.

2.1.3 Radiation Pattern

What distinguishes one antenna from another is the distribution of power radiated into different directions or bearings. There is no such thing as an antenna that radiates isotropically (equally in all directions) at any instant in time. The radiation pattern of an antenna is a diagram (usually in polar coordinates) showing where the RF power is concentrated. Radiation patterns are sometimes represented in 3-D renderings or, more commonly, using planar cuts. By convention, radiation patterns are normalized to a maximum value of unity. They may be drawn with linear or logarithmic axes, although care must be taken with scaling in the latter case to avoid problems near the nulls in the pattern.

The average power density (2.3) for the Hertzian dipole can be expressed as

$$P_r(\theta,\phi) = \frac{Z_\circ}{8}\left(\frac{Idl\sin\theta}{r\lambda}\right)^2 \; (W/m^2). \tag{2.4}$$
$$\propto \sin^2\theta$$

Cuts through $P_r(\theta,\phi)$ are plotted in Figure 2.2. The two cuts are taken through the E and H planes, respectively, the planes in which the electric and magnetic lines of force lie.

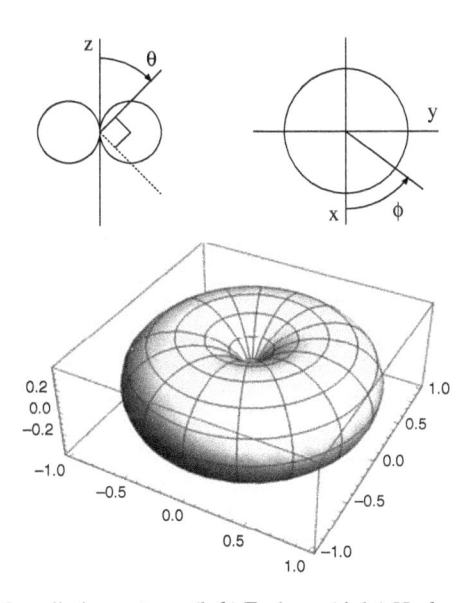

Figure 2.2 Hertzian dipole radiation pattern: (left) E-plane, (right) H-plane, (bottom) 3-D rendering. The overall pattern is a torus.

The radial distance from the origin to the surface that defines the pattern is proportional to $\sin^2(\theta)$ in either case. The patterns have been plotted here using linear scaling.

Bulbs in the radiation pattern are referred to as beams or lobes, whereas zeros are called nulls. Some lobes have larger peak amplitudes than others. The lobe with the greatest amplitude is called the main lobe, and smaller ones are called sidelobes. If there are more main lobes than one, the pattern is said to have grating lobes. The nomenclature can be ambiguous at times. In this example, there is essentially a single lobe in the H-plane that wraps around all azimuths.

Several measures of the width of the lobes in the patterns are in common use. These include the half-power full beamwidth (HPFB or HPBW), which gives the range of angles through which the radiation pattern is at least half its maximum value in a specified plane. The HPBW is found by equating $\sin^2 \theta$ with 1/2, yielding $\theta = 45°, 135°$ in this case. The difference between these limits is the HPBW, which is exactly 90° here. Note the importance of correct normalization in this calculation. Discussing the HPBW in the H-plane is not meaningful in the case of a dipole.

Another measure is the beamwidth between first nulls (BWFN), which is usually about twice the HBPW. (The easiest way to estimate the latter may be to calculate the former and divide by two.) Since the Hertzian dipole radiation pattern has nulls at $\theta = 0°$ and $180°$, the BWFN is 180° and exactly twice the HPBW in this case.

While the widths of planar cuts through an antenna beam can be described by angles, the overall beam width can be described by the solid angle (see Fig. 2.3). Solid angles are to conical shapes in a volume what angles are to sectors in a plane. Just as the arc length swept out on a circle of radius r by a sector with a differential angle $d\theta$ is $r\,d\theta$, the differential area swept out on a sphere of radius r by a cone with differential solid angle Ω is $dA = r^2 d\Omega$. Thus, we define $d\Omega = dA/r^2$. In spherical coordinates, we can readily see that $d\Omega = \sin\theta\, d\theta\, d\phi$. The units of solid angle are steradians (Str.) The total solid angle enclosed by any closed surface is 4π Str. Narrow antenna beams have solid angles much smaller than 4π Str.

A radiation pattern can be characterized by its solid angle. For example, if the pattern is conical, uniform radiation emitted within the cone and no radiation outside, the solid angle of the beam would be the solid angle of the cone. A formal definition of the solid angle of an arbitrary beam shape will be given below after the related concept of directivity has been introduced.

There are a number of approximations that are used to relate the solid angle of a radiation patter to its half-power beamwidth. If the pattern has a rectangular shape, as is the case for many phased-array antennas, then the area illuminated by the antenna at range r can be written in terms of the product of the half-power beamwidths in the E and H planes, i.e.,

$$r^2 \Omega \approx (r \cdot \text{HPBW}_E)(r \cdot \text{HPBW}_H)$$
$$\Omega \approx \text{HPBW}_E \cdot \text{HPBW}_H, \qquad (2.5)$$

where the half-power beamwidths are measured in radians. If the beam shape is instead circular, then a better approximation for the relationship would be

2.1 Hertzian Dipole

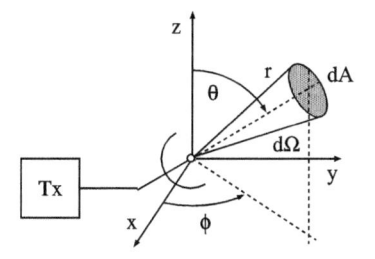

Figure 2.3 Antenna radiation pattern characterized by its solid angle $d\Omega$.

$$r^2\Omega \approx \pi(r \cdot \text{HPBW}/2)^2$$
$$\Omega \approx (\pi/4)(\text{HPBW})^2, \quad (2.6)$$

which is slightly smaller by virtue of the removal of the corners of the radiation pattern.

A specific subcase of a beam with a circular shape is is a Gaussian beam, i.e., $P_r \propto \exp(-\theta^2/2\sigma^2)$, where θ is the polar angle measured from the antenna boresight and σ describes the width of the pattern. Such a function is often used as a convenient proxy for a more complicated beam shape. In terms of this parameter, the half-power full beamwidth of this pattern is $\text{HPBW} = \sqrt{8\ln 2}\sigma \approx 2.35\sigma$. Below, we will see that the exact value for the solid angle of a Gaussian beam is $[\pi/4\ln(2)](\text{HPBW})^2$. This is somewhat larger than the approximation given immediately above, reflecting the gradually sloping sides of the Gaussian function.

Note that none of these approximations can be applied in the case of the ideal dipole since the HPBW in the H-plane is not defined. The directivity and the beam solid angle can instead be calculated exactly using definitions provided below.

2.1.4 Total Radiated Power

We calculate the total power radiated by the antenna by integrating the Poynting flux through any closed surface surrounding the antenna, a sphere or arbitrary radius being the natural choice. The computation frequently necessitates either integral tables and variable transformations or numerical computation. The Hertzian dipole case is elementary, however:

$$\begin{aligned}
P_t &= \oint_S P_r\, dS \\
&= \oint_S P_r r^2\, d\Omega \\
&= \underbrace{\int_0^{2\pi} d\phi}_{2\pi} \int_0^{\pi} d\theta \frac{Z_\circ}{8}\left(\frac{Idl}{r\lambda}\right)^2 r^2 \sin^3\theta \\
&= \frac{2\pi Z_\circ}{8}\left(\frac{Idl}{\lambda}\right)^2 \int_0^{\pi} \sin^3\theta\, d\theta.
\end{aligned}$$

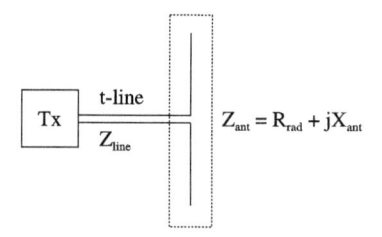

Figure 2.4 Antenna viewed as circuit element with complex load impedance.

Notice that the result does not depend on r, as it must not. We frequently must calculate definite integrals of powers of trigonometric functions. In this case, $\int \sin^3\theta d\theta = \int \sin\theta(1-\cos^2\theta)d\theta = -[\cos\theta - \cos^3\theta/3]_0^\pi = 4/3$. Finally,

$$P_t = Z_\circ \frac{\pi}{3}\left(\frac{Idl}{\lambda}\right)^2 (W),$$

where the reader is reminded that $dl \ll \lambda$ by definition.

2.1.5 Radiation Resistance

Since an antenna dissipates power, it must have an associated resistance. (Antennas are also reactive in general, although one generally tries to design purely resistive antennas in practice so as to facilitate impedance matching with feedlines – see Fig. 2.4). We associate the *radiation resistance* R of an antenna with the average flow of power into the antenna. For many practical antennas, this resistance is typically tens or hundreds of Ohms. Radiation resistance is usually unrelated to the Ohmic resistance associated with the conductors from which the antenna is constructed. In most (but not all) circumstances, Ohmic losses can be neglected. This is tantamount to equating gain with directivity.

Equating the total radiated power for this antenna with the average power dissipated by an equivalent resistive circuit element according to Ohm's law yields

$$P_t = I_{rms}^2 R_{rad} = \frac{1}{2}|I|^2 R_{rad}$$
$$R_{rad} = 80\pi^2 (dl/\lambda)^2 (Ohms).$$

Whereas the leading coefficient in this expression is of the order of 1000, the ratio (dl/λ) is again small by definition, and the radiation resistance of a Hertzian dipole is apt to be very small. Consider an AM car radio antenna operating at a frequency of ~ 1 MHz, for example. Here, the wavelength $\lambda \sim 300$ m, whereas the antenna length $dl \sim 1$m. In this case, the radiation resistance is predicted to be $R_{rad} \sim 0.01\Omega$, a virtual short circuit. Such electrically short antennas pose practical problems for impedance matching and are likely to be inefficient. Radio engineering at very low frequencies (VLF and ULF), where antennas are necessarily short compared to the wavelength, is consequently very challenging, and exotic solutions have been proposed, including the use of superconducting wires, long wires trailed from aircraft, and even modification of the conductivity of the ionosphere in such a way that it can function as an enormous antenna.

Returning to the AM car radio example, despite the poor efficiency and impedance matching problems, an electrically short antenna provides acceptable performance in this application. What ultimately matters in a communications (or radar) link is the signal-to-noise ratio rather than the absolute signal power, since the latter can always be boosted with amplification. Antenna inefficiency and mismatches attenuate the signal and atmospheric or sky noise equally, and so long as the noise ultimately delivered by the antenna to the receiver is large compared to the noise generated within the receiver, the link is not compromised. Background sky noise at AM radio frequencies is very intense, so antenna performance is not a critical issue. Note that this reasoning does not apply to antennas used for transmission, however! Transmitting antennas must be efficient and readily matched and are generally physically large ($\lambda/4$ or longer).

2.1.6 Directivity

Another measure of the degree of confinement is the directivity. The directivity, $D(\theta, \phi)$, is defined as the power density in some direction, normalized to the average power density in all directions,

$$\begin{aligned} D(\theta, \phi) &\equiv \frac{P_r(\theta, \phi)|_r}{\langle P_r \rangle} \\ &= \frac{P_r(\theta, \phi)|_r}{P_t/4\pi r^2}, \end{aligned}$$

where r is the range at which the directivity is evaluated.

In Chapter 4 of the text, we will see that the gross performance of an antenna is different at large distances (in the so-called "far field") than it is close by (in the near field). In the far field, all the rays connecting a point in space to different points on the antenna may be regarded as being parallel for purposes of determining distances, measured in wavelengths. This approximation does not hold in the near field. The separatrix between the near and far fields depends on the physical size of the antenna and on the wavelength. (A rule of thumb places the separatrix at a distance $r = 2d^2/\lambda$, where d is the antenna size and λ is the wavelength.) In the majority of practical radar applications, the object being observed is comfortably in the far field. In the case of the Hertzian dipole antenna, there is no near field since the size of the antenna is infinitesimal.

The directivity defined above depends on range. In the far field, however, the directivity of all antennas is invariant with range, and so it is expedient in that case to express directivity in terms of the power density per unit solid angle ($P_\Omega(\theta, \phi)$) which is also invariant with range. Recognizing that a per-unit-area quantity multiplied by r^2 becomes a per-unit-solid-angle quantity, we can define directivity as

$$D(\theta, \phi) \equiv 4\pi \frac{r^2 P_r(\theta, \phi)|_r}{P_t}$$
$$= 4\pi \frac{P_\Omega(\theta, \phi)}{P_t}.$$

A single number is sometimes quoted for the directivity, referring to the directivity in the direction of the main beam, i.e., $D = D(\theta, \phi)_{\max}$. For example, revisiting the example of the exactly conical beam with solid angle Ω, $P_\Omega = P_t/\Omega$ within the beam, and we find that $\Omega = 4\pi/D$. This is in fact a general result for any beam shape (i.e., not just conical), as will be demonstrated shortly below. Since Ω can be no more than 4π, D can be no less than unity. In fact, it is never less than 1.5.

Combining the definition of the directivity with the approximate relationship between solid angle and HPBW for a rectangular beam shape gives

$$D \approx \frac{4\pi}{\text{HPBW}_E \text{HPBW}_H},$$

where the half-power beamwidths are to be evaluated in radians. For the case of the circular beam shape, the factor of π in the numerator can be replaced with a factor of 4. Either formula provides a rough estimate of the directivity on the basis of nothing more than the half-power beamwidths in two planes. In fact, they are apt to yield overestimates, since they neglect radiation emitted through sidelobes in the radiation pattern. A generally more accurate approximation is given by this rule of thumb,

$$D \approx \frac{26{,}000}{\text{HPBW}_E \text{HPBW}_H},$$

where the beamwidths are given in degrees this time and where an "efficiency" factor of about 64 percent has been applied to represent different factors that reduce antenna directivity in practice. Exact means of calculating directivity will follow, but the approximations given here are sufficient for many purposes.

For the special case of a Gaussian beam shape, the directivity can be calculated directly. Neglecting the constant of proportionality, the total radiated power is given by

$$P_t = \int_{-\pi}^{\pi} \int_0^{\pi} P_\Omega \, d\Omega$$
$$= \int_{-\pi}^{\pi} \int_0^{\pi} e^{-\theta^2/2\sigma^2} \sin\theta \, d\theta \, d\phi$$
$$\approx 2\pi \int_0^{\pi} e^{-\theta^2/2\sigma^2} \theta \, d\theta$$
$$= 2\pi\sigma^2,$$

in which $\sin\theta$ was replaced with the first term in its Taylor-series expansion under the assumption that the Gaussian pattern is very narrow and nonzero only at small angles. This makes the directivity

$$\begin{aligned}D &= 4\pi\frac{P_\Omega}{P_t}\\ &= \frac{2}{\sigma^2}\\ &= \frac{16\ln(2)}{(\text{HPBW})^2}.\end{aligned}$$

This is an exact result by virtue of the fact that the radiation pattern was specified explicitly. The approximation lies in taking the radiation pattern to have a narrow Gaussian shape to begin with.

For the Hertzian dipole, we calculate the exact result,

$$\begin{aligned}D(\theta,\phi) &= 4\pi\frac{(Z_\circ/8)(Idl\sin\theta/\lambda)^2}{Z_\circ(\pi/3)(Idl/\lambda)^2}\\ &= 1.5\sin^2\theta.\end{aligned}$$

In the case of the Hertzian dipole, $D = 1.5$, and the directivity is a function neither of wavelength nor of range. Note that this antenna is an idealization and that these characteristics are not universal.

2.1.7 Gain

Gain G (sometimes called directive gain) and directivity differ only for antennas with Ohmic losses. Whereas the directivity is defined in terms of the total power transmitted, the gain is defined in terms of the power delivered to the antenna by the transmitter. Not all the power delivered is radiated if the antenna has significant Ohmic losses.

$$G \equiv \frac{P_r(\theta,\phi)|_r}{P_{\text{del}}/4\pi r^2}.$$

Note that $G \le D$. In many applications, antenna losses are negligible, and the gain and directivity are used interchangeably. Both directivity and gain are dimensionless quantities, and it is often useful to represent them in decibels (dB).

$$D \approx G \;=\; 10\log_{10}\frac{P_{max}}{P_{avg}}.$$

In some cases, Ohmic losses in antennas cannot be neglected, and directivity and gain differ. This is likely in cases involving electrically short antennas, for example.

2.1.8 Beam Solid Angle

We have already stated that the beam solid angle is always given by $\Omega = 4\pi/D$ regardless of the beam shape. This can be proven by expressing the total radiated power as an integral and invoking the definition of directivity:

$$P_t = \int P_\Omega(\theta,\phi)\,d\Omega$$
$$= \int D(\theta,\phi)\frac{P_t}{4\pi}d\Omega.$$

Here, P_t is a constant (the total radiated power) that can be canceled from both sides of the equation, leaving the identity:

$$4\pi = \int D(\theta,\phi)d\Omega.$$

The formal definition for the beam solid angle, meanwhile, is the integral of the differential solid angles in the radiation pattern, weighted by the normalized power density,

$$\Omega \equiv \int (D(\theta,\phi)/D)\,d\Omega,$$

where the fact that the directivity and power per unit solid angle are proportional has been utilized. Substituting above gives the desired result:

$$\Omega = \frac{4\pi}{D}.$$

In general, the solid angle of a radiation pattern is the solid angle of a uniform sector pattern (cone) with the same directivity as the pattern. The beam solid angle of the ideal dipole is $8\pi/3$ Str. The solid angle of a Gaussian beam is $[\pi/4\ln(2)](\text{HPBW})^2$.

2.1.9 Antenna Effective Area

We have already defined the effective area of an antenna as the ratio of the power it delivers to a matched load (i.e., no reflected power) to the incident power density, i.e.,

$$P_{rx}(W) = P_{inc}(W/m^2)A_{\text{eff}}(m^2).$$

Whereas antenna directivity and gain are natural parameters for characterizing transmission, effective area may be more intuitive for characterizing reception. However, directivity and effective area are intimately related, and high-gain antennas will be found to have large effective areas.

Effective area has an intuitive interpretation for aperture antennas (reflectors, horns, etc.), and we will see that the effective area is always less than or equal to (less than, in practice) the physical area of the aperture. The effective area of a wire antenna is unintuitive. Nevertheless, all antennas have an effective area, which is governed by the *reciprocity theorem*. It will be shown below that

$$D = \frac{4\pi}{\lambda^2} A_{\text{eff}},$$

where λ is the wavelength of the radar signal. High gain implies a large effective area, a small HPBW, and a small solid angle. For wire antennas with lengths L large compared to a wavelength, we will see that HPBW $\approx \lambda/L$ in general.

2.2 Reciprocity Theorem

The analysis to this point emphasized the performance of antennas used for transmitting radiation. Antennas of course also receive radiation, and their performance as transmitter and receiver are related by the reciprocity theorem. Reciprocity is best illustrated by a kind of thought experiment, where we consider what happens when the same antennas are used interchangeably for transmission and reception.

Refer to Figure 2.5. Consider first the diagram on the left, which shows a communication link with a transmitter feeding power to a dipole antenna and a receiver collecting power from an aperture antenna. The particular kinds of antennas used are not important. We know that the power received at site 2, the aperture antenna, due to radiation from site 1, the dipole, is given by

$$P_{12} = P_t \frac{D_1}{4\pi r^2} A_{\text{eff} 2}, \tag{2.7}$$

where the subscripts refer to the antennas at the two sites. Consider next the equivalent circuit describing the link. If the intervening material between the antennas is linear, the link can be regarded as a two-port linear network, with

$$\begin{pmatrix} Z_{11} & Z_{12} \\ Z_{21} & Z_{22} \end{pmatrix} \begin{pmatrix} I_1 \\ I_2 \end{pmatrix} = \begin{pmatrix} V_1 \\ V_2 \end{pmatrix}.$$

The lower portion of Figure 2.5 shows the corresponding Thevenin equivalent circuit. For this discussion, we suppose that the transmitting and receiving antennas are sufficiently distant to neglect the coupling term Z_{12} that would otherwise appear as an additional source in the transmitter circuit. The total power delivered by the transmitter (the generator) to the transmitting antenna is

$$P_t = \frac{1}{2}|I_1|^2 R_1,$$

where $R_1 = \Re Z_{11}$ is the transmitting antenna radiation resistance. Coupling cannot be neglected in the receiver circuit, of course, since that is the only thing driving a current.

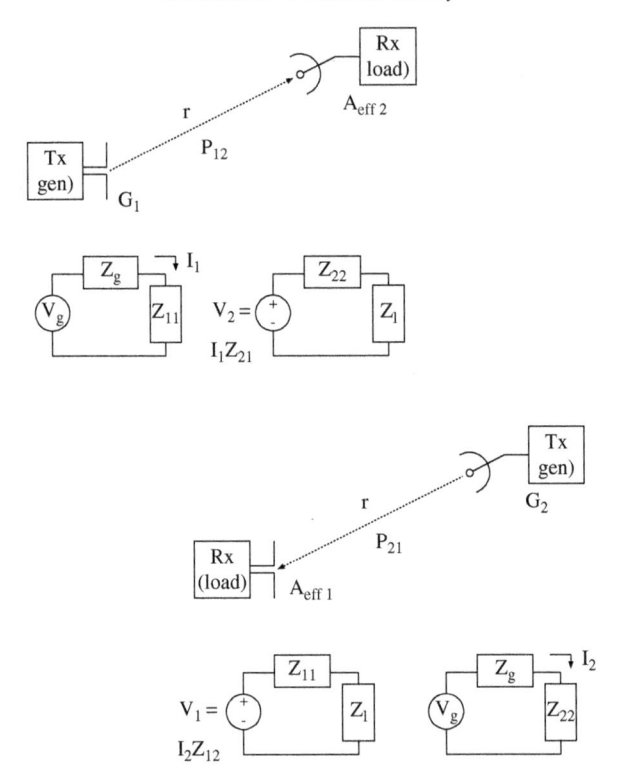

Figure 2.5 Communication links illustrating the reciprocity theorem. Two different configurations of a communications link are shown in which the roles of transmitter and receiver are reversed. Immediately below each configuration is its Thevenin equivalent circuit. The subscripts "g" and "l", refer to the generator (transmitter) and load (receiver) circuit characteristics, respectively. The coupled antennas are represented by the impedance matrix Z. $V_{1,2}$ is the open-circuit voltage of the given receiving antenna. $I_{1,2}$ are the currents driving the given transmitting antenna.

The power delivered to the receiver (the load) under matched load conditions, $Z_l = Z_{22}^*$, is given by

$$P_{12} = \frac{1}{2} \left| \frac{V_2}{Z_{22} + Z_l} \right|^2 R_l$$
$$= \frac{|I_1 Z_{21}|^2}{8R_2},$$

where $R_2 = \Re Z_{22} = \Re Z_l$ is the radiation resistance of the receiving antenna. This gives the ratio

$$\frac{P_{12}}{P_t} = \frac{|Z_{21}|^2}{4R_1 R_2}, \qquad (2.8)$$

which must be consistent with the result from (2.7).

2.2 Reciprocity Theorem

Now, reverse the roles of the sites so that transmission occurs from site 2 and reception at site 1, as shown on the right side of Figure 2.5. This time, the power received at site 1 due to transmission at site 2 under matched load conditions is given by

$$P_{21} = P_t \frac{D_2}{4\pi r^2} A_{\text{eff} 1}, \qquad (2.9)$$

where the transmit power is kept the same as before. In terms of equivalent circuit theory, we find that

$$\frac{P_{21}}{P_t} = \frac{|Z_{12}|^2}{4R_1 R_2}, \qquad (2.10)$$

which must be consistent with P_{21}/P_t from (2.9).

In the Appendix B, it is shown that for linear, isotropic (but not necessarily homogeneous) intervening media, $Z_{12} = Z_{21}$. This behavior is sometimes referred to as *strong reciprocity*. In view of Equations (2.8) and (2.10), one consequence of strong reciprocity is that $P_{12} = P_{21}$, which is a somewhat less general condition referred to as *weak reciprocity*. We can restate this condition using (2.7) and (2.9) as

$$P_t \frac{D_1}{4\pi r^2} A_{\text{eff} 2} = P_t \frac{D_2}{4\pi r^2} A_{\text{eff} 1},$$

which demands that

$$\frac{D_1}{D_2} = \frac{A_{\text{eff} 1}}{A_{\text{eff} 2}}.$$

This fundamental result shows that the ratio of the directivity for any antenna to its effective area is a universal constant (that may depend on λ, which is held fixed in the swap above). That constant can be found by performing a detailed calculation for some specific antenna configuration. It can also be found using the principle of detailed balance which is examined later in Chapter 5. The results can be stated as

$$D(\theta, \phi) = \frac{4\pi}{\lambda^2} A_{\text{eff}}(\theta, \phi). \qquad (2.11)$$

The factor of λ^2 is to be expected; directivity is a dimensionless quantity and can only be equated with another dimensionless quantity, such as the effective area measured in square wavelengths. Note also that the antenna effective area has been generalized and is now a function of bearing, just like the directivity. When quoting a single figure for the effective area, the maximum value is given.

We can consider the S-band planetary radar system at the Arecibo Radio Telescope as an extreme example. This radar operates at a frequency of 2.38 GHz or a wavelength of 12 cm. The diameter of the Arecibo reflector is 300 m. Consequently, we can estimate the directivity to be about 6×10^7 or about 78 dB, equating the effective area with the physical area of the reflector. In fact, this gives an overestimate, and the actual gain of the system is only about 73 dB (still enormous). We will see how to calculate the effective area of an antenna more accurately later in the text.

Two important points should be recognized:

1. For aperture antennas, we will see in Chapter 4 that the effective area is related to the physical area but that $A_{\text{eff}} \leq A_{phys}$. In practice, the former is always less than the latter.
2. All antennas have an effective area even if they have no obvious physical area. For the Hertzian dipole antenna, we found that $D = 1.5$. Consequently, $A_{\text{eff}} = 1.5\lambda^2/4\pi$ independent of the length of the antenna! For the AM car radio example, $A_{\text{eff}} = 1 \times 10^4 \, \text{m}^2$!

As stated earlier, the reciprocity theorem assumes that the antennas are matched electrically to the transmission lines, which we take to be matched to the transmitters and receivers. If not, then reflections will occur at the antenna terminals, and power will be lost to the communication budgets. The reflection coefficient at the terminals is given by

$$\rho = \frac{Z_l - Z_t}{Z_l + Z_t},$$

where the l and t subscripts denote the antenna (load) and transmission line impedances, respectively. Using 50-Ohm transmission line, the fraction of power reflected is then given by $|\rho|^2$. In the AM car radio example, $|\rho|^2 \sim 0.9992$.

The *Friis* transmission formula extends the reciprocity theorem to the case of potentially lossy and improperly matched antennas,

$$P_{rx} = \left(1 - |\rho_r|^2\right) \underbrace{\frac{\lambda^2}{4\pi} G_r(\theta_r, \phi_r)}_{A_{\text{eff} \, r}} \left(1 - |\rho_t|^2\right) \frac{P_t}{4\pi r^2} G_t(\theta_t, \phi_t),$$

where the r and t subscripts refer to the receiver and transmitter, respectively. Note that the formula neglects the effects of polarization, which are discussed later in the chapter. Mismatches between the polarization of the signals for which the transmit and receive antennas were designed cause additional power loss. In Chapter 9, we will discuss Faraday rotation. Note also that this and other phenomena occurring in anisotropic media can defeat reciprocity.

2.3 Practical Wire Antennas

The Hertzian dipole antenna is an idealization, but practical wire antennas can be viewed as assemblies of these idealized elements. This provides an approach for analyzing wire antennas that are physically long compared to the wavelength. In principle, one must know the current distribution on the wires to solve for the radiation fields. In practice, it is often acceptable to guess the distribution and then calculate the fields using superposition. Keeping track of the phases of the contributions from the radiating elements is the

2.3 Practical Wire Antennas

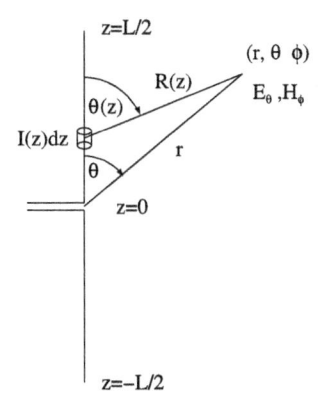

Figure 2.6 Long-wire antenna geometry. The observation point is (r, θ, ϕ).

main difficulty. Rigorous computational solutions agree reasonably well with the solutions presented below and with measurements. We consider the far field only in what follows.

We will analyze the long-wire (length L) dipole antenna in Figure 2.6, using the results from the Hertzian dipole analysis. We expect only E_θ and H_ϕ field components for the radiative fields. It is sufficient to solve for the former, since the latter is related by the impedance, since $E_\theta/H_\phi = Z_o$. From a Hertzian dipole $I(z)dz$, we expect the differential contribution

$$dE_\theta = \frac{jZ_o k}{4\pi} I(z) dz \sin\theta(z) \frac{e^{-jkR(z)}}{R(z)}.$$

Integrating the contributions from all the elemental dipoles that constitute the long-wire antenna will then yield the total observed electric field. Note that the harmonic time dependence $exp(j\omega t)$ is implicit.

We next make a number of assumptions to simplify the line integral:

1. Take $r \gg \lambda, L$. The first inequality restates the radiative-field limit. The second is being introduced for the first time here. This is the "far field" assumption. Its implications will be considered in more detail later in the text.
2. We can consequently replace $\theta(z)$ with θ.
3. In the denominator, we can let $R(z) \to r$.
4. More care must be taken with the phase term, since even small variations in R (comparable to a wavelength) with z can produce significant changes. Regarding all the rays from the antenna to the observation point as parallel permits us to replace $R(z) \to r - z\cos\theta$ in the phase term.

The differential contribution from each elemental dipole to the observed electric field then becomes

$$dE_\theta = \frac{jZ_o k}{4\pi r} I(z) dz \sin\theta e^{-jkr} e^{jkz\cos\theta}.$$

What is needed next is a specification of the current distribution on the wire antenna, $I(z)$. Calculating this self consistently is very difficult, even computationally. Intuition can provide a reliable guide, however. We expect the current to nearly vanish at the ends of the wire in view of the demands of current continuity. Elsewhere on the wire, we may expect a standing wave pattern. A distribution that satisfies these conditions is

$$I(z) = I_\circ \sin\left(k[L/2 - |z|]\right).$$

This turns out to be a reasonably good approximation for the actual current distribution on a long-wire antenna, except that the actual distribution never exactly vanishes at the feedpoint.

Substituting $I(z)$ and integrating yields the following expression for the observed electric field:

$$\begin{aligned}
E_\theta &= \frac{jkZ_\circ \sin\theta I_\circ}{4\pi r} e^{-jkr} \left[\int_{-L/2}^{0} \sin k(L/2+z) e^{jkz\cos\theta} dz + \int_{0}^{L/2} \sin k(L/2-z) e^{jkz\cos\theta} dz \right] \\
&= \frac{jkZ_\circ \sin\theta I_\circ}{2\pi r} e^{-jkr} \left[\frac{\cos(k\frac{L}{2}\cos\theta) - \cos(k\frac{L}{2})}{k\sin^2\theta} \right] \\
&= \frac{jZ_\circ I_\circ}{2\pi r} e^{-jkr} \left[\frac{\cos(k\frac{L}{2}\cos\theta) - \cos(k\frac{L}{2})}{\sin\theta} \right] \\
&= Z_\circ H_\phi.
\end{aligned}$$

The Poynting flux can then be calculated.

$$\begin{aligned}
P_r &= \frac{1}{2}\Re(\mathbf{E} \times \mathbf{H}^*) \cdot \hat{\mathbf{r}} = \frac{1}{2}\Re(E_\theta H_\phi^*) \\
&= \frac{Z_\circ}{2} \frac{|I_\circ|^2}{(2\pi r)^2} \left[\frac{\cos(k\frac{L}{2}\cos\theta) - \cos(k\frac{L}{2})}{\sin\theta} \right]^2.
\end{aligned}$$

Like the Hertzian dipole, this antenna does not radiate power in the direction of the wire despite the $\sin^2\theta$ term in the denominator. However, it is no longer the case that the main beam of the radiation pattern need lie in the equatorial plane.

2.3.1 Half-Wave Dipole

An illustrative and practical case is that of a half-wave dipole antenna with $L = \lambda/2$. Now, $kL/2 = \pi/2$ and the second cosine term in the numerator vanishes. This leaves

$$P_r = \frac{|I_\circ|^2 Z_\circ}{8\pi^2 r^2} \left[\frac{\cos(\frac{\pi}{2}\cos\theta)}{\sin\theta} \right]^2.$$

2.3 Practical Wire Antennas

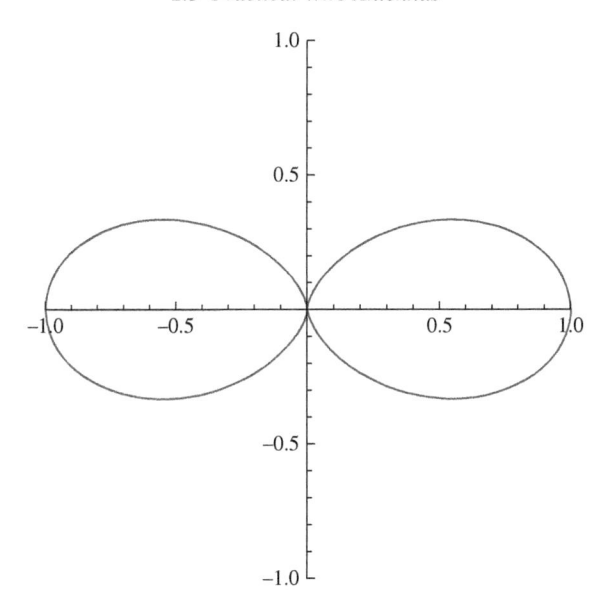

Figure 2.7 Radiation pattern of a half-wave dipole. The vertical axis here is the polar ($\hat{z}$) axis with which the antenna is aligned.

We will see that trigonometric functions of trigonometric functions are rather common expressions in antenna theory. While the shape of this particular expression is not immediately obvious, it is actually not very different from $\sin^2 \theta$ (see Figure 2.7). Note that the term in square brackets is already normalized to unity. Note also that the radiated power density decreases as the square of the range such that the power density per unit solid angle is invariant with range. This is true for all antennas in the radiative far field.

The total power radiated by the half-wave dipole antenna is found by integrating the Poynting flux over the surface of a sphere of any radius or, equivalently, integrating P_Ω over all solid angles,

$$P_t = \frac{|I_\circ|^2 Z_\circ}{8\pi^2} \int_0^{2\pi} \underbrace{\int_0^{\pi} \left[\frac{\cos(\frac{\pi}{2} \cos\theta)}{\sin\theta} \right]^2 \sin\theta\, d\theta\, d\phi}_{1.219}$$

$$\approx 36.6 I_\circ^2 \ (W),$$

where the last result was obtained using $Z_\circ = 120\pi\, \Omega$ and with the help of numerical integration. (Without the aid of numerical integration, the theta integral can be expressed in terms of the cosine integral function $C_i(x) = -\int_x^\infty \cos u/u\, du$, which is a tabulated function.)

As noted above, the radiation pattern for this antenna is not very different from that of a Hertzian dipole. The HPBW in the E-plane is 78° and therefore only slightly narrower than the Hertzian dipole. The BWFN is still 180°. A more meaningful statistic is the directivity:

$$D(\theta,\phi) = \frac{P_\Omega}{P_t/4\pi}$$

$$= \frac{\frac{Z_\circ |I_\circ|^2}{8\pi^2}\left[\frac{\cos(\frac{\pi}{2}\cos\theta)}{\sin\theta}\right]^2}{36.6|I_\circ|^2/4\pi}$$

$$\approx 1.64\left[\frac{\cos(\frac{\pi}{2}\cos\theta)}{\sin\theta}\right]^2. \qquad (2.12)$$

The overall directivity is therefore 1.64, only slightly more than 1.5 for a Hertzian dipole, since the quotient in (2.12) is unity when $\theta = 0$.

The most important difference between the half-wave dipole and the Hertzian dipole is its radiation resistance. Equating

$$P_t = \frac{1}{2}|I_\circ|^2 R_{rad}$$
$$\rightarrow R_{rad} \approx 73.2\,\Omega.$$

The higher impedance of the half-wave dipole permits a practical match using commercially available 75 Ω transmission line. In fact, the impedance of the half-wave dipole is slightly reactive. Shortening the antenna slightly below $\lambda/2$ makes it purely resistive and reduces the radiation resistance to a figure closer to 70 Ω, permitting a satisfactory match using standard 50 or 75 Ω transmission line. Furthermore, since this radiation resistance is likely to be much greater than the Ohmic resistance of the antenna, efficiency will be high, and we are justified in neglecting Ohmic losses and equating directivity with gain. These qualities make the half-wave dipole practical. The gain of the antenna is, however, still quite modest.

2.3.2 Driving-Point Impedance

As an aside, we consider here briefly the calculation required to estimate the imaginary part of the complex driving-point impedance of a practical wire antenna. The calculation is essentially more difficult than the one leading to the radiation resistance evaluated above. The first reason for this is that the reactive components of the electromagnetic field of the Hertzian dipole have to be considered and retained, since it is these that determine the reactance of the antenna. Secondly, it will be necessary to evaluate the electromagnetic field from the antenna in the near field, meaning that the simplifying assumptions incorporated in the integral in Section 2.3 are no longer valid. Nor are the electric and magnetic fields related simply by a factor of $Z_\circ$.

The normal procedure is to calculate the magnetic vector potential **A** for a finite-size wire with a prescribed current distribution $I(z)$ through careful integration of (2.2), calculate the magnetic field **H** from the curl of **A**, and then calculate the electric field **E** using Faraday's law. It will be expedient to perform the calculations in a cylindrical coordinate system aligned with the wire antenna. The $\hat{z}$ component of the electric field for a wire antenna of length L can be shown in this way to be

$$E_z(\rho,z) = -j\frac{Z_\circ I_\circ}{4\pi}\left[\frac{e^{-jkr_-}}{r_-} + \frac{e^{-jkr_+}}{r_+} - 2\cos\left(\frac{kL}{2}\right)\frac{e^{-jkr}}{r}\right], \qquad (2.13)$$

where $r^2 \equiv \rho^2 + z^2$ and $r_\pm^2 \equiv \rho^2 + (z\pm L/2)^2$ and where the current distribution is again taken to be $I(z) = I_\circ \sin(k[L/2 - |z|])$. Notice that the cosine term above vanishes when the length of the antenna is an integral number of half wavelengths.

Finally, an integral over a surface surrounding the antenna must be carried out. The surface of interest is not a sphere in the radiative far field but instead a cylinder in the reactive near field, on the surface of the wire antenna itself. This implies a consideration of the wire diameter, something heretofore neglected. In fact, the reactance of a wire antenna depends on the wire thickness. Thick dipole antennas have a smaller reactance than thin ones and will therefore be useful over a broader range of frequencies surrounding the resonant frequency where the impedance is purely resistive. The so-called "fat dipole" or "squirrel cage" antennas constructed from wire frames are often employed in broadband applications.

The formula for estimating the driving-point impedance of a long-wire dipole antenna in isolation is

$$Z_{in} = -\frac{1}{I^2}\int_{-L/2}^{L/2} H_\phi^*(a,z)E_z(a,z)\,ad\phi dz \qquad (2.14)$$

$$= -\frac{1}{I^2}\int_{-L/2}^{L/2} I(z)E_z(a,z)dz, \qquad (2.15)$$

where $I = I_\circ \sin(kL/2)$ is the feedpoint current. This can be understood in the following way. The right side of (2.14) is the outward complex Poynting flux integrated over the surface of the wire. Whereas the real part of this quantity is associated with the driving-point resistance as before, the imaginary point can be associated with the driving-point reactance. In order to calculate the magnetic field, the boundary condition at the surface of the wire equating H_ϕ with the surface current density can be used. Since the electric and magnetic fields are axisymmetric about the wire, the integral in azimuth of the surface current density just produces a factor of the current $I(z)$, which is assumed to flow in a sheet very near the wire surface.

Using the electric field specification given by (2.13) together with the same current distribution assumed previously in (2.15) yields estimates for the driving-point resistance and reactance expressed in terms of the cosine and sine integrals, C_i and S_i:

$$R_{rad} = \frac{I_\circ^2}{I^2}\frac{Z_\circ}{2\pi}\left\{C + \ln(kL) - C_i(kL) + \frac{\sin(kL)}{2}[S_i(2kL) - 2S_i(kL)] \right. \qquad (2.16)$$
$$\left. + \frac{\cos(kL)}{2}[C + \ln(kL/2) + C_i(2kL) - 2C_i(kL)]\right\}$$

$$\chi = \frac{I_o^2}{I^2}\frac{Z_o}{4\pi}\left\{2S_i(kL) + \cos(kL)\left[2S_i(kL) - S_i(2kL)\right]\right. \tag{2.17}$$

$$\left. - \sin(kL)\left[2C_i(kL) - C_i(2kL) - C_i\left(\frac{2ka^2}{L}\right)\right]\right\},$$

where, $C = 0.5772$ is Euler's constant. The first of these formulas implies the same value of the radiation resistance as the previous calculation based on the average Poynting flux evaluated in the far field. The second gives the reactance at the antenna feedpoints. Evaluating the expression shows that the reactance vanishes for dipoles with approximately half-integer wavelengths, i.e., $L \sim \lambda/2, 3\lambda/2, 5\lambda/2, \cdots$. The reactance is negative (positive) for slightly shorter (longer) antennas, and the transition is more gradual, the larger the wire radius a. Resonances occur at intermediate wavelengths where the reactance changes sign.

The complex driving-point impedance of long wire antennas calculated according to (2.16) and (2.17) is depicted in Figure 2.8. The wire radius was taken to be 0.001 times the length for these calculations. Antennas with half-integer wavelength sizes are practical to feed.

2.3.3 Antenna Effective Length

Just as all antennas have an effective area, they also have an effective length. This is related to the physical length in the case of wire antennas, but there is no intuitive relationship between effective length and antenna size for aperture antennas. **h** antennas have large effective lengths as a rule.

Formally, the effective length is that quantity which, when multiplied by the electric field in which the antenna is immersed, gives the open circuit voltage appearing across the antenna terminals. The concept of effective length is useful for calculating the power received by an antenna without invoking the reciprocity theorem. There is a simple, intuitive interpretation of effective length that applies to wire antennas, which is described below.

We've seen that the electric fields for a half-wave dipole and a Hertzian dipole are given respectively by:

$$E_\theta = jZ_o I_o \frac{2\cos((\pi/2)\cos\theta)}{\sin\theta}\frac{e^{-jkr}}{4\pi r}$$

$$E_\theta = jZ_o I dl k \sin\theta \frac{e^{-jkr}}{4\pi r}.$$

Let us rewrite the second of these as

$$\mathbf{E} = \frac{jZ_o k e^{-jkr}}{4\pi r}I\mathbf{h} \tag{2.18}$$

$$\mathbf{h} = dl\sin\theta\hat{\theta},$$

2.3 Practical Wire Antennas

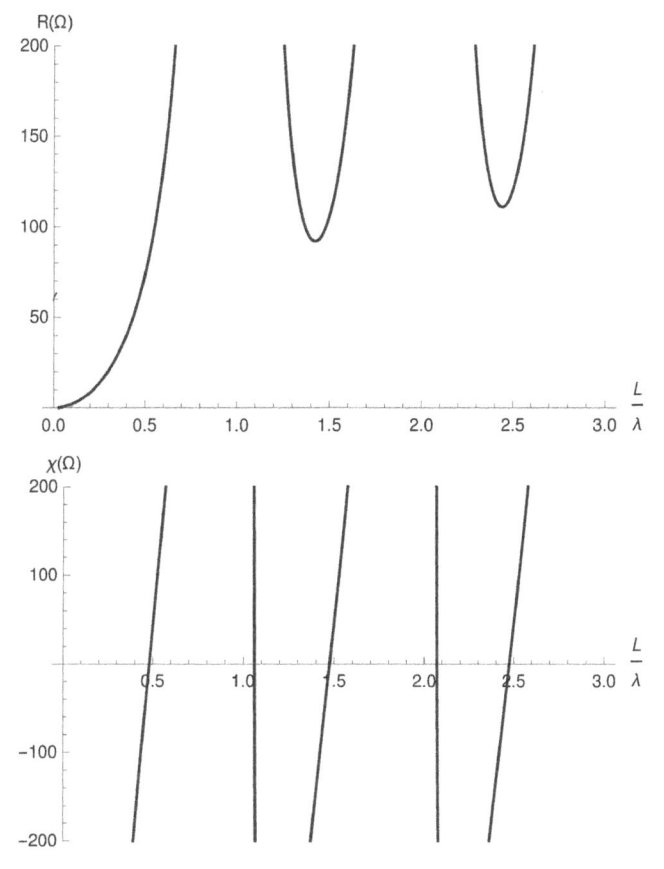

Figure 2.8 Driving-point resistance R (above) and reactance χ (below) vs. length for long-wire dipole antennas.

where **h** represents the effective length of the Hertzian dipole. It obviously has the apparent physical length of the Hertzian antenna as viewed by an observer. We assert that the effective and physical lengths of the Hertzian dipole are identical since the current is uniform, meaning that the entire length dl is fully utilized.

By extension, if we can cast the radiation field of any antenna in the form of (2.18), the effective length of that antenna is whatever becomes of **h**. In the case of the half-wave dipole, we have

$$\mathbf{h} = \frac{\lambda}{\pi} \frac{\cos((\pi/2)\cos\theta)}{\sin\theta} \hat{\boldsymbol{\theta}}.$$

Note that $|\mathbf{h}| \leq \lambda/\pi \leq \lambda/2$. The effective length of the half-wave dipole is less than its physical length, reflecting the fact that the current distribution on the wire is tapered and that the wire is therefore not fully utilized for transmission and reception. Similar results

2.3.4 Full-Wave Dipole

Another seemingly obvious choice is the full-wave dipole ($L = \lambda$). Following the procedures above, one can show that the full-wave dipole has a narrower beam with an HPBW of 48°. However, note that the standing wave pattern assumed for the current distribution has a null at the feedpoint for a full-wave dipole. This implies a very high driving-point impedance in practice, making the antenna impractical for most applications.

Note also that directivity does not increase with increasing wire length as a rule. Wires with increasing lengths have increasingly narrow main beams, but these come at the expense of a multiplicity of sidelobes. The highest directivity that can be achieved with a dipole is approximately $D = 2$ for a wire with a length of about $L = 1.22\lambda$. The utility of long-wire dipoles by themselves for radar work is limited.

2.3.5 Quarter-Wave Monopole

A common antenna configuration is illustrated in the left side of Figure 2.9. Here, a vertical monopole antenna is excited from a feedpoint at the level of a ground plane (earth), which we regard as perfectly conducting. The monopole antenna could be a large tower of the kind used for AM radio transmission. The current distribution on the monopole antenna can be assumed to have the same form as one "leg" of the long-wire dipole studied earlier. Surface charges and currents induced in the ground plane by the radiation from the antenna will themselves emit, effecting the net radiation. Calculating the effect would seem to necessitate first calculating the sources induced on the ground plane and then their

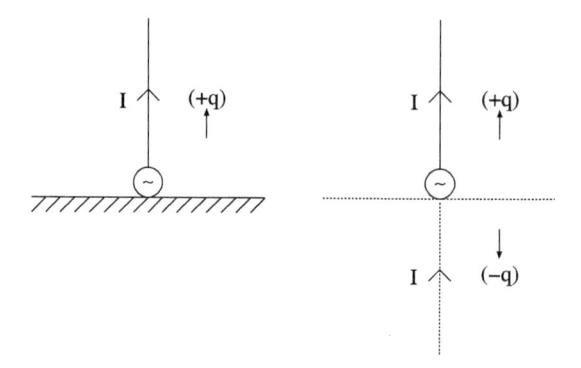

Figure 2.9 Monopole antenna above a perfectly conducting ground plane. We can think of the current in the monopole being carried by positive charge carriers moving in the direction of the current. The corresponding image charges are negative and move in the opposite direction, and the image current is consequently in the same direction as the original current.

2.3 Practical Wire Antennas

contributions to the total radiation. While the aforementioned method is certainly tractable, a more expedient alternative is offered by the method of images.

There will obviously be no electromagnetic fields below the ground plane, in the lower half space. We concentrate therefore on the upper half space, the region bounded below by the plane just above the ground plane and above by a hemispherical surface of infinite radius centered on the antenna feedpoint.

Within this "region of interest," the uniqueness of solution of differential equations guarantees that only one solution for the electromagnetic field is consistent with Maxwell's equations, the sources in the region, and the conditions at the boundary encompassing the region. We are meanwhile free to replace the sources outside the upper half space, including the sources on the ground plane, with other sources of our choosing. This obviously modifies neither the sources in the upper half space, which are on the antenna itself, nor Maxwell's equations. So long as it does not alter the conditions at the boundaries, the electromagnetic fields in the upper half space will be unchanged from the original specification. The goal is to choose replacements for the sources on the ground plane carefully so as to simplify the calculations as much as possible.

The boundary conditions to be maintained in the upper half space immediately adjacent to the ground plane are the absence of tangential electric field and normal magnetic field components, since neither can exist at the surface of a perfect conductor. Both of these conditions are met by the introduction of a hypothetical monopole antenna into the lower half space that is the mirror image of the original. If we envision the current density being carried by hypothetical positive charge carriers in the original antenna, the current in the image should be carried by negative charge carriers moving in the opposite direction. The sense of the current directions is therefore the same in both monopoles, and the shape of $I(z)$ the same. Overall, the configuration becomes the long-wire dipole antenna.

Since neither the original monopole nor its twin produces a magnetic field in the direction normal to the ground plane at ground-plane height, that component of the boundary condition is satisfied. The tangential components of the electric fields due to the original monopole and its image are equal and oppositely directed at ground-plane height, so that component of the boundary condition is satisfied as well. This is true not only for the radiative but also the reactive field components. Evidently, no new calculations are necessary; the electromagnetic field in the upper half space due to a monopole above a perfectly conducting ground plane is the same as that due to a vertical dipole antenna with the same feedpoint current.

We can now characterize the behavior of the monopole compared to that of the dipole almost by inspection. By comparison to the dipole antenna with identical feedpoint current,

- $P_\Omega(\theta, \phi)$ is unchanged in the upper-half space and has a maximum at broadside;
- P_t is halved (no radiation below ground plane);
- R_{rad} is halved (from Ohm's law);

- $D = 4\pi P_\Omega/P_t$ is doubled;
- HPBW and beam solid angle are halved.

In practice, the Earth's surface behaves more like a lossy dielectric than a perfect conductor. Ground wires are often run radially above or in the ground to better simulate an ideal ground plane. In the absence of such wires, the results presented above must be regarded as approximations. Their accuracy depends on frequency, soil type, and weather conditions. Imperfect ground planes can give rise to rather severe losses.

2.3.6 Folded Dipole Antenna

Another common antenna is the half-wave folded dipole. You may be able to find one of these on the back of your FM stereo. The folded dipole offers a simple means of increasing the radiation resistance of an ordinary half-wave dipole. Folded dipoles are therefore often incorporated in parasitic arrays which tend otherwise to have very low radiation resistance. Folded dipole antennas can offer a good match with several kinds of inexpensive, twin-lead feedline and can, in fact, be made from such feedline.

To understand how this works, imagine constructing one from a piece of two-wire transmission line one-half wavelength long. Figure 2.10 shows the sense of the current during one particular phase of the wave. The end is shorted and therefore corresponds to a current maximum. A current node exists one-quarter wavelength closer to the transmitter, and another current maximum exists at the feedpoint. We suppose that deforming the line into the folded dipole antenna at the right leaves the sinusoidal current distribution unchanged. The feedpoint remains a current maximum, the endpoints current nodes. Clearly, the two wires comprising the folded dipole have identical current distributions that are also identical to that of an ordinary half-wave dipole.

Since there are two wires carrying current, the total current is doubled compared to an ordinary half-wave dipole. Consequently, for some input current,

- the electric field is doubled
- the Poynting flux and total radiated power increase by a factor of 4
- since the feedpoint currents are the same, the radiation resistance must have increased by a factor of 4 (to approximately 300 Ω, which is the intrinsic impedance for one common kind of twin-lead feedline)
- the radiation pattern is meanwhile unchanged from that of a half-wave dipole
- the directivity, effective area, HPBW, and beam solid angle are all unchanged

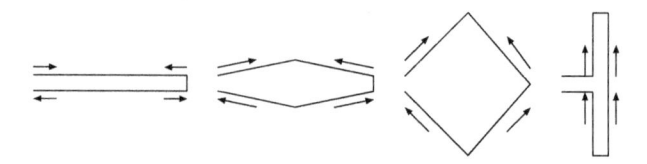

Figure 2.10 Constructing a folded dipole antenna from a two-wire transmission line.

2.3.7 Short Dipole

It is often necessary or desirable to use wire antennas that are very short compared to the wavelength. This situation comes about when space is limited or when the wavelength is very long, as with VLF and ULF communications. A short antenna is not a Hertzian dipole. The latter is an idealization and requires that the current on the antenna be uniform along its length. In practice, the current vanishes at the ends of a short wire antenna, just as for long wires.

Nothing about the calculations in Section 1.5 necessitated that the wire antenna be long, and those results are completely applicable to the short dipole. However, we can simplify things considerably with a few well justified assumptions. The first is that the distances measured along the wire antenna are so small (compared to λ) that the sinusoidal current distribution we assumed earlier can be replaced by its argument (i.e., the first term if its Taylor series expansion):

$$I(z) \approx I_\circ \left(1 - \frac{|z|}{L/2}\right).$$

The current distribution is approximately triangular on any sufficiently short wire. Furthermore, we can approximate

$$dE_\theta = \frac{jZ_\circ k}{4\pi r} I(z) dz \sin\theta e^{-jkr} \underbrace{e^{jkz\cos\theta}}_{\sim 1}$$

again, since all the distances measured along the wire antenna are small compared to a wavelength. The integration over dz now involves only $I(z)$ and is tantamount to computing the average current on the wire, i.e., $L\langle I(z)\rangle = I_\circ L/2$. Finally, we have

$$E_\theta = \frac{jZ_\circ k}{4\pi r} I_\circ \frac{L}{2} \sin\theta e^{-jkr}$$

for the electric field due to a short dipole antenna, precisely the same result as for the Hertzian dipole, only with dl replaced by $L/2$. This means that the effective length of a short dipole is $L/2$ (times the factor of $\sin\theta$), half its visible physical length. The reduction is associated with the incomplete utilization of the wire length due to the tapered current distribution.

By comparison to a Hertzian dipole with the same physical length, the short dipole generates electric fields half as intense and radiates one-quarter the Poynting flux and the total power. Consequently, its radiation resistance is also reduced by a factor of 4 (from an already small figure). The radiation pattern is identical, however, meaning that the directivity, HPBW, solid angle, etc., are also the same.

2.3.8 Small Current Loop

Another important antenna is a small current loop with radius $a \ll \lambda$, as shown in Figure 2.11. Here, the observer is at location (r, θ, ϕ) in spherical coordinates. The sources

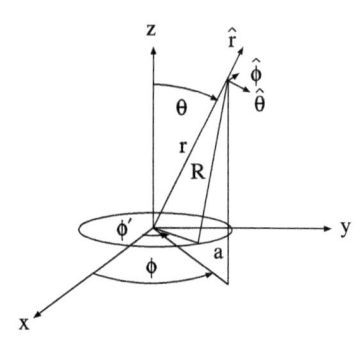

Figure 2.11 Radiation from a small current loop of radius *a* carrying current *I*.

of radiation are differential line segments $Idl = Ia d\phi'$ at locations $(a, \pi/2, \phi')$ in the x-y plane. (As elsewhere, primed and unprimed coordinates are used to represent source and observation points here.) The direction of the source current anywhere on the loop is given by $\hat{\phi}' = -\hat{x}\sin\phi' + \hat{y}\cos\phi'$. The vector potential at the observation point can then be expressed in terms of a line integral of the source current elements on the loop,

$$\mathbf{A} = \frac{\mu_\circ}{4\pi} Ia \int_0^{2\pi} \frac{e^{-jkR}}{R} \left(-\hat{x}\sin\phi' + \hat{y}\cos\phi'\right) d\phi',$$

where R is the distance between the source and observation points. In the limit $r \gg a$, R is approximately equal to r minus the projection of the vector from the origin to the source on the vector from the origin to the observer:

$$R \sim r - a\sin\theta\cos(\phi - \phi')$$
$$= r - a\sin\theta(\cos\phi\cos\phi' + \sin\phi\sin\phi').$$

Since we are interested in calculating the far field, we may replace R with r in the denominator of the vector potential expression. More care is required in the exponent in the numerator, however, where we instead note that $ka \ll 1$ and so use the small-argument approximation:

$$e^{-jkR} = e^{-jkr}e^{jka\ldots}$$
$$\approx e^{-jkr}\left(1 + jka\sin\theta(\cos\phi\cos\phi' + \sin\phi\sin\phi')\right).$$

The vector potential therefore becomes

$$\mathbf{A} = \frac{\mu_\circ Ia}{4\pi r} e^{-jkr} \int_0^{2\pi} \left(1 + jka\sin\theta(\cos\phi\cos\phi' + \sin\phi\sin\phi')\right)\left(-\hat{x}\sin\phi' + \hat{y}\cos\phi'\right) d\phi'.$$

The only terms that survive the ϕ' integration are those involving $\sin^2\phi'$ and $\cos^2\phi'$. Performing the necessary integrals yields

$$\mathbf{A} = jk\mu_\circ I\pi a^2 \sin\theta \frac{e^{-jkr}}{4\pi r}\hat{\phi}.$$

The resulting magnetic field is obtained by taking the curl of the vector potential. The electric field in the far-field region is normal to the magnetic field and to the direction of propagation (radially away), with an amplitude governed by the impedance of free space,

$$\mathbf{H} = \frac{1}{\mu_\circ}\nabla \times \mathbf{A}$$
$$= -Mk^2 \sin\theta \frac{e^{-jkr}}{4\pi r}\hat{\theta}$$
$$\mathbf{E} = -Z_\circ \hat{\mathbf{r}} \times \mathbf{H}$$
$$= MZ_\circ k^2 \sin\theta \frac{e^{-jkr}}{4\pi r}\hat{\phi},$$

where $M = I\pi a^2$ is the magnetic moment of the current loop.

The radiation pattern for the small current loop is given by $P_r = \sin^2\theta$, just like the Hertzian electric dipole. The total radiated power and radiation resistance can be calculated in the same way, leading to the result $R_a \propto (a/\lambda)^4$. In view of the small radius limit assumed here, the radiation resistance of the small current loop is necessarily very small, making the antenna an inefficient one. The radiation resistance of an antenna made of multiple (n) current loops grows as n^2, which can improve the efficiency to some degree. In some situations, such as the short dipole AM car radio antenna discussed earlier, receiving antennas with poor efficiency represent acceptable engineering solutions, as we will see after analyzing the effects of system noise on signal detectability. Poor efficiency is seldom acceptable for transmitting antennas, however.

2.3.9 Duality

The similarity in performance of the ideal electric dipole antenna and the small current loop is no accident, but results from underlying isomorphism in Maxwell's equations. Consider Faraday's and Ampère's laws for a region free of sources:

$$\nabla \times \mathbf{E} = -j\omega\mu\mathbf{H}$$
$$\nabla \times \mathbf{H} = j\omega\varepsilon\mathbf{E}.$$

Notice that the equations remain unchanged with the replacement of $\mathbf{E}$ by $\mathbf{H}$, $\mathbf{H}$ by $-\mathbf{E}$, μ by ε, and ε by μ. Consequently, any general solution to the source-free equations remains a solution after undergoing such an exchange. Put another way, any general solution to the equations is the solution to two distinct problems: the original and its dual. This is also true of a particular solution, although the boundary conditions in the original problem may not map to physically realizable boundary conditions in the dual problem.

The source for the small current loop is a magnetic dipole $M = I\pi a^2$. For the Hertzian dipole, it is the current element Idl. More precisely, it is the time-varying electric dipole $p = qdl$, with I being the time rate of change of the charge q, making $p = Idl/j\omega$. Replacing p in the field equations in Section 2.1.2 with μM and performing the other substitutions given above produces the results:

$$E_\phi = -jMZk\frac{1}{4\pi}\sin\theta\left(\frac{jk}{r} + \frac{1}{r^2}\right)e^{j(\omega t - kr)}$$

$$H_r = jM\frac{1}{2\pi}\cos\theta\left(\frac{k}{r^2} + \frac{1}{jr^3}\right)e^{j(\omega t - kr)}$$

$$H_\theta = -M\frac{1}{4\pi}\sin\theta\left(\frac{k^2}{r} - \frac{jk}{r^2} - \frac{1}{r^3}\right)e^{j(\omega t - kr)}.$$

The radiative field components given here agree with the results for the small current loop given above. In fact, the new results are accurate in the radiative and reactive regions and were derived expediently on the basis of the known results for the Hertzian dipole.

Isomorphism of Maxwell's equations would seem to break down when sources are considered, since there are no magnetic currents in Faraday's law to match the electric currents in Ampère's law and no magnetic charges (monopoles) to match electric charges in Gauss' law. Indeed, no such sources have been found in nature at the time of this writing. Such charges could nevertheless exist, and there is nothing to prevent their introduction in Maxwell's equations where it is expedient to do so. Consideration of hypothetical magnetic charges and currents is a useful approach to analyzing aperture antennas and will be explored later in Chapter 4.

2.4 Polarization

This chapter would be incomplete without a discussion of electromagnetic wave polarization. The antennas discussed above all transmit and receive waves with a specific polarization, and polarizations must be "matched" in order to optimize communication and radar links. Below, the basic ideas behind wave polarization are introduced.

The *polarization* of a wave refers to variations in the direction of the wave's electric field. We restrict the discussion to transverse electromagnetic (TEM) plane waves, in which surfaces of constant phase are planes oriented normally to the direction of propagation. Taking the direction of propagation to be $\hat{\mathbf{z}}$, we can express the electric field of the wave as

$$\mathbf{E} = \left(E_1\hat{\mathbf{x}} + E_2 e^{j\psi}\hat{\mathbf{y}}\right)e^{j(\omega t - kz)},$$

where E_1, E_2, and ψ are taken to be real constants. The plane wave nature of the wave is contained in the trailing exponential, which is assumed if not always explicitly written. The

ψ term controls the relative phase of the vector components of the transverse electric field. Depending on the ratio of E_1 to E_2 and the value of ψ, the wave can be linearly, elliptically, or circularly polarized.

2.4.1 Linear Polarization

If $\psi = 0$, the two transverse components of **E** are in phase, and in any $z =$const plane, they add to give a vector with the same direction specified by $\alpha = \tan^{-1} E_y/E_x = \tan^{-1} E_2/E_1$. Since the electric field always lies in the same plane, we refer to such waves as linear or plane polarized. Examples are vertically ($\alpha = \pi/2$) and horizontally ($\alpha = 0$) polarized waves. Note that one can speak of "plane-polarized plane waves," meaning linearly polarized waves with an $\exp(j(\omega t - kz))$ phase dependence.

2.4.2 Circular Polarization

Another important special case is given by $E_1 = E_2$, $\psi = \pm\pi/2$ so that

$$\mathbf{E} = (\hat{\mathbf{x}} \pm j\hat{\mathbf{y}}) E_1 e^{j(\omega t - kz)}. \tag{2.19}$$

In instantaneous notation, this is

$$\begin{aligned}\mathbf{E}(z,t) &= \Re\left[(\hat{\mathbf{x}} \pm j\hat{\mathbf{y}})E_1 e^{j(\omega t - kz)}\right] \\ &= E_1\left[\hat{\mathbf{x}}\cos(\omega t - kz) \mp \hat{\mathbf{y}}\sin(\omega t - kz)\right].\end{aligned}$$

The electric field vector therefore traces out a circle with a radius E_1 such that $E_x^2 + E_y^2 = E_1^2$ and an angle with respect to $\hat{\mathbf{x}}$ given by $\alpha = \tan^{-1} E_y/E_x = \tan^{-1} \mp \sin(\omega t - kz)/\cos(\omega t - kz) = \mp(\omega t - kz)$.

Imagine holding the spatial coordinate z fixed. At a single location, increasing time decreases (increases) α given a plus (minus) sign in (2.19). In the example in Figure 2.12, which illustrates the plus sign, the angle α decreases as time advances with z held constant, exhibiting rotation with a left-hand sense (left thumb in the direction of propagation, fingers curling in the direction of rotation). The plus sign is therefore associated with left circular polarization in most engineering and RF applications, where the polarization of a wave refers to how the electric field direction varies in time. This text follows this convention.

Consider, though, what happens when time t is held fixed. Then the angle α increases (decreases) with increasing z given the plus (minus) sign in the original field representation (2.19). Referring again to Figure 2.12, the angle α increases as z advances with t held constant, exhibiting rotation with a right-hand sense (right thumb in the direction of propagation, fingers curling in the direction of rotation). In many physics and optics texts, polarization refers to how the electric field direction varies in space, and the plus sign is associated with right circular polarization. *This is not the usual convention in engineering, however.*

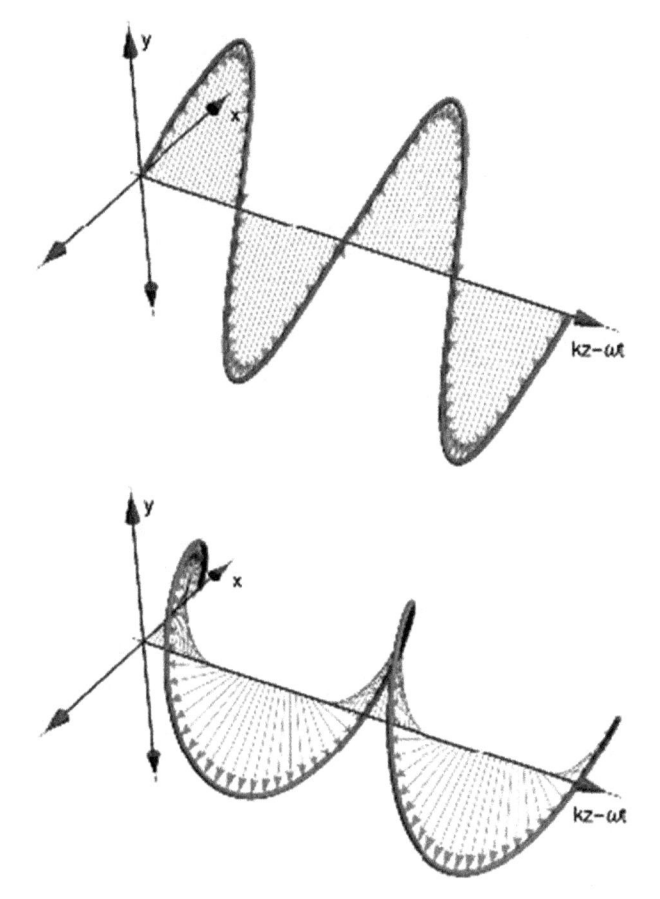

Figure 2.12 Illustration of linear (above) and circular (below) polarized waves. The direction or propagation ($\hat{\mathbf{z}}$) is toward the right. Arrows indicate electric-field directions. In the case of the linearly polarized wave shown, $\alpha = 45°$. In the case of circularly polarized waves, α varies in space and time (see text). Both cases are combinations of vertically and horizontally polarized waves with equal amplitudes.

2.4.3 Elliptical Polarization

This is the general case and holds for $\psi \neq \pm\pi/2$ or $E_1 \neq E_2$. We are left with

$$\mathbf{E}(z,t) = \Re\left[(E_1\hat{\mathbf{x}} + E_2 e^{\pm j\psi}\hat{\mathbf{y}})e^{j(\omega t - kz)}\right]$$
$$= \hat{\mathbf{x}} E_1 \cos(\omega t - kz) + \hat{\mathbf{y}} E_2 \cos(\omega t - kz \pm \psi).$$

At fixed $z = 0$, for example, this is

$$\mathbf{E}(z = 0, t) = \hat{\mathbf{x}} E_1 \cos(\omega t) + \hat{\mathbf{y}} E_2 \cos(\omega t - \pm\psi),$$

2.4 Polarization

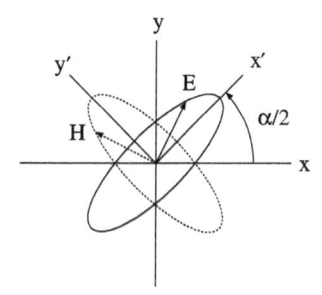

Figure 2.13 Elliptically polarized TEM wave.

which is a parametric equation for an ellipse. With a little effort, one can solve for the major and minor axes and tilt angle of the ellipse in terms of $E_{1,2}$ and ψ. One still has left and right polarizations, depending on the sign of ψ.

The following principles are important for understanding free-space wave polarization (see Fig. 2.13):

- **H** is in every case normal to **E**, with $\mathbf{E} \times \mathbf{H}$ in the direction of propagation and $|\mathbf{E}|/|\mathbf{H}| = Z_\circ$.
- A wave with any arbitrary polarization can be decomposed into two waves having orthogonal polarizations. These two waves may be linearly, circularly, or elliptically polarized. For example, a circularly polarized wave can be decomposed into two linearly polarized waves with polarizations at right angles. This construction is sometimes how circularly polarized waves are generated.
- All antennas transmit and receive signals with a specific polarization. Waves with polarizations orthogonal to the antennas are invisible to them. Efficient communication and radar links require matched polarizations.
- Reflection/scattering can affect the polarization of a signal, as can propagation in certain interesting materials (crystals, plasmas). Signals reflected off a surface at or near to the Brewster angle will retain only one linear polarization, for example. In space plasmas and birefringent crystals, waves spontaneously decompose into orthogonal polarizations (called the X and O modes) which then propagate differently.
- *Polarimetry* involves monitoring the polarization of a signal scattered from the ground to detect surface properties. As depicted in Figure 2.14, the sense of rotation of circularly polarized waves reverses upon reflection. It can reverse multiple times upon scattering from rough surfaces. Revealing imagery can be constructed by monitoring the fraction of the power in one circular polarization of a scattered wave.

The generic expression for an elliptically polarized wave is arguably difficult to visualize. However, its meaning can be clarified with a couple of revealing variable transformations that make use of some of the principles stated above. The first of these involves decomposing the wave into left- and right-circularly polarized components, i.e.,

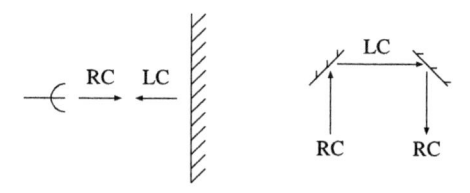

Figure 2.14 Illustration of polarization changes occurring upon reflection or scattering.

$$(E_1\hat{\mathbf{x}} + E_2 e^{j\psi}\hat{\mathbf{y}}) = E_l(\hat{\mathbf{x}} + j\hat{\mathbf{y}}) + E_r(\hat{\mathbf{x}} - j\hat{\mathbf{y}}),$$

where we take E_1 and E_2 to be real constants. Applying a little algebra reveals that

$$E_r = \frac{1}{2}\left[E_1 + jE_2 e^{j\psi}\right]$$

$$E_l = \frac{1}{2}\left[E_1 - jE_2 e^{j\psi}\right].$$

Now, the complex amplitudes E_r and E_l can always be expressed as real amplitudes E_R and E_L modifying a complex phase term involving symmetric and antisymmetric phase angles ϕ and α, i.e.,

$$E_r = E_R e^{j\alpha/2 + j\phi/2}$$

$$E_l = E_L e^{-j\alpha/2 + j\phi/2}.$$

It is left as an exercise to solve for the new constants. In terms of them, the total electric field becomes

$$\mathbf{E} = E_L e^{-j\alpha/2}(\hat{\mathbf{x}} + j\hat{\mathbf{y}})e^{j(\omega t - kz + \phi/2)} + E_R e^{j\alpha/2}(\hat{\mathbf{x}} - j\hat{\mathbf{y}})e^{j(\omega t - kz + \phi/2)}.$$

Define the total phase angle to be $\phi' = \omega t - kz + \phi/2$. Then, after taking the real part of the electric field, we find

$$\Re(\mathbf{E}(z,t)) = \cos\phi' \underbrace{\left(\hat{\mathbf{x}}\cos\frac{\alpha}{2} + \hat{\mathbf{y}}\sin\frac{\alpha}{2}\right)}_{\hat{\mathbf{x}}'} \underbrace{(E_L + E_R)}_{E_+} - \sin\phi' \underbrace{\left(\hat{\mathbf{y}}\cos\frac{\alpha}{2} - \hat{\mathbf{x}}\sin\frac{\alpha}{2}\right)}_{\hat{\mathbf{y}}'} \underbrace{(E_L - E_R)}_{E_-}$$

$$= \Re\left\{(E_+\hat{\mathbf{x}}' + jE_-\hat{\mathbf{y}}')e^{j\phi'}\right\}.$$

This clearly has the form of an elliptically polarized wave with a semimajor (semiminor) axis $\hat{\mathbf{x}}'$ ($\hat{\mathbf{y}}'$), amplitude E_+ (E_-), and phase angle ϕ'. As time t or position z advances, the electric field vector sweeps out the path of the ellipse. The sense of rotation depends on the signs of E_+ and E_- and ultimately on the original value of ψ.

2.4.4 Stokes Parameters

The arbitrarily polarized wave discussed above was parametrized first in terms of the complex amplitudes $(E_1, E_2 e^{j\phi})$, together called the Jones vector, and the Cartesian basis vectors $\hat{\mathbf{x}}$ and $\hat{\mathbf{y}}$. This was converted into a parametrization involving the semi-major and semi-minor axes of an ellipse with a certain orientation and sense of rotation. This parametrization is suitable for graphing. Still another way to parametrize the wave involves the so-called Stokes parameters I, Q, U, and V. Together, these parameters form the Stokes vector. The Stokes vector constitutes an intuitive means of quantifying observations of polarized light.

The elements of the Stokes vector are defined in terms of the expectations of the wave amplitudes in different coordinate systems:

$$I \equiv \langle E_x^2 \rangle + \langle E_y^2 \rangle = \langle E_a^2 \rangle + \langle E_b^2 \rangle = \langle E_l^2 \rangle + \langle E_r^2 \rangle$$
$$Q \equiv \langle E_x^2 \rangle - \langle E_y^2 \rangle$$
$$U \equiv \langle E_a^2 \rangle - \langle E_b^2 \rangle$$
$$V \equiv \langle E_l^2 \rangle - \langle E_r^2 \rangle.$$

In these formulas, the subscripts refer to different coordinate systems. The x and y subscripts refer to a Cartesian coordinate system, and the a and b subscripts refer to another Cartesian coordinate system rotated from the first by $45°$. Finally, the l and r subscripts refer to left- and right-circular coordinate systems, respectively, i.e., $\hat{\mathbf{l}} = (\hat{\mathbf{x}} + j\hat{\mathbf{y}})/\sqrt{2}$ and $\hat{\mathbf{r}} = (\hat{\mathbf{x}} - j\hat{\mathbf{y}})/\sqrt{2}$. The brackets imply the average of the relevant quantity or the expectation in cases where the polarization is a random variable, such as in the case of unpolarized light.

Clearly, I is the intensity of the wave, and Q, U, and V measure the degree of wave polarization in the given coordinate system. For example, written as column vectors, $(1\,1\,0\,0)$, $(1\,-1\,0\,0)$, $(1\,0\,0\,1)$ and $(1\,0\,0\,-1)$ imply horizontally, vertically, left-circularly, and right-circularly polarized waves, respectively.

A number of useful mathematical properties and relationships involving the Stokes vector makes it a useful tool for describing how polarization changes in different propagation contexts. The components Q, U, and V, for example, are points on a sphere (Poincaré sphere) with azimuth and zenith angles in spherical coordinates related to angles inscribed by the polarization ellipse. Stokes parameters are often used in radar polarimetry, where they provide an expedient means of sorting backscatter data from intricate spatial targets.

2.4.5 Helical Antennas

The antennas considered so far all emit linearly polarized signals. Elliptical and circular polarization can be achieved by combining two or more linearly polarized antennas and exciting them out of phase. Crossed dipoles driven $90°$ out of phase are often used to transmit and receive circularly polarized signals, for example. Similarly, crossed Yagi antennas (parasitic arrays of dipoles discussed in the next chapter) can be used to achieve higher gain. Hybrid networks facilitate the feeding of the antennas. However, there are also antennas that are elliptically or circularly polarized natively. One of these is the helical antenna or helix.

Figure 2.15 shows a section of a helical antenna. The total length of an antenna with n turns is nS. The circumference C of the turns is πD. The behavior of the helical antenna depends on whether its dimensions are large or small compared to a wavelength. When $D, nS \ll \lambda$, the helix is electrically small and operates in the so-called "normal" mode. When $C, S \sim \lambda$, the helix operates in the "axial" mode. While the radiation fields can be determined by calculating the vector potential directly as in previous examples, experience has shown that the performance of the helix is well approximated by some simple models.

In the normal mode, each turn of the antenna behaves like the superposition of a Hertzian electric dipole and an elemental magnetic dipole antenna, as shown in Figure 2.15. Since the antenna is electrically small, the current is uniform along the length of the antenna. Since the turns are closely spaced compared to a wavelength, each contributes equally to the far field, and it is sufficient to consider the contribution from a single turn. The contributions from the electric and magnetic dipoles are, respectively,

$$\mathbf{E}_e = jZ_\circ kIS \sin\theta \frac{e^{-jkr}}{4\pi r} \hat{\theta}$$
$$\mathbf{E}_m = \frac{\pi}{4} Z_\circ k^2 ID^2 \sin\theta \frac{e^{-jkr}}{4\pi r} \hat{\phi},$$

with the total far field being the vector sum of the two. Both components share the same $\sin\theta$ spatial dependence, and so the radiation pattern is broadside and wide. The two components are, however, 90° out of phase, and the polarization of the signal is consequently elliptical and the same in all directions. The polarization becomes circular in all directions when the ratio of the amplitudes

$$\frac{|E_\theta|}{|E_\phi|} = \frac{2\lambda S}{C^2}$$

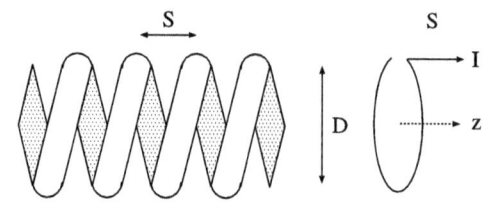

Figure 2.15 Section of a helical antenna with diameter D and turn spacing S. To the right is a normal mode model of a turn of the helix.

is equal to 1. This occurs when the circumference of the helix is the geometric mean of the wavelength and the turn spacing, $C = \sqrt{2\lambda S}$.

If the circumference and turn spacing of the helix are increased until both are comparable to a wavelength, the antenna operates in axial mode. When the circumference of a loop antenna is approximately one wavelength, the currents on opposing points on the loop have the same direction, and constructive interference occurs along the normal axis, which is the endfire pointing axis in this case. The element factor for such a loop is approximated well by $\cos\theta$. Along the endfire direction, the polarization is circular, with a sense determined by the sense of the turns of the helix. The axial mode helical antenna can further be modeled as a uniform linear array of n such loop antennas, with the contributions from the different loop elements adding constructively if they are also separated by about a wavelength. Very high gain is possible when there are a large number of turns, making the antenna a simple and practical means of transmitting and receiving circularly polarized signals. They are, consequently, frequently used for satellite communications, for reasons that will be made clear later in the text. Further analysis of the axial mode helical antenna requires an investigation of antenna arrays.

2.5 Balanced and Unbalanced Transmission Lines and Balun Transformers

At VHF frequencies and below, antennas are usually coupled to transceivers by transmission lines. (Waveguides are usually used at UHF frequencies and above, but not always.) Two types of transmission lines – balanced and unbalanced – are used. Balanced lines are symmetric in the sense that neither of the two conductors differs fundamentally in its construction from the other. An example is twin-lead or ladder line (or "window"), composed of two identical, parallel wires with uniform spacing. The near-absence of a dielectric material between the conductors means that the lines can have very low losses. An example of an unbalanced line is coaxial cable, where one conductor runs inside the other and is thus physically distinguishable. Some kind of dielectric material is generally used to secure the inner conductor. Coaxial cables are popular with RF engineers because they isolate the signals they carry from the environment and are also easy to install and maintain.

Dipole antennas and other antennas based on them are balanced since the two conductors involved are constructed in the same way. Monopole antennas and others based on them are unbalanced. Most transmitters and receivers are unbalanced, with one antenna terminal usually being connected to the internal electronics and the other to the grounded chassis.

Along with balanced/unbalanced transmission lines, antennas, transmitters, and receivers, one can also distinguish between balanced/unbalanced signals or modes. A balanced mode is one in which the currents on the transmission line are equal and opposite everywhere on the line. This was the assumption for all the analyses considered in this chapter. Since the net current everywhere along the transmission line is zero, there can be no radiation from the line. This is normally desirable, since transmission lines are not designed or optimized to radiate. In contrast, the currents in an unbalanced mode are equal and not opposite and do not cancel. This means that the transmission line itself will radiate, becoming effectively part of the antenna, and that the two poles of the dipole antenna will be driven asymmetrically, violating the premise underlying their analysis to this point and altering the performance in some way.

An arbitrary signal can be decomposed into balanced and unbalanced modes, and a balance mismatch on either end of the transmission line is likely to lead to excitation of an unbalanced mode. The unbalanced current component may be supplied by the earth ground, in which case the ground wire itself will radiate. Such a current may be unintentionally rectified, leading to the generation of harmonic spurs. For these reasons and the ones mentioned above, unbalanced modes are normally undesirable and require suppression. This can be accomplished with a balanced/unbalanced transformer or a "balun."

Baluns come in a number of configurations. Figure 2.16 shows three examples. In the first, a transformer is used to match an unbalanced transceiver to a balanced transmission line. In the second, an RF low-pass filter (represented by a ferrite ring) is added where the antenna meets the transmission line. The filter suppresses any net RF current on the transmission line associated with the unbalanced mode without affecting the balanced mode. The third configuration employs a half-wavelength section of transmission line to

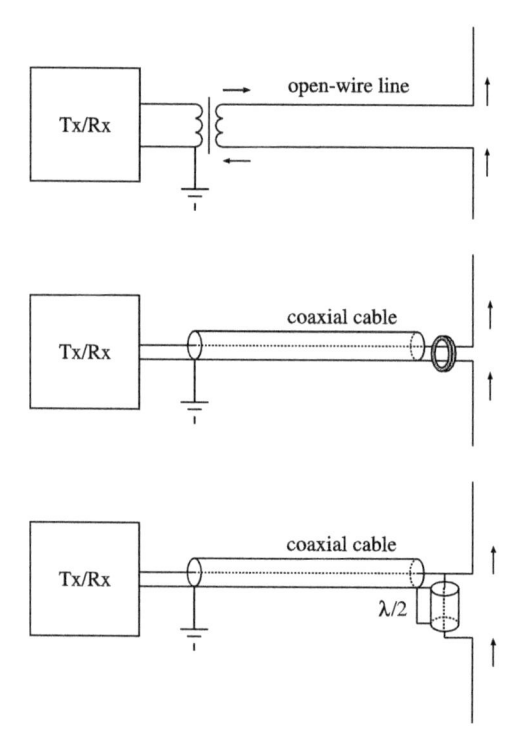

Figure 2.16 Examples of a dipole antenna connected to a transceiver via balanced and unbalanced transmission lines.

excite both poles of a dipole antenna from a common drive point. Different balun configurations may also have differing input and output impedances, which may be useful for simultaneous balance and impedance matching.

2.6 Notes and Further Reading

Two of the most popular introductory texts on electrodynamics are those by D. K. Cheng and D. J. Griffiths. (The former is favored by engineers, and the latter by physicists.) More advanced texts include those by J. D. Jackson, L. D. Landau and E. M. Lifshitz, R. F. Harrington, and W. K. H. Panofsky and M. Phillips.

The basic principles of antenna theory are well known and have been covered in quite a few textbooks, including the very well-known texts by R. E. Collin and W. L. Stutzman and G. A. Thiele. Another truly encyclopedic volume on the subject is the textbook by C. A. Balanis. This text in particular has an extensive treatment of mutual coupling between antenna elements. Another recently updated classic text on antennas is the very comprehensive one by J. D. Kraus and R. J. Marhefka. Practical information about antennas can be found in the ARRL Handbook for Radio Communications, published annually by the American Radio Relay League.

A text that spans electromagnetic theory and antenna theory is "Electromagnetic Wave Theory," by J. A. Kong. This text includes extensive treatments of the field equivalence principle and of reciprocity.

C. A. Balanis. *Theory: Analysis and Design*, 4th ed. John Wiley & Sons, Hoboken, NJ, 2016.

D. K. Cheng. *Field and Wave Electromagnetics*, 2nd ed. Addison-Wesley, Reading, MA, 1989.

R. E. Collin. *Antennas and Radiowave Propagation*. McGraw-Hill, New York, 1985.

D. J. Griffiths. *Introduction to Electrodynamics*, 3rd ed. Prentice Hall, Upper Saddle River, NJ, 1999.

R. F. Harrington. *Time-Harmonic Electromagnetic Fields*, 2nd ed. Wiley-IEEE Press, New York, 2001.

J. D. Jackson. *Classical Electrodynamics*, 3rd ed. John Wiley & Sons, New York, 1975.

J. A. Kong. *Electromagnetic Wave Theory*. EMW Publishing, Cambridge, MA, 2008.

J. D. Kraus and R. J. Marhefka. *Antennas*, 3rd ed. McGraw-Hill, New York, 2001.

L. D. Landau and E. M. Lifshitz. *The Classical Theory of Fields*, 3rd ed. Pergamon Press, New York, 1971.

W. K. H. Panofsky and M. Phillips. *Classical Electricity and Magnetism*, 2nd ed. Dover Publications, New York, 2005.

W. L. Stutzman and G. A. Thiele. *Antenna Theory and Design*, 3rd ed. John Wiley & Sons, New York, 2012.

Exercises

2.1 Starting with the formula given in the text for the magnetic vector potential of a Hertzian dipole aligned with the z axis,

$$\mathbf{A} = \frac{\mu_\circ I dl}{4\pi r} e^{j(\omega t - kr)} \hat{\mathbf{z}},$$

derive the expressions for the corresponding electric and magnetic fields. You will need to use the relationships between the fields and the potentials

$$\mathbf{B} = \nabla \times \mathbf{A}$$
$$\mathbf{E} = -\nabla \phi - \frac{\partial \mathbf{A}}{\partial t}$$

together with the Lorenz gauge condition,

$$\nabla \cdot \mathbf{A} + \frac{1}{c^2} \frac{\partial \phi}{\partial t} = 0.$$

Retain the radiation and storage field components. In order to do this, you will have to compute the time derivative, curl, divergence, and gradient of a function in spherical coordinates. To get started, note that $\hat{\mathbf{z}} = \hat{\mathbf{r}} \cos\theta - \hat{\boldsymbol{\theta}} \sin\theta$. The absence of a phi-dependence in the vector potential will greatly simplify things. Make use of the algebraic relationships between ω, k, c, and $Z_\circ$.

2.2 Using the *full* equations for the **E** and **H** fields generated by an elementary dipole antenna, show by direct integration of the Poynting vector (which is $(1/2)\Re(\mathbf{E} \times \mathbf{H}^*)$ in phasor notation) that the total average power passing through a spherical surface of *any* radius r is

$$40\pi^2 |I|^2 (dl/\lambda)^2 \text{watts}$$

in the near zone as well as in the far zone. This must be true if the power is not to accumulate somewhere. Note that this result does not hold for instantaneous power flow, the difference being due to the oscillating storage fields. Note also that $\mathbf{E} \times \mathbf{H}^* \neq |\mathbf{E}|^2/Z_\circ \hat{\mathbf{r}}$ in general, but only in the far field.

2.3 Suppose the radiated power density from an antenna is proportional to $\cos^n(\theta)$ for $0 < \theta < \pi/2$, where n is an integer, and is zero elsewhere.
   a) Find the expression for the directivity of the antenna. This should be a pencil-and-paper calculation.
   b) What is the minimum value of n giving a maximum directivity of at least 20?

2.4 We saw that the radiation pattern of a half-wave dipole antenna has the functional form

$$f(\theta) = \left[\frac{\cos(\frac{\pi}{2}\cos\theta)}{\sin\theta} \right]^2,$$

where θ is the usual polar angle measured from the z axis, which is aligned with the dipole.

Suppose that the dipole is now reoriented so that it is aligned with the x axis. Express the radiation pattern for this configuration in polar coordinates (θ, ϕ). Repeat the process for a dipole aligned with the y axis. Note that you *cannot* just exchange θ and φ, for example. Your results will be helpful when we consider arrays of dipole antennas that are not aligned with the z axis.

Hint: note that $\cos(\theta)$ is another way to write $\hat{z} \cdot \hat{r}$ and that $\sin^2(\theta) = 1 - \cos^2(\theta)$. You can therefore replace the trigonometric terms in f with factors of $\hat{z} \cdot \hat{r}$. Next, replace $\hat{z}$ with $\hat{x}$ and $\hat{y}$, and rewrite these products in terms of polar coordinates to find the desired answers.

2.5 Return to the radiation pattern from Problem 1. Perform the following steps computationally. Use the value of n you found in part (b).
  a) Plot the radiation pattern in polar coordinates in the plane. Make the plot have a dB scale. Since the radiated power density in this problem varies from 0 to 1, the base-10-logarithm will vary from $-\infty$ to 0. To make your plot, add an arbitrary number (say 30 dB) and set a floor for plotting (say 0 dB). Put labels on the plot and make sure it is intelligible.
  b) Compute the directivity of the pattern and make sure your answer satisfies part (a) of the problem.

2.6 Sometimes we need to resolve waves into orthogonally polarized pairs. Suppose a plane wave propagating in free space with a frequency of 250 MHz is described by

$$\mathbf{E} = E_\circ(3\hat{x} + 2e^{j\pi/4}\hat{y})e^{j(\omega t - kz)},$$

where $E_\circ = 1$ mV/m.
  a) How much power would a half-wave dipole, oriented parallel to the $\hat{y}$ axis, receive, assuming it is connected to a matched receiver?
  b) How much power would be received by an antenna designed to receive left circularly polarized radiation if its effective area is 20 m^2?

2.7 Starting with the results given for a dipole antenna of arbitrary length, find the expression for the total radiated power of a dipole of total length 1.5λ and compute the directivity function $D(\theta, \phi)$, the HPBW in the E-plane, the effective area for reception, and the radiation resistance. Be careful to use the correct form for the driving current. Compare the maximum directivity, HPBW, and radiation resistance with that of a half-wave dipole. Try to account for the differences qualitatively.

You will have to perform a numerical integral to get the directivity and the radiation resistance. Finding the HPBW is easily done by trial and error, although you can find this numerically as well if you wish. It will probably be helpful to plot the radiation pattern to make sure that you've calculated the HPBW correctly.

2.8 Demonstrate the reciprocity theorem with a simple example. Consider a plane wave in free space incident on a short dipole antenna from a distant source. The amplitude of the wave's electric field is 1 V/m. Using the reciprocity theorem, calculate the power received by the antenna and delivered by a matched load. Next, repeat the calculation, this time without using the reciprocity theorem. Using the effective length of the antenna, calculate the open-circuit voltage induced across the antenna terminals. Then, calculate the power delivered to a matched load using circuit theory. It will be helpful to draw an equivalent circuit for the antenna connected to the matched load. The two results should agree.

3
Antenna Arrays

None of the antennas considered so far has the potential for very high gain, as is required for radar. There are two practical ways to achieve high gain. One involves using large-aperture antennas, a subject covered in the next chapter. The other is to combine a number of small, low-gain antennas together into arrays. Not only does this permit very high gain, it also serves as the foundation for electronic beam steering techniques widely used in radar applications. Of course, large-aperture antennas can also be organized into arrays, increasing the gain still further.

The basic idea is to consider the radiation from an assembly of n identical elemental antennas, with the ith element located at $\mathbf{r}_i$ as shown in Figure 3.1. Let $\hat{\mathbf{r}}$ be the unit vector from the origin to the observer. The vector $\mathbf{R}_i$ is the displacement from the ith element to the observer. If the distance to the observer is large compared to any of the $\mathbf{r}_i$, then all of the $\mathbf{R}_i$ are parallel, with lengths given by $R_i = r - \hat{\mathbf{r}} \cdot \mathbf{r}_i$, i.e., the distance from the observer to the origin minus the projection of the element location on $\hat{\mathbf{r}}$. This is essentially the same "far-field" limit adopted for analyzing long-wire antennas in the previous chapter. In this limit, we expect the directivity of the antenna to be invariant with range. Be aware that certain large wire and array antennas may well violate this limit, in which case performance will have to be re-evaluated. Some of the consequences will be considered in the next chapter of the text.

Suppose that the electric field produced by an elemental radiator placed at the origin and excited with unity amplitude is given by

$$\mathbf{E}(\mathbf{r}) = f(\theta, \phi) \frac{e^{-jkr}}{4\pi r},$$

where $f(\theta, \phi)$ is the element factor. Then the radiation seen by an array of such elements with complex excitations given by $C_i e^{j\alpha_i}$ is given by superposition:

$$\mathbf{E}(\mathbf{r}) = \sum_{i=1}^{n} C_i e^{j\alpha_i} f(\theta, \phi) \frac{e^{-jkr + jk\hat{\mathbf{r}} \cdot \mathbf{r}_i}}{4\pi r}$$

$$= f(\theta, \phi) \frac{e^{-jkr}}{4\pi r} \sum_{i=1}^{n} C_i e^{j\alpha_i + jk\hat{\mathbf{r}} \cdot \mathbf{r}_i}.$$

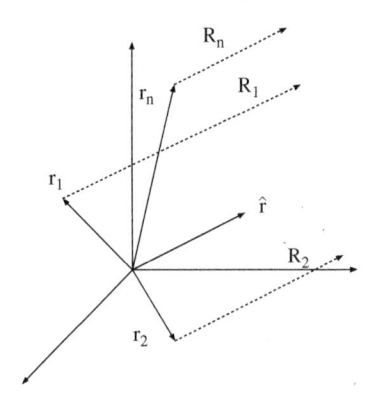

Figure 3.1 Antenna array configuration.

Note that we have substituted R_i in the exponential (phase) terms, but used a common factor of r in the denominator for all the elements in accordance with the assumption that $r_i \ll r$. This allows us to express the radiation field in terms of a common factor of $e^{-jkr}/4\pi r$, the signature for spherically expanding wavefronts. We were also able to bring the common element factor $f(\theta,\phi)$ outside the summation. The remaining summation is unique to the particular antenna array in question. This term is called the array factor $F(\theta,\phi)$, and determines how the array configuration influences the radiation:

$$\mathbf{E}(\mathbf{r}) = f(\theta,\phi) F(\theta,\phi) \frac{e^{-jkr}}{4\pi r}$$

$$F(\theta,\phi) = \sum_{i=1}^{n} C_i e^{j\alpha_i + jk\hat{\mathbf{r}} \cdot \mathbf{r}_i}.$$

The Poynting flux and directivity of the antenna array are proportional to the square of the modulus of the electric field:

$$D(\theta,\phi) \propto |f(\theta,\phi)|^2 |F(\theta,\phi)|^2, \tag{3.1}$$

where $|f(\theta,\phi)|^2$ and $|F(\theta,\phi)|^2$ are known as the element pattern and array pattern, respectively, and are usually normalized to unity by convention. Whereas the former of these is typically broad, the latter can be made very narrow using a large number of widely separated antenna elements.

Furthermore, (3.1) implies an important property of antenna arrays known as the *principle of pattern multiplication*. This says that the radiation pattern of an array of identical antennas is given by the product of the radiation pattern of the antenna elements and the array pattern. The principle is recursive. One can conceive of building complicated structures that are arrays of arrays, or arrays of arrays of arrays, in which case the principle of pattern multiplication gives a recipe for simple calculation of the overall radiation pattern.

3.1 Uniform Linear Arrays

Although the sums expressed above can be calculated readily with the aid of a computer, they can be tedious and do not lend themselves to manual calculation and analysis. Fortunately, there are a number of important classes of antenna arrays that lend themselves to considerable simplification. One of these is the uniform linear array. These arrays are made of a number of identical, evenly spaced elements driven at uniform amplitude and with a linear phase progression. They are easy not just to analyze but also to build and feed.

Refer to Figure 3.2. The element indices run from 0 to N by convention, meaning that there are $N+1$ elements. The nth of these has constant amplitude $C_n = I_\circ$ and phase $\alpha_n = n\alpha d$, where we interpret α as the phase progression per unit length from one element to the next. With ψ being the included angle between the array direction (here $\hat{x}$) and the direction to the observer $\hat{r}$, the product $\hat{r} \cdot \mathbf{r}_n = nd\cos\psi$.

Consequently, the array factor becomes

$$F(\theta,\phi) = I_\circ \sum_{n=0}^{N} e^{j\alpha nd + jknd\cos\psi}$$

$$= I_\circ \sum_{n=0}^{N} e^{jn\xi}$$

$$\xi \equiv kd\cos\psi + \alpha d.$$

Remarkably, the sum can be expressed in closed form. Consider differencing the following two equations to produce the third:

$$F = I_\circ \sum_{n=0}^{N} e^{jn\xi}$$

$$Fe^{j\xi} = I_\circ \sum_{n=1}^{N+1} e^{jn\xi}$$

$$F\left(1 - e^{j\xi}\right) = I_\circ \left(1 - e^{j(N+1)\xi}\right)$$

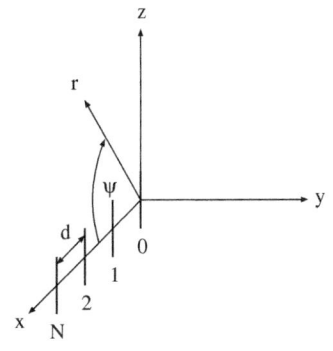

Figure 3.2 Uniform linear array of dipole antennas.

and then solving for F,

$$F(\theta,\phi) = I_o \frac{1 - e^{j(N+1)\xi}}{1 - e^{j\xi}}$$
$$= I_o \frac{1 - e^{j(N+1)(k\cos\psi+\alpha)d}}{1 - e^{j(k\cos\psi+\alpha)d}}.$$

This can be further simplified with the aid of the Euler theorem and some trigonometric identities:

$$F(\theta,\phi) = I_o \underbrace{e^{j\frac{N}{2}(k\cos\psi+\alpha)d}}_{\text{phase factor}} \frac{\sin\left(\frac{N+1}{2}(k\cos\psi+\alpha)d\right)}{\sin\left(\frac{1}{2}(k\cos\psi+\alpha)d\right)}.$$

The phase factor above does not affect the array pattern, which depends only on the modulus of F,

$$|F(\theta,\phi)| = I_o \left|\frac{\sin\left(\frac{N+1}{2}(k\cos\psi+\alpha)d\right)}{\sin\left(\frac{1}{2}(k\cos\psi+\alpha)d\right)}\right|$$
$$= I_o \left|\frac{\sin\left(\frac{N+1}{2}(u+u_o)\right)}{\sin\left(\frac{1}{2}(u+u_o)\right)}\right|,$$

where the auxiliary variables $u = kd\cos\psi$ and $u_o = \alpha d$ have been defined.

Figure 3.3 shows an example array factor (modulus) for a uniform linear array. Note that the array pattern is often plotted in polar coordinates, often on a dB scale. When plotting it in Cartesian coordinates like this, it is important to remember that, while the array factor is a periodic function of its arguments, u can only take on values between $\pm kd$ since $\cos\psi$ can only take on values between ± 1. The interval of allowed values for u is called the *visible region*; only those features of the beam shape falling within the visible region are actually expressed in space.

A little analysis reveals the following properties of the array factor:

- The maximum value of $|F|$ is $I_o(N+1)$ prior to normalization, i.e., the wave amplitude is multiplied by the number of elements.

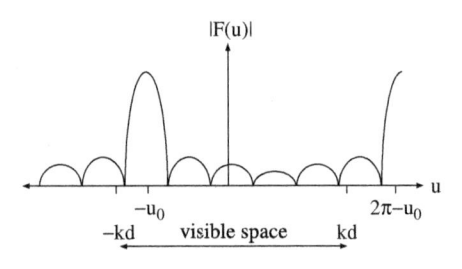

Figure 3.3 Array factor for a ULA.

- Major lobes occur where $u = -u_\circ$ (main lobe) and $(u+u_\circ)/2 = m\pi$ (grating lobes).
- There are $N-1$ minor lobes (sidelobes) between major lobes.
- The largest sidelobes occur where $(u+u_\circ)(N+1)/2 \approx \pm 3\pi/2$. (The $\pm\pi/2$ points lie within the main lobe.)
- Nulls occur in between the sidelobes where $(u+u_\circ)(N+1)/2 = \pm\pi$.
- The location of the main beam is controlled by the phase progression α.

To repeat – among the features itemized above, only those falling within the visible region will be part of the actual radiation pattern.

3.1.1 Broadside Array

The broadside array is an important special case in which the main beam of the radiation pattern is directed normal to the array line. It occurs when the phases of the array elements are uniform, so that $u_\circ = \alpha d = 0$. The main beam then falls where $u = kd\cos\psi = 0$ or $\psi = \pi/2$. This is the equation of a plane normal to the array line. The main beam of a broadside array is toroidal.

Other major lobes (grating lobes) occur where $u = kd\cos\psi = 2\pi m$, where m is a nonzero integer of either sign. Grating lobes are suppressed if this condition is met only outside the visible region. This is the case when $d < \lambda$.

Nulls in the radiation pattern occur where $kd\cos\psi(N+1)/2 = \pm\pi$ or

$$\cos\psi = \frac{\pm\lambda}{(N+1)d}.$$

The corresponding BWFN can be estimated with the help of a few approximations. Taking N to be a large number, we expect $\cos\psi$ to be small, such that $\psi \sim \pi/2$. Expanding $\psi = \pi/2 \pm \Delta\psi$ allows us to approximate $\cos\psi \sim \pm\sin\Delta\psi \sim \pm\Delta\psi$. The BWFN is then given by $2\Delta\psi$ or about $2\lambda/L$, where $L \equiv Nd$ is the array length. This illustrates the need for long arrays (measured in wavelengths) to achieve narrow main beams.

Suppose now that the broadside array is composed of half-wave dipoles spaced by a half wavelength as shown in Figure 3.4. The principle of pattern multiplication specifies what the radiation pattern will look like. In this example, the array pattern has nulls along the x axis, whereas the element pattern has nulls along the z axis. The combined pattern will therefore have nulls along both sets of axes and main beams along the remaining $\pm y$ axes. The E-plane radiation pattern is entirely controlled by the half-wave dipole pattern, and the E-plane HPBW is consequently $\sim 78°$ or 1.36 rad. The H-plane pattern is controlled by the array pattern and may have considerably narrower beams, depending on the number of array elements.

An exact calculation of the directivity requires numerical integration of the combined radiation pattern. We can estimate the directivity, however, using

$$D \approx \frac{4\pi}{\text{HPBW}_E \, \text{HPBW}_H}.$$

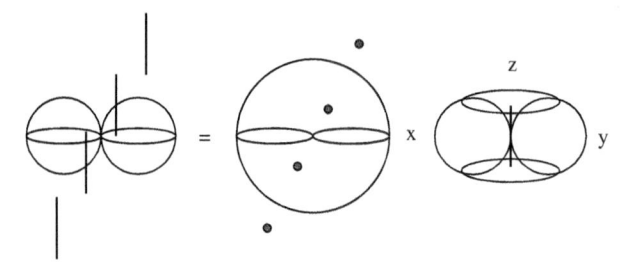

Figure 3.4 Pattern multiplication example for a broadside uniform linear array composed of half-wave dipoles spaced by a half wavelength. The array and element patterns are figures of revolution about the x and z axes, respectively.

The E-plane beamwidth was already found to be 1.36 rad. We determine the H-plane HPBW by finding the angle ψ that makes the array pattern half its maximum value.

$$\frac{\sin^2\left(\frac{N+1}{2}u\right)}{\sin^2\left(\frac{1}{2}u\right)} = \frac{1}{2}(N+1)^2$$
$$\sin^2\left(\frac{N+1}{2}u\right) = \frac{1}{2}(N+1)^2\left(\frac{u}{2}\right)^2, \qquad (3.2)$$

where we assume that N is large, the main beam is narrow, and u is sufficiently small near the main beam to permit the replacement of $\sin(u/2)$ with its argument. The product $(N+1)u/2$ is too large to permit this approximation for the sine term in which it appears. However, retaining higher-order terms in the Taylor expansion gives us:

$$\sin\frac{N+1}{2}u \approx \frac{N+1}{2}u - \frac{1}{6}(N+1)^3\left(\frac{u}{2}\right)^3.$$

We can equate this expansion with the square root of the right side of (3.2) and solve the resulting quadratic equation for the value of u corresponding to a half-power point in the array pattern.

$$u_{1/2} = \pm\frac{2.65}{N+1} = kd\cos\left(\pi/2 \pm \Delta\psi_{1/2}\right)$$
$$\approx \pm kd\Delta\psi_{1/2},$$

where ψ has once again been expanded as $\pi/2 \pm \Delta\psi$. Finally,

$$\text{HPBW}_H = 2\Delta\psi_{1/2}$$
$$= \frac{2.65\lambda}{(N+1)\pi d}$$
$$\approx \frac{2.65\lambda}{\pi L},$$

3.1 Uniform Linear Arrays

which is again proportional to $(L/\lambda)^{-1}$ and slightly less than half the BWFN. Altogether, the estimate for the directivity of the array of half-wavelength dipoles is:

$$D = \frac{4\pi}{2 \cdot 1.36 \cdot \text{HPBW}_H} \approx 5.48 \frac{L}{\lambda}. \tag{3.3}$$

Note that the expression for the estimate of D based on the E- and H-plane HPBW assumes the presence of only a single beam. The factor of 2 in the denominator here reflects the fact that this particular array has two equal beams, halving the directivity.

This is the first high-gain antenna we have studied; D can be made arbitrarily large, given sufficiently large L. For example, consider a 21-element ($N = 20$) array of half-wave dipoles spaced $d = 0.9\lambda$ apart to avoid grating lobes. According to (3.3), the directivity for this antenna should be 103, or about 20 dB. The H-plane pattern for this array is shown in Figure 3.5. Numerical calculations show that the directivity is actually about 77, or about 18.9 dB. The discrepancy is rooted in the sidelobes of the pattern, which were neglected in the analysis above.

3.1.2 Endfire Array

The endfire array is another special case in which the excitation of the array elements is set so that the main beam is in the direction of the array line. This is accomplished by setting

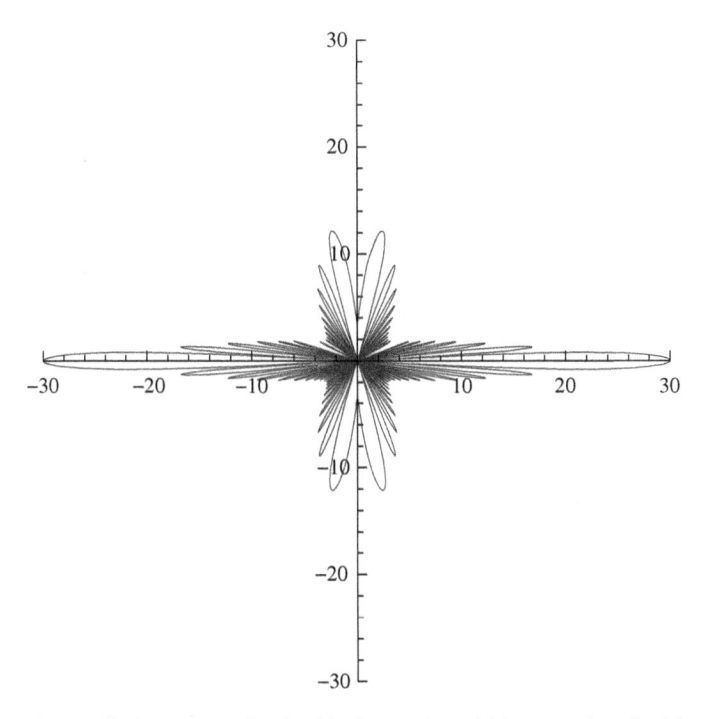

Figure 3.5 H-plane radiation pattern for the 21-element broadside array described in the text. The polar plot reflects a log scale spanning 30 dB of dynamic range.

$u_\circ = -kd$ so that the main beam occurs where $u = kd\cos\psi = -u_\circ = kd$ or where $\psi = 0$. (Taking $u_\circ = kd$ gives a main beam in the opposite direction.)

Now, grating lobes will occur where $u + u_\circ = kd(\cos\psi - 1) = 2m\pi$, where m is a nonzero integer. These can be excluded from the visible region only if $d < \lambda/2$ – half the distance necessary to exclude them for a broadside array. Closer element spacing is required, implying lower directivity for the same number of elements.

For an endfire array,

$$|F| = I_\circ \left| \frac{\sin([(N+1)/2]kd(\cos\psi - 1))}{\sin((kd/2)(\cos\psi - 1))} \right|.$$

Let us once again estimate the BWFN from this expression. The nulls just beside the main beam occur where $kd(\cos\psi - 1)(N+1)/2 = \pm\pi$. For a large array, we assume $\psi \sim \Delta\psi$ is a small quantity, and we can therefore expand $\cos\Delta\psi \sim 1 - (\Delta\psi)^2/2$ this time, leading to

$$\frac{(\Delta\psi)^2}{2} = \frac{2\pi}{(N+1)kd}.$$

This time, BWFN $= 2\Delta\psi = 2\sqrt{2\lambda/L}$, which is only proportional to the *square root* of the reciprocal array length in wavelengths. (For the broadside array, it is proportional to the reciprocal of the array length in wavelengths.) While this is not as narrow as the broadside array, note that narrowing is occurring in both planes and not just one. Overall, the directivity increases with increasing array length about as quickly for broadside and endfire arrays (see below). Note that, for long arrays, the element pattern is less likely to contribute to the directivity of an endfire array than to that of a broadside array.

Neglecting the element pattern and repeating the kind of analysis performed for the broadside array, we find

$$\Delta\psi_{1/2} = 1.64 \left[\frac{\lambda}{\pi(N+1)d} \right]^{1/2}$$

$$\text{HPBW} \approx 1.83 \sqrt{\frac{\lambda}{L}},$$

which is again a little more than half the BWFN. The HPBW is the same in the E and H plane in this analysis, which neglected the contribution from the element pattern.

Let us estimate the directivity of the endfire array in another way by beginning with an estimate of the beam solid angle. We can regard the main beam as occupying a cone with an angular width of $\Delta\psi_{1/2}$. The solid angle is therefore

$$\Omega = \int_0^{2\pi} \int_0^{\Delta\psi_{1/2}} \sin\theta \, d\theta \, d\phi$$
$$= 2\pi(1 - \cos\Delta\psi_{1/2})$$
$$\approx \pi(\Delta\psi_{1/2})^2.$$

3.1 Uniform Linear Arrays

Therefore, we can estimate $D = 4\pi/\Omega \approx 4.73L/\lambda$. While this is somewhat less than the figure for a broadside array of the same length, recall that it neglects the contribution from the element pattern and so represents an underestimate.

3.1.3 Hansen–Woodyard Array

It turns out surprisingly that the directivity of an endfire array can be increased somewhat by taking $|u_\circ| = |\alpha d| > kd$ so as to actually exclude the peak of the main lobe from the visible region. The Hansen–Woodyard condition for maximum directivity is given by

$$\alpha d = u_\circ = -kd - \frac{p}{N+1},$$

where p is a number close to 3 that depends on the number of elements in the array. This phasing gives a more peaked main beam at the expense of elevated sidelobes. The correction becomes less significant the greater the number of elements in the array.

A practical example of a Hansen–Woodyard array is given by an axial mode helical antenna, which was introduced at the end of Chapter 2. A helical antenna is an example of a traveling wave antenna, where a traveling current wave propagates along the antenna wire (versus the usual standing wave discussed so far), providing the source of radiation. The traveling wave propagates somewhat slower than the speed of light because of interactions with the material that makes up the antenna. This is the distinguishing feature of so-called "slow-waves" that propagate on transmission lines, waveguides, and guiding structures of all kinds. Remarkably, the additional phase delay introduced in this case can be enough to satisfy the Hansen–Woodyard condition.

Recall that the turns of the axial-mode helix could be regarded as elements of an endfire array, each with an element factor approximated by $\cos\theta$. Assuming uniform amplitude along the helix, the combined element-array factor is then

$$F(\psi) = \cos\theta \frac{\sin((N+1)\psi/2)}{(N+1)\sin(\psi/2)},$$

where $N+1$ is now the number of turns, $\psi = kS\cos\theta + \alpha S$, $d \to S$ is the turn spacing, and αS is the phase progression of the driving current from turn to turn. Note that the phase delay from turn to turn can be written as $\alpha S = -\beta C$, where C is the circumference of a turn and β is the propagation constant for the current waves traveling on the wire. This is not the same as the propagation constant for waves in free space $k = \omega/c = 2\pi/\lambda$, and the difference introduces the required phase delay. Is it enough?

$$-\beta C \stackrel{?}{=} -kS - \frac{\pi}{N+1} - 2\pi.$$

Let us use this equation to calculate the velocity factor for the slow waves, $f = v/c$, where v is their phase velocity. Note that we are always free to add a factor of 2π to the $u_\circ$ without modifying the problem; doing so at the end of the equation above admits physically reasonable velocity factors for a typical antenna configuration. Writing $\beta = \omega/v$, we can find:

$$f = \frac{C/\lambda}{S/\lambda + \frac{N+3/2}{N+1}}.$$

Since $C \sim \lambda$ and $S < \lambda$ typically, this formula predicts velocity factors somewhat less than unity, which are what are expected to occur in practice. Helical antennas can therefore satisfy the Hansen–Woodyard condition if properly constructed.

3.2 Binomial Arrays

The arrays considered so far employed uniform amplitudes for excitation. This situation is both the easiest to analyze and the easiest to construct in the field. However, there are times when it is worthwhile to shape or taper the excitation of the array elements, for instance in the pursuit of a very specific beam shape. Tapering, the gradual reduction of element excitation with distance from the physical center of the array, is also a means of reducing sidelobes, at the cost of reduced effective area and directivity. A particularly simple means of eliminating sidelobes altogether is the binomial array.

A binomial array is one with element excitations specified by the coefficients of a binomial expansion $(a+b)^n$ as given by Pascal's triangle. A three-element binomial array has element amplitudes in the proportion 1:2:1, etc. The amplitudes are thus tapered, and in a special way:

Table 3.1 *First few rows of Pascal's triangle.*

```
                1
            1       1
        1       2       1     ←
     1      3       3       1
  1      4       6       4       1
```

Suppose those three elements are uniformly spaced by an incremental distance d. Notice that this three-element array can be viewed as the superposition of two collinear two-element arrays spaced also by a distance d. The two inside array elements overlap, giving rise to a doubling of the amplitude in the center element of the net array. What we have is, in effect, a two-element array of two-element arrays. We can therefore use the principle of pattern multiplication.

A two-element linear array with element separation d has the array pattern

$$|F(\theta)|^2 = \cos^2(kd\cos(\theta)/2),$$

where θ is measured off the array line. Using the principle of pattern multiplication, a three-element binomial array therefore has the array pattern

$$|F(\theta)|^2 = \cos^4(kd\cos(\theta)/2).$$

Continuing on, a four-element binomial array can be viewed as a two-element array of three-element arrays, with the total array pattern given by cosine to the sixth power. In general, the array pattern goes as cosine to the nth power, $n = 2(m-1)$, with m the number of elements in the array. The element factor can then be incorporated into the overall radiation pattern of the array using pattern multiplication as usual.

Since the two-element linear array has no sidelobes for $d \leq \lambda$, neither will a binomial array of any total length, so long as $d \leq \lambda$. Tapering reduces sidelobe levels in general; in this case, it eliminates them entirely. The tradeoff comes in the form of a broadened main beam and reduced directivity compared to a uniform linear array.

3.3 Inter-Element Coupling and Parasitic Arrays

Chapter 2 dealt with the properties of simple antennas used for transmission, where current imposed on the antenna produced electromagnetic radiation, and reception, where electromagnetic radiation in the environment induced current on the antenna. In either case, the antenna was considered in isolation. When another conducting or dielectric body is introduced into the picture, antenna performance changes. "Parasitic" current will be induced on the body by ambient electromagnetic radiation, and the body will radiate in turn. As a consequence, new current will appear on the antenna, leading to additional radiation, etc. An equilibrium situation will develop in which both the radiation pattern of the antenna and the current and voltage distribution (and the impedance) are altered from the isolated case.

Now consider the fact that every element of an antenna array is such a body insofar as the other elements are concerned. On transmission and reception, currents will appear on every element in response to currents on all the other elements. Such inter-element coupling has been neglected throughout the discussion of antenna arrays, but is not generally insignificant. Solving for the impedances of the driven elements in an array and the overall radiation pattern is usually performed computationally. The Numerical Electromagnetic Code (NEC) produced at the Lawrence Livermore National Laboratory is widely available and employed for such purposes.

In some antennas, inter-element coupling is exploited as a key design feature. Consider the case of two dipole antennas, only one being driven, aligned side-by-side. If the parasitic element is cut to the right length and placed sufficiently close to the driven element for strong coupling to occur, an array is formed by the two antennas with possibly high directivity. The directivity can be increased further with the introduction of more parasitic elements. Such parasitic arrays are practical to build since they can be driven from a single point, obviating the need for complicated feed networks.

We consider here the simplest case of a two-element array – one driven, and one parasitic. Examples of suitable elements are wire dipoles, as shown in Figure 3.6. Such an array can be viewed generally as a two-port linear network, with

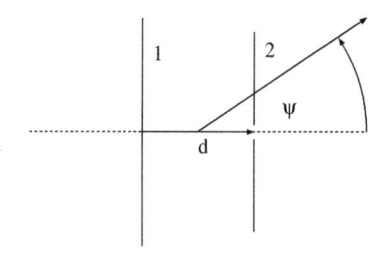

Figure 3.6 Diagram of a two-element parasitic array. Elements 1 and 2 are the parasitic and driven elements here, respectively.

$$\begin{pmatrix} Z_{11} & Z_{12} \\ Z_{21} & Z_{22} \end{pmatrix} \begin{pmatrix} I_1 \\ I_2 \end{pmatrix} = \begin{pmatrix} 0 \\ V_2 \end{pmatrix},$$

where the subscripts refer to the first (parasitic) and second (driven) element in the pair, V, I, and Z refer to the element feedpoint voltages, currents, and impedances, and the parasitic element feedpoint is short-circuited so that $V_1 = 0$. The diagonal or self-impedance terms Z_{11} and Z_{22} are the driving-point impedances of the two dipoles in isolation estimated in Section 2.3.2. The off-diagonal, cross, or mutual impedance terms Z_{12} and Z_{21} represent changes in the driving-point impedance due the presence of the neighboring antenna. They can be calculated in a manner similar to the self-impedances, except considering complex Poynting fluxes arising from currents in both antenna elements, i.e., the magnetic field from one and the electric field from the other.

Solving for the feedpoint currents gives (with $Z_{12} = Z_{21}$)

$$\begin{pmatrix} I_1 \\ I_2 \end{pmatrix} = \frac{V_2}{Z_{11}Z_{22} - Z_{12}^2} \begin{pmatrix} -Z_{12} \\ Z_{11} \end{pmatrix},$$

which specifies the complex ratio of the element amplitudes. The corresponding normalized array factor is then

$$|F(\psi)| = |1 - |Z_{12}/Z_{11}| e^{j\alpha d - jkd\cos\psi}|,$$

where the angle ψ is measured off the array line as usual and where αd is the phase angle of the quotient Z_{12}/Z_{11}.

Ideally, we might want $|F|$ to have a maximum along the line of fire ($\psi = 0$) and a minimum in the opposite direction ($\psi = \pi$). This requires strong inter-element coupling so that $|Z_{12}/Z_{11}|$ is a substantial fraction of unity. The first and second conditions also imply $\alpha d - kd \sim \pm \pi$ and $\alpha d + kd \sim 0, 2\pi$, respectively. Strong coupling will only occur for $d \leq \lambda/4$. Assuming $d = \lambda/4$, the two conditions can be satisfied if $\alpha d = -\pi/2$. In this case, the parasitic element will act as a reflector.

Equally well, we might want $|F|$ to have maxima and minima in the $\psi = \pi$ and $\psi = 0$ directions. In that case, and by the same arguments, the corresponding conditions are satisfied by $\alpha d = \pi/2$. This time, the parasitic element acts as a director.

Whether a parasitic element acts like a reflector or a director depends on the phase angle αd, which depends mainly on the negative of the phase angle of the parasitic element impedance Z_{11}. (The phase angle of Z_{11} varies much more with antenna length than that of Z_{12} near resonance.) Elements longer than (shorter than) a half wavelength were already found to be inductive (capacitive), having a positive (negative) phase angle and therefore contributing to negative (positive) values of αd. Hence, long parasitic elements behave like reflectors, whereas short elements behave like directors.

So-called "Yagi-Uda" or just "Yagi" antennas have both a reflector and a director. Antennas with multiple directors and sometimes more than one reflector can be constructed, although the increasing distance of the additional elements from the driven element leads to diminishing returns. A 20-element Yagi antenna can have up to about 19 dBd of gain, and antennas with more elements than this are seldom used. The optimal inter-element spacing decreases with increasing numbers of elements, and the directors get smaller with distance from the driven element. Design tables for Yagi antennas optimized for different objectives (high gain, low sidelobes, high front-to-back ratios, etc.) have been determined empirically and computationally and are available in the literature.

Yagi antennas have two shortcomings. For one, they tend to have feedpoint impedances significantly smaller than that of the driven element in isolation. This problem can be combated by using a folded dipole as the driven element. Most practical Yagi antennas also have matching networks built into the feed system to facilitate matching to common transmission lines. A second shortcoming is that Yagi antennas tend to be rather narrow-banded, with bandwidths of only a few percent of the operating frequency. Strategies for broadening the bandwidth attempt to divorce the design of the antenna from strict adherence to spatial scaling laws. This can be done, for example, by bending the elements into "V" shapes and shaping them to look like fat bow ties rather than slender wires. Element lengths and spacings can also be made to vary logarithmically rather than linearly, giving rise to the so-called "log-periodic" broadband antenna.

3.4 Uniform Two-Dimensional Arrays

We next consider the uniform 2-D antenna array, which can be viewed as a uniform linear array of uniform linear arrays, in view of the principle of pattern multiplication. We assume identical elements with equal excitations, only allowing for a linear, progressive phase shift across the elements of the form $\exp(jn\alpha d + jm\beta d)$, where n and m are index counters and α and β are the phase shifts per unit length in the two cardinal directions of the array. For simplicity, we take the inter-element spacing d to be the same in both directions. The total number of elements is $(N+1)(M+1)$ (see Fig. 3.7).

Previously, we defined $\cos\psi$ as the dot product between unit vectors in the directions of the array line and the observer, e.g., $\cos\psi = \hat{\mathbf{r}} \cdot \hat{\mathbf{x}}$, referring back to Figure 3.2. However, we need two angles to specify the bearing with respect to a 2-D array. In this example, where the array lines are aligned with the x and z axes, we can make use of polar coordinates and write $\cos\theta = \hat{\mathbf{r}} \cdot \hat{\mathbf{z}}$ and $\sin\theta\cos\phi = \hat{\mathbf{r}} \cdot \hat{\mathbf{x}}$. We consequently define the auxiliary variables,

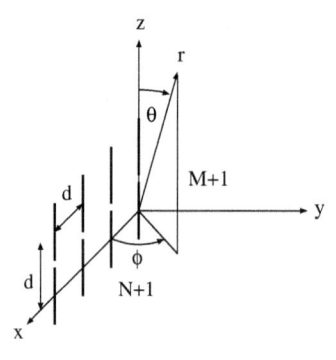

Figure 3.7 Uniform two-dimensional (2-D) array of dipole antennas.

$$u = kd\sin\theta\cos\phi, \quad u_\circ = \alpha d$$
$$v = kd\cos\theta, \quad v_\circ = \beta d$$

and proceed as before. Invoking the principle of pattern multiplication allows us to write

$$|F| = I_\circ \left| \frac{\sin\{[(N+1)/2](u+u_\circ)\}\sin\{[(M+1)/2](v+v_\circ)\}}{\sin[(u+u_\circ)/2]\sin[(v+v_\circ)/2]} \right|.$$

Some comments about the array factor above are in order:

- The maximum value of $|F|$ prior to normalization is at most $I_\circ(N+1)(M+1)$.
- The main beam occurs where $u = -u_\circ$, $v = -v_\circ$. If $\alpha = \beta = 0$, this is normal to the plane of the array. The beam may be steered by adjusting α and β, at the expense of the directivity. This is the idea behind phased-array radar (vs. mechanically steered radar).

The results for the 1-D broadside uniform linear array can be applied to an estimate of the directivity of the 2-D array,

$$\text{HPBW}_{xy} = \frac{2.65\lambda}{(N+1)\pi d}$$

$$\text{HPBW}_{yz} = \frac{2.65\lambda}{(M+1)\pi d}$$

$$D \approx \frac{4\pi}{2 \cdot \text{HPBW}_{xy} \cdot \text{HPBW}_{yz}}$$

$$= \frac{8.83(N+1)(M+1)d^2}{\lambda^2}$$

$$\approx 8.83 \frac{A_{phys}}{\lambda^2},$$

where the extra factor of 2 in the denominator again reflects the presence of two main beams in the $\pm\hat{y}$ direction. Note that the directivity will be reduced if the beam is steered away from broadside.

3.5 Three-Dimensional Arrays

The 3-D generalization of the problem should be obvious by now. The radiation pattern for a 3-D array has the form

$$\begin{pmatrix} \text{power} \\ \text{pattern} \end{pmatrix} = \begin{pmatrix} \text{element} \\ \text{pattern} \end{pmatrix} \begin{pmatrix} \text{array} \\ \text{pattern} \end{pmatrix}_x \begin{pmatrix} \text{array} \\ \text{pattern} \end{pmatrix}_y \begin{pmatrix} \text{array} \\ \text{pattern} \end{pmatrix}_z,$$

where care must be taken to make the geometry of each component consistent. Each component will depend on the polar coordinates θ and ϕ in a different way, depending on the direction of the given array line. The notation can be complicated, but the idea is straightforward.

One of the terms above could be associated with locating an antenna array above ground. The antenna and its image below ground form a two-element array with its own array pattern. The relative phasing of the antenna and its image depends on the orientation of the antenna elements with respect to ground. Figure 3.8 shows vertically and horizontally polarized dipole antennas located a distance h above a perfectly conducting ground plane. Arrows show the sense of the current that flows in a snapshot in time. Imagining that the current in the vertical antenna is carried by positive charge carriers moving upward, the current in its image would be carried by negative charge carriers moving downward. In either case, the direction of the current is upward, and the antenna and its image are in phase. For the horizontal antenna, however, the charge carriers all move to the right, and the antenna and its image are consequently out of phase by 180°.

The two-element array composed of an antenna and its image has the array factor

$$|F| = \left| \frac{\sin(v+v_\circ)}{\sin((v+v_\circ)/2)} \right|$$

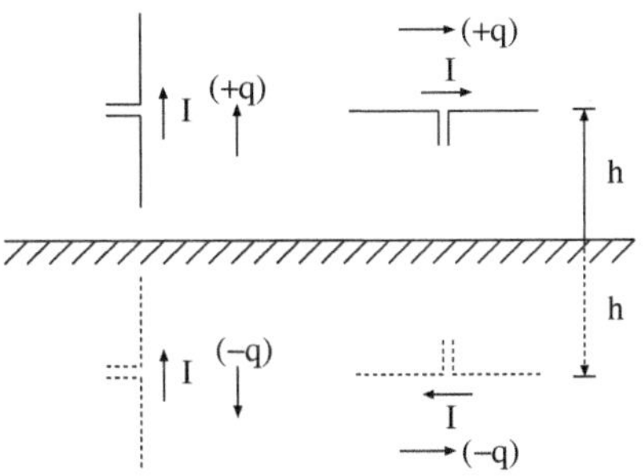

Figure 3.8 Vertical and horizontal dipole antennas above ground with their images.

(note that $N+1 = 2$ here). For the case of horizontally polarized antennas placed a quarter-wavelength above ground, we have $v_o = \pm\pi$ and $d = 2h = \lambda/2$. Taking the array line to lie in the z direction, this gives $v + v_o = \pi(\cos\theta \pm 1)$. The resulting array factor consequently has a maximum in the vertical, as it must for a reflecting ground.

In the case of very large arrays, both the element factor and the factor arising from image theory may have a negligible effect on the overall shape of the main antenna beam, but the fact that no radiation is directed downward into the ground results in an immediate doubling of the directivity.

3.6 Very Long Baseline Interferometry (VLBI)

Much of the discussion above assumed that grating lobes are a feature of antenna arrays to be avoided. In some applications, however, antenna arrays with long spatial separations and multiple grating lobes can be tremendously valuable. One of these is very long baseline interferometry, used in radio astronomy to observe distant sources. Synthetic antenna arrays formed from widely separated receivers may be able to discern the motion and even resolve the width of the objects if the grating lobes are sufficiently narrow.

Imagine forming a two-element antenna array using the Arecibo S-band telescope and a comparable system located in Scandinavia, a distance of some 10,000 km away. Both telescopes could be trained on the same region of space. The radiation pattern of the array would consist of a large number of narrow grating lobes, the width of each being $\lambda/L \sim 10^{-8}$ radians or about 7×10^{-7} degrees. The power received from a "point" radio source, a pulsar for example, moving through the radiation pattern would nearly reproduce the pattern. The finite angular width of the pulsar would broaden the grating lobes slightly, however, and the width could be deduced from the broadening. Together, the telescopes constitute a sensitive radio interferometer. The beamwidth of the grating lobes is the diffraction limit of the interferometer.

If the pulsar experiment was repeated every few years, subtle differences in the positions of the grating lobes would be observed. This is due to the slight relative motion of the two telescopes caused by plate tectonics and continental drift. The displacement need only be a fraction of a wavelength or a few cm to be noticeable. Such experiments were an important means of measuring plate motion prior to the advent of GPS and other satellite-based navigation instruments.

3.7 Preliminary Remarks on SAR

We will investigate synthetic aperture radar (SAR) principles later in the text (see Chapter 12). For now, it is sufficient to consider how it is possible for airborne and space-borne radar systems carrying modest antennas to make images of distant targets with such remarkably good spatial resolution. From orbit, SAR radars on satellites can resolve features with dimensions as small as about 10 cm. Using conventional side-looking radar tactics, this would necessitate using antennas with beamwidths of the order of 10^{-5}

3.7 Preliminary Remarks on SAR

degrees. Such antennas would be enormous and impractical to fly. In fact, it turns out that flying small antennas is actually advantageous for SAR.

The basic idea behind SAR involves regarding samples taken at different points along the vehicle trajectory as having come from a virtual or synthetic array. The target must remain fixed throughout data acquisition for this to work, and the data storage and processing required is formidable. The size of the virtual array can be enormous, however, affording SAR radars fine spatial resolution.

Figure 3.9 illustrates the geometry for an SAR data acquisition pass. The arc of the vehicle trajectory, which is the length of the virtual array, is L. In order for the target (a surface feature here) to remain visible across the entire arc, the beamwidth of the antenna mounted on the vehicle must be sufficiently large. If the antenna has a physical size d, then its beamwidth will be approximately λ/d, meaning $L \leq R\lambda/d$, where R is the altitude of the vehicle.

The beamwidth of a virtual array of length L is then given by

$$\text{HPBW} = \frac{\lambda}{L} \geq \frac{d}{R},$$

and is bounded by the altitude of the vehicle relative to the size of the radar antenna. Finally, the spatial resolution on the ground is given by the product of the HPBW and the altitude, or $\delta x \geq d$. Amazingly, the resolution is limited by the size of the portable antenna (rather than its inverse)! While it may seem advantageous to fly the smallest radar antenna possible, signal detectability will suffer if the antenna is too small, and some compromise must be reached.

As mentioned above, extracting images from SAR data requires considerable data processing. The target must remain fixed during the observation, since Doppler information (presumed to come entirely from the vehicle motion) is used in constructing the images. Three-dimensional images are possible with the incorporation of interferometry, increasing the computational burden further. If interferometric data are acquired using successive

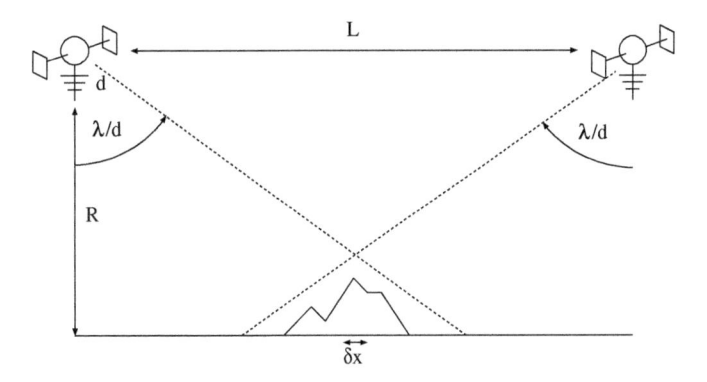

Figure 3.9 Figure illustrating the geometry behind synthetic aperture radar (SAR).

orbital passes, then the target is further required to remain stationary from one orbit to the next. So-called InSAR data are most useful for studying slowly changing geologic features rather than rapidly evolving situations (e.g., tracking). It is used extensively to study features of Earth's surface as well as the surfaces of other planets (notably Venus and Mars).

3.8 Beam Synthesis

The preceding discussion has concerned the analysis of antenna arrays where the element excitation has been chosen for the sake of expediency. However, there exist several techniques for beam synthesis, where the excitations are chosen for the purpose of custom designing a radiation pattern. The excitation of array elements with different phases and amplitudes especially poses practical challenges, is likely to drive system costs, and imposes a penalty in aperture efficiency, since the uniformly excited array is the most efficient one. However, there are times when these penalties may be acceptable, for example when it is necessary to exclude certain zones from the radiation pattern to avoid radar clutter or interference from or to other services. Below, we briefly explore two of the simplest synthesis strategies for the case of uniformly spaced linear antenna arrays.

The simplest and most intuitive of all synthesis methods is the Fourier synthesis method. Recall that the array factor for a linear array can be expressed as

$$F(u) = \sum_{n=-N/2}^{N/2} I_n e^{j\alpha_n + jnu}$$

$$= \sum_{n=-N/2}^{N/2} C_n e^{jnu},$$

where $u = kd\cos\psi$ as before but where the amplitude I_n and phase α_n associated with each element are now arbitrary. Note also that the indexing convention has been changed here so that the $n = 0$ element is at the center of the array. The second expression above is written in such a way that it is clear that the array factor can be interpreted as a truncated Fourier series expansion, where the $C_n = I_n \exp(j\alpha_n)$ are the Fourier series coefficients. The Fourier series permits a band-limited representation of $F(u)$ on the interval $-\pi \leq u \leq \pi$. Standard Fourier series theory gives a recipe for calculating the coefficients that best represent $F(u)$ in a least-squares sense on that interval:

$$C_n = \frac{1}{2\pi} \int_{-\pi}^{\pi} F(u) e^{-jnu} du.$$

The more antenna elements and the more terms in the summation, the finer the detail in the pattern that can be accurately reconstructed.

In the event that the element spacing is $d = \lambda/2$, the interval where $F(u)$ is reconstructed corresponds to the visible region. This is the ideal situation for Fourier series synthesis, since the element excitations C_n are guaranteed to give the most accurate representation possible of the desired array factor for the given number of elements available. For other choices of d, however, the visible region will not coincide with the interval covered by the Fourier series reconstruction of $F(u)$, and so its representation will be suboptimal.

More flexibility for arrays with inter-element spacing unequal to half a wavelength is offered by the Woodward-Lawson sampling method. In this simple method, we regard the element excitation distribution as the superposition of $N+1$ different distributions, each with a real amplitude A_n and a linear phase progression α_n, both of which need to be determined. The resulting array factor is a superposition of $2N+1$ known array factors,

$$F(u) = \sum_{n=-N/2}^{N/2} A_n e^{i\phi_n} \frac{\sin\left[\frac{N+1}{2}(u-u_n)\right]}{(N+1)\sin\left[\frac{1}{2}(u-u_n)\right]},$$

where the $u_n = -\alpha_n d$ (note the minus sign) are chosen to be uniformly distributed over the visible space with

$$u_n = \frac{2\pi n}{N+1}.$$

Now, the sampling locations (where $u = u_n$) fall on the nulls of all but one of the terms in the sum and on the main lobe of the remaining one. Consequently, we have $|F(u_n)| = A_n$ at those locations, providing a recipe for setting the A_n for the desired array factor.

Having found the amplitudes A_n corresponding to the phase progressions α_n, the element excitations C_m are given by the appropriate superposition:

$$C_m = \frac{1}{N+1} \sum_{n=-N/2}^{N/2} A_n e^{-m\alpha_n d}.$$

Other synthesis techniques designed specifically to minimize sidelobes or the width of the main beam have also been developed. All of these techniques are non-adaptive, in that optimization takes place just at the time of design. In order to cope with changing conditions, engineers turn to adaptive optimization techniques, which constantly update the array excitations in response to the properties of the received signals. These are especially useful for coping with a noisy, dynamic environment.

3.9 Adaptive Arrays

Consider the following not-so-hypothetical situation. A large antenna used for sensitive radio astronomy experiments (e.g., the Arecibo telescope) is receiving strong interference from one or more moving sources. The interference is strong enough to be received by several nearby, low-gain antennas as well. It would seem possible to estimate the interference from the low-gain antenna signals and then subtract it (with the proper phasing) from the signal on the high-gain antenna, leaving behind just the desired astronomical signal. The summing could be performed with analog or digital equipment, although allowances would have to be made for the potentially large dynamic range of the signals involved, given that the interference can be many tens of dB stronger than the astronomical signal.

As illustrated in Figure 3.10, what is being described is the creation of an adaptive antenna array. By weighting and summing all of the signals correctly, it is possible to synthesize an array pattern with a main lobe in the direction of the desired signal and nulls in the direction of the interference sources. The more antennas, the more nulls are possible. The antennas could be homogeneous (for example, in the case of the VLA array) or heterogeneous (the Arecibo case).

In the case of moving and changing (appearing, vanishing) interference sources, the problem becomes one of determining the optimal weights quickly on the basis of the signal characteristics. This is an important application for adaptive arrays in which the weights must be computed and applied "on the fly."

Suppose $x_i(t)$ are the signals coming from $i = 1 \cdots n$ antennas and w_i are the complex weights applied to each. We can represent the signals and weights as column vectors and make use of linear algebra to express the array output in time,

$$y(t) = w^\dagger x(t),$$

where $w^\dagger$ is the Hermitian transpose (complex conjugate transpose) of w. The output power is the expectation of $|y|^2$, which we write as

$$\begin{aligned}\langle |y|^2 \rangle &= \langle w^\dagger x x^\dagger w \rangle \\ &= w^\dagger \langle x x^\dagger \rangle w \\ &= w^\dagger R w.\end{aligned}$$

Here, R is the covariance matrix, an $n \times n$ matrix with terms $R_{ij} = \langle x_i x_j^* \rangle$ that can be tabulated as data are collected. The covariance matrix is a complete specification of the statistical properties of the data, which are assumed to behave like Gaussian random variables. Specialized equipment (correlators) exists to compute R rapidly.

Many different schemes exist for adjusting the weights to optimize the array output. One of these is called linear constrained minimum variance (LCMV), which seeks to minimize

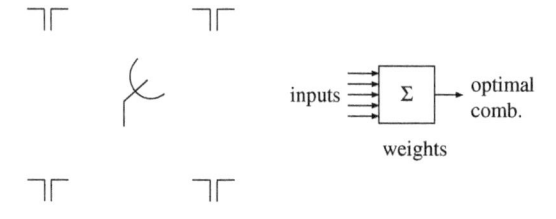

Figure 3.10 Schematic representation of an adaptive array.

the output power while maintaining constant gain in a specified direction. The idea is that interference always adds to the output power, which should therefore be minimized, but not at the expense of reduced gain in the direction of interest.

Let the column vector e represent the array output for a signal arriving from a direction specified by the wavevector $\mathbf{k}$, prior to weighting, i.e.,

$$e^\dagger = \begin{bmatrix} e^{j\mathbf{k}\cdot\mathbf{d}_1} & e^{j\mathbf{k}\cdot\mathbf{d}_2} & \cdots & e^{j\mathbf{k}\cdot\mathbf{d}_n} \end{bmatrix}, \tag{3.4}$$

where $\mathbf{d}_i$ is the vector position of antenna element i. The LCMV problem is to minimize the output power while constraining the desired signal level $e^\dagger w$ to be constant (unity). This is accomplished by minimizing the functional

$$L(w,\gamma) = w^\dagger R w + \gamma(e^\dagger w - 1),$$

where γ is a Lagrange multiplier with the purpose of enforcing the constraint it multiplies. The functional is minimized by setting all of its derivatives (with respect to the unknown weights and also γ) to zero. Doing so, we find

$$w = R^{-1} e (e^\dagger R^{-1} e)^{-1}$$
$$\langle |y^2| \rangle = \frac{1}{e^\dagger R^{-1} e}.$$

Note that the weights never actually have to be calculated. Instead, the array output is just expressed in terms of the inverse covariance matrix and the vector e, which is known for any desired target. The vector e can be fixed on a particular target or varied, affording an interference-free view of all the points in the sky. This last strategy is one means of radio/radar imaging, where the array output is constrained to be minimally influenced by radiation from undesired locations. Radar imaging will be discussed in more detail in Chapter 12.

3.10 Acoustic Arrays and Non-Invasive Medicine

The preceding discussion concerned constructing an adaptive array for the purpose of optimizing the signal-to-noise or signal-to-interference ratio of a received signal. The technique has imaging overtones, since the ratio can be optimized one bearing at a time in the process of mapping the sky. A related problem exists in medicine, where physicians make images of targets within living tissue. Here, acoustic pulses replace radio signals, and piezoelectric transducers replace antennas. Inhomogeneous tissue replaces free space, so that the acoustic waves travel neither along straight lines nor at constant speed. Scattering and absorption of the signal also can occur along its propagation path, further complicating things. The problem of identifying the target from the signals it scatters is nevertheless similar to the adaptive beamforming problem.

The medical problem does not stop at imaging diagnostics, however. An array of transducers is also capable of transmitting acoustic signals of sufficient strength to modify the tissue (e.g., shatter a kidney stone) if they can be made to arrive at the target simultaneously and in phase. The problem then becomes one of finding the signals that must be transmitted through each of the transducers to most efficiently focus the acoustic power on the target on the basis of signals collected with those same sensors. Such focusing not only leads to finer imagery but can also be used for non-invasive surgery. One kind of array that performs this feat is known as a time reverse mirror (TRM).

Time reverse focusing is a space-time generalization of the matched filter theorem, which, as will see in Chapter 5, governs the optimal response of a linear filter to an input signal in time. The output of a filter with an impulse response function $h(t)$ is maximized by a signal of the form $h(T-t)$, since the output $h(t) \otimes h(T-t)$, which is the autocorrelation function, is necessarily a maximum at $t = T$. The output of any linear filter is therefore maximized by a signal that is the time-reverse of the filter response to an impulse. TRM generalizes this notion to filtering in space and time, where the spatial domain is the volume enclosed by the sensor array and containing the target.

In the notation of TRM, we define the usual temporal impulse response function for the ith transducer in the array, $h_i(t)$. We also define a forward spatio-temporal impulse response function, $h_i^f(\mathbf{r}_i, t)$. This represents the acoustic signal present at the ith transducer at time t and position $\mathbf{r}_i$ in response to an impulsive acoustic pulse launched by the target, which resides at $\mathbf{r} = 0$, $t = 0$. A reverse spatio-temporal impulse response function $h_i^r(\mathbf{r}_i, t)$ can also be defined as the signal present at the target at time t due to an impulse at the ith transducer at time 0. According to the reciprocity theorem, however, $h_i^f(\mathbf{r}_i, t) = h_i^r(\mathbf{r}_i, t)$, which is crucial. (Henceforth, the superscripts will not be written.) Note also that reciprocity holds for inhomogeneous as well as homogeneous media.

The first step in the TRM process involves illuminating the inhomogeneous medium with the signal from one transducer. The signal will scatter off the target (and perhaps from other targets), which behaves like an impulsive acoustic source. Different signals will be received by the transducers at different times, reflecting different propagation paths. Each transducer will consequently record

$$h_i(t) \otimes h_i(\mathbf{r}_i, t).$$

Altogether, the recordings can be regarded as the output of a spatio-temporal filter to an impulsive input injected at the origin. In the next step of the process and as a generalization of matched filtering, each transducer is driven by an electrical signal equal to the time-reverse of the one it recorded:

$$h_i(T-t) \otimes h_i(\mathbf{r}_i, T-t).$$

Each signal must pass back through the transducer and then through the medium on its way to the target. This implies additional convolutions with the temporal and space-time impulse response functions. The acoustic pulse arriving back at the target due to the ith transducer will therefore be

$$h_i(T-t) \otimes h_i(\mathbf{r}_i, T-t) \otimes h_i(\mathbf{r}_i, t) \otimes h_i(t),$$

and the total pulse arriving at the target is the sum of the contributions from all the transducers

$$\sum_i h_i(T-t) \otimes h_i(\mathbf{r}_i, T-t) \otimes h_i(\mathbf{r}_i, t) \otimes h_i(t).$$

The main point is that, by virtue of the matched filter theorem, all of the individual signals will go through maxima at the time $t = T$ at the target and will therefore interfere constructively and be in focus there.

If there are multiple targets in the medium, focusing on any one of them will be imperfect initially. The strongest scatterer will tend to be in better focus than the others, however, so scatter from it will dominate the signals received by the transducers in successive pulses. Iteration tends to improve the focus of the most important target, making the others less visible. The process is therefore not only adaptive but self-focusing. In the final step of the process, the amplitude of the signals from the transducers is increased dramatically until the target shatters. The process can then be repeated as necessary.

The TRM principle relies on the reciprocity theorem, the field equivalence principle, and the reversibility of the acoustic waves. Just as scattering from the target produces a diverging wavefront that intercepts the transducer array in different places and times, time-reversed forcing applied at the transducer array produces a converging wavefront that coalesces on the target in a time-reversed duplicate process. In fact, the process is lossy and not entirely reversible, and TRM is an imperfect procedure. Multiple-scattering events, movement of the target and the medium, coarse transducer sampling in space, and non-ideal transducer behavior can limit the efficacy of the process further. Improving the performance by incorporating sophisticated signal processing and modeling methods is a very active area of contemporary research.

3.11 Notes and Further Reading

The references cited at the end of Chapter 2 are also relevant for the material in this chapter. In addition, several good texts focus on phased-array antennas in particular. These include the texts by R. C. Hansen and R. L. Haupt. A contemporary treatment of all manner of antennas including phased arrays can be found in the text by Drabowitch et al.

The Numerical Electromagnetics Code is described in a three-part series of technical reports from the Lawrence Berkeley Laboratory authored in 1981 by G. J. Burke and A. J. Poggio (UCID-18834), which may be obtained from the National Technical Information Service. A number of cost-free and commercial versions of the NEC are available today.

The foundational treatment of the LCMV method for adaptive beamforming was given by J. Capon. A fairly recent review of VLBI in geodesy and astronomy was published by

O. J. Sovers et al. The general topic of interferometry in the context of radio astronomy is treated in the text by A. R. Thompson, J. M. Moran, and G. W. Swenson, Jr.

For more information about synthetic aperture radar, consult the references at the end of Chapter 12.

J. Capon. High-resolution frequency-wavenumber spectrum analysis. *Proc. IEEE*, 57:1408, 1969.
S. Drabowitch, A. Papiernik, H. D. Griffiths, J. Encinas, and B. L. Smith. *Modern Antennas, 2nd ed.* Springer, Dordrecht, The Netherlands, 2005.
R. C. Hansen. *Phased Array Antennas*. John Wiley & Sons, New Jersey, 1998.
W. W. Hansen and J. R. Woodyard. A new principle in directional antenna design. *Proc. IRE*, 26(3):333–345, 1938.
R. L. Haupt. *Antenna Arrays*. John Wiley & Sons, New Jersey, 2010.
O. J. Souvers, J. L. Fanselow, and C. S. Jacobs. Astrometry and geodesy with radio interferometry: experiments, models, results. *Rev. Mod. Phys.*, 70:1393–1454, 1998.
A. R. Thompson, J. M. Moran, and G. W. Swenson. *Interferometry and Synthesis in Radio Astronomy*. John Wiley & Sons, New York, 1986.
H. Yagi and S. Uda. Projector of the sharpest beam of electric waves. *Proc. Imp. Acad. Japan*, 2(2):49–52, 1926.

Exercises

3.1 Fill in the details leading to the expression for the half-power full beamwidth of an endfire array given in Section 3.1.2. This should be a fairly brief derivation.

3.2 Consider a three-element array of antennas spaced by a distance d on the $\hat{x}$ axis. Find values of α and d such that there is no radiation in the $-\hat{x}$ direction but maximum radiation in the $+\hat{x}$ direction. Sketch the radiation pattern in the $x-y$ plane. You can consider the antenna elements themselves to be isotropic radiators for this problem. If there is more than one solution to the problem, give the general rule, and sketch a few representative solutions.

3.3 The directivity of endfire arrays can be increased (at the cost of increased sidelobes) by taking $|\alpha d| > k_\circ d$. In particular, we can take

$$u + u_\circ = kd(\cos\psi - 1) - \frac{x}{n+1}.$$

The notes suggest that taking $x \approx 3$ gives the maximum directivity. In fact, this is just a rule of thumb. The optimum value of x depends on the size of the array.

Consider arrays with element spacing $d = 0.4\lambda$. To keep the calculation simple, consider the elements in the arrays to be *isotropic* radiators. To work this problem, you will need to write routines to evaluate the array pattern of an endfire array with an arbitrary number of elements $n+1$ and phase offset x. You will also need a routine to calculate the directivity of said array.

a) Consider arrays with 5, 10, and 20 elements. For each one, plot the endfire directivity as a function of x. Identify the value of x for which the directivity is a maximum in each case. For each number of elements, find the directivity of the conventional endfire array ($x = 0$) and the enhanced directivity array (with x optimized).

b) For the 20-element case, plot the radiation pattern for a conventional endfire array and for the enhanced directivity array. Comment on the relative main beamwidths and sidelobe levels. Try to combine the two plots in the same figure, and make sure they are normalized appropriately.

c) Again for the 20-element case, find the precise HPBW for the conventional endfire array and the enhanced directivity array. Compare these with the value estimated by the generic endfire formula given in class. Also, compare the HPBWs you find with what you would estimate solely on the basis of the directivities.

3.4 The text used the example of a 21-element array of dipole antennas spaced by 0.9λ to estimate the performance of broadside antenna arrays. Do the calculations exactly using numerical analysis.

a) Calculate the HPBW in the E and H planes. Show plots for both planes (in dB so that the sidelobes show).

b) Estimate the gain of the array on the basis of what you found in (a).

c) Now, make an exact calculation of the gain. This will require 2D numerical integration over the entire pattern to find the total radiated power. Compare the result with (b) and also with the estimate from class.

3.5 Design an endfire antenna array made up of half-wavelength dipole elements oriented parallel to the $\hat{y}$ axis and spaced $\lambda/4$ in the $\hat{z}$ axis. As usual, θ is measured from the $\hat{z}$ axis, and ϕ is measured from the $\hat{x}$ axis in the $x - y$ plane.

a) Using trial and error or any means you like, find the smallest number of elements that will give a HPBW of 33° in the $x - z$ plane (the $\phi = 0$ plane, also called the H-plane in this case).

b) For this number of elements, calculate the HPBW in the $x - z$ and $y - z$ planes. The array factors are the same in both cases, but the results will be slightly different because of the element factor. Show plots for both planes (in dB so that the sidelobes show).

c) Estimate the antenna gain on the basis of the results from b).

d) Now, make an exact calculation of the gain. This will require a 2D numerical integration over the entire pattern to find the total radiated power. Compare c) and d). The results will be different but should not be *too* different.

3.6 Suppose you have a 6-by-6 rectangular array of 36 half-wave dipoles in the $x - y$ plane. The separation between dipole centers is a half wavelength in both directions, and the dipoles are oriented parallel to the $\hat{x}$ axis. The array is located a

96 Antenna Arrays

quarter-wavelength above a perfectly conducting ground, and all the dipoles are excited in phase with a current $I_\circ \cos \omega t$.

a) Derive a complete expression for the radiated power vs. θ and ϕ in the far field in terms of $I_\circ$. Be sure to include the correct element factor and the effect of the ground plane. We have here essentially a three-dimensional array ($6 \times 6 \times 2$). Express the results in standard polar coordinates as usual.
b) Find, numerically, the HPBW in the $x-z$ ($\phi = 0$) and the $y-z$ ($\phi = \pi/2$) planes.
c) Make plots of the radiation patterns in these two planes, and make sure that they are consistent with your HPBW calculations. For example, both patterns should have maxima at $\theta = 0$.
d) How would you phase the array to shift the beam maximum to $\theta = 25°$, $\phi = 90°$?
e) Plot the pattern for d) in the $y-z$ plane and recompute the HPBW in this plane. Compare the results with those from b) and c). Do the results make sense?

3.7 The HAARP facility in Alaska is a transmitter designed to emit radiation in the HF band nearly vertically into the upper atmosphere. The antenna is a 12×15 array of crossed dipoles. The antennas are spaced a distance 24.4 m and elevated 11.7 m above a conducting ground screen. The lengths of the dipoles are 21.6 m.

The antenna is meant to function between frequencies of about 2.5–10 MHz.

For the purposes of this problem, suppose emission takes place just with the dipole elements oriented in the direction in which the array is wider.

a) At what frequency is the directivity of the antenna looking straight up precisely 27 dB?
b) For this frequency, plot the E- and H-plane radiation patterns in a manner that clearly illustrates the shape of the main beam and the first sidelobes.
c) Find the beamwidth in the E- and H-planes.
d) How far can the radiation pattern be steered in the direction in which the array is wider before the peak directivity drops to 26 dB? What is the E-plane HPBW now?

Check your work by applying the appropriate approximations used in this chapter.

3.8 Arrays with random antenna spacings can have lower sidelobe levels than uniform arrays. Consider a 1-D broadside array, for example. Recall from class that the array factor can be written as

$$F = \frac{1}{N+1} \sum_{n=0}^{N} e^{ikd_n \cos \psi}.$$

Suppose the element spacing is not quite uniform and that $d_n = nd + \Delta_n$, where Δ_n is a small shift and a Gaussian random variable. This means

$$F = \frac{1}{N+1} \sum_{n=0}^{N} e^{iknd \cos \psi} e^{ik\Delta_n \cos \psi}.$$

Exercises

Show that the expectation of the array pattern, $\langle |F|^2 \rangle$, is the pattern of a regular broadside array, only a) multiplied by a function that reduces the sidelobe level and b) added to by a small function that fills in the nulls in the pattern a little.

In order to compute the expectation, note the following statistical characteristics of Δ_n:

$$\langle \Delta_n \rangle = 0$$
$$\langle \Delta_n^2 \rangle = \sigma^2$$
$$\langle \Delta_n \Delta_m \rangle = \delta_{m,n} \sigma^2.$$

You cannot make use of the compact formulas for ULAs given in the text. Rather, just start with

$$FF^* = \frac{1}{(1+N)^2} \sum_{n=0}^{N} \cdots \sum_{m=0}^{N} \cdots$$

and go from there. Recall that the expectation of a sum of terms is the sum of their expectations. Recall also that we can expand $e^\varepsilon \sim 1 + \varepsilon + \varepsilon^2/2$ for small values of ε. You will need to retain terms up to order ε^2 for this problem.

4
Aperture Antennas

High gain can also be achieved with the use of aperture antennas. Aperture antennas are characterized by a large collecting area, at least several wavelengths in each dimension, through which radiation passes during transmission and reception. The simplest example is an open-ended waveguide. The waveguide aperture can be flared for impedance matching purposes or the radiation can pass through tuned slots. To increase the size of the aperture, waveguides are often used in conjunction with reflectors or RF lenses, which then behave like secondary apertures. In any case, we assume the electromagnetic field across the antenna aperture is somehow known. The problem becomes one of calculating the resulting radiation fields.

Treating aperture antennas demands an appreciation of diffraction theory. Diffraction is a wave phenomenon where the rays that describe the wave deviate from rectilinear propagation paths. Ray bending can also occur in media where the index of refraction varies gradually in space. In that case, the phenomenon is called refraction. Ray bending in homogeneous media that include obstacles, including the boundaries of an aperture, is called diffraction.

The first explanation for diffraction was provided in the seventeenth century by Christian Huygens, who considered each point on a wavefront to be a kind of disturbance and a source of secondary spherical wavefronts. If the initial wavefront is an infinite plane, the superposition of all the spherical waves will be another planar wavefront, displaced from the original in the direction of propagation. If the initial wavefront has finite extent, however, the spherical waves will combine constructively in some places and destructively in others, producing an interference pattern with lobes and nulls like the radiation patterns we have already seen. Huygens' principle can explain, for example, how radiation bends around corners and why the radiation from a finite aperture is not entirely like a "searchlight" pattern.

Throughout the eighteenth century, little attention was paid to Huygens' principle, as the corpuscular theory of light espoused by Newton gained favor. During this time, the theory of geometric optics was developed. This theory is still used to describe refraction, and will be discussed later in the text. In the nineteenth century, however, the wave picture of light was revisited. Gustav Kirchhoff combined Fresnel's diffraction pattern calculations based on Huygens' principle with Maxwell's identification of light with electromagnetic waves to advance a complete theory of scalar wave diffraction. In that theory, each component of the electromagnetic field is assumed to obey the same scalar wave equation with the same boundary conditions. In particular, each field quantity and its derivative is assumed to vanish on and within any obstacles. Outside of obstacles, the field quantities are assumed to behave as if the obstacles were not there.

Despite the success of scalar diffraction theory in accounting for a great number of observed phenomena, the assumptions at its foundations are flawed, since the obstacles perturb the field quantities nearby and since the fields inevitably penetrate some distance into the obstacles. At the end of the nineteenth century, Poincaré and Sommerfeld showed that the assumptions are incompatible with Maxwell's equations. A more complete vector diffraction theory was subsequently developed. More tractable approaches to the diffraction problem continued to emerge throughout the twentieth century.

In what follows, the Kirchhoff–Fresnel scalar diffraction theory is described rather than the more exact Rayleigh–Sommerfeld vector theory or its more contemporary successors. The scalar theory can be regarded as a first approximation to the vector theory. So long as the aperture in question is very large in terms of the wavelength, the approximation is a very good one. A more careful treatment of diffraction is presented in Chapter 9.

4.1 Simple Apertures

The top row of Figure 4.1 shows a waveguide terminating in an aperture. We presume that a source of radiation, a dipole antenna perhaps, exists inside the waveguide and that the electromagnetic field across the aperture is known. The waveguide problem is analogous to that of a plane wave passing through a perforated barrier. The sources that contribute to radiation to the right of the barrier lie on the aperture (i.e., not on any barriers).

The second row of Figure 4.1 illustrates the field equivalence principle, the second important idea (after Huygens' principle) for analyzing aperture antennas. The lower left corner of the figure shows a volume enclosed by a surface containing sources of electromagnetic radiation. We again suppose that the fields on the surface are somehow known. The field equivalence principle applies to calculations of the electromagnetic field outside

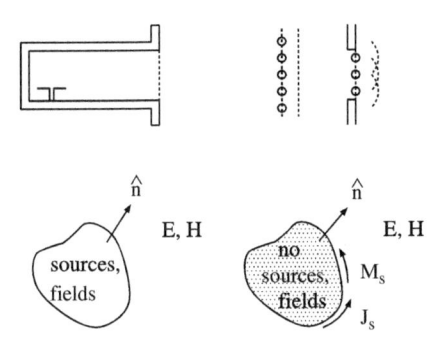

Figure 4.1 Illustration of Huygens' principle (top row) and field equivalence (bottom row).

the surface. It says that we can replace the region enclosed by the surface with a source-free and field-free one with equivalent sources on the surface. So long as those sources reproduce the surface fields, they reproduce the field everywhere outside the surface. This principle, which follows from the uniqueness of a solution of Maxwell's equations, given specific boundary conditions, simplifies our calculations greatly.

What constitutes suitable equivalent sources? Ampère's law can be manipulated to show that the magnetic field tangent to a field-free region must be supported by a surface current density,

$$\mathbf{J}_s = \hat{\mathbf{n}} \times \mathbf{H},$$

where $\hat{\mathbf{n}}$ is the outward unit normal. Specifying the tangential component of $\mathbf{H}$ allows the specification of the equivalent surface current density, which is one of two *Huygens sources*.

The problem is incomplete, however, as the electric field at the surface has not yet been accounted for. In fact, Faraday's law specifies that there is no electric field tangential to a field-free region, which would seem to make it impossible to reproduce an arbitrary surface field. The problem is remedied with the introduction of a fictional magnetic surface current density, $\mathbf{M}_s$, carried by hypothetical magnetic monopoles. Inclusion of this second Huygens source in Faraday's law leads to another boundary condition analogous to the first:

$$\mathbf{M}_s = \hat{\mathbf{n}} \times \mathbf{E}.$$

The reader should not be concerned about the physical non-existence of magnetic surface current. Neither electric nor magnetic current actually flows on the aperture. If they did, however, they would generate the specified aperture fields, and by the uniqueness theorem, the associated radiation fields. So long as we do not introduce new sources (fictional or otherwise) into the region of interest, the region illuminated by the aperture, we have violated no laws of electrodynamics. Note that when we account for the tangential components of the aperture fields, the normal components are automatically accounted for by Maxwell's equations.

Finally, it can be shown that the radiation fields arising from $\mathbf{J}_s$ and $\mathbf{M}_s$ are identical, and we can therefore work entirely with one of the Huygens sources and simply double the result. Moreover, we will not be concerned with constants of proportionality or vector directions in the scalar diffraction calculations that follow for the moment, since they affect neither the radiation pattern nor the directivity.

The strategy for analyzing aperture antennas is as follows:

1 Establish the electromagnetic fields specified on the aperture surface. For simplicity, we will consider planar surfaces.
2 Derive from these the equivalent Huygens source(s).
3 Calculate the radiation fields from the Huygens source(s).

For the last step, we utilize the phasor form of the retarded vector potential equation (see Appendix B),

4.1 Simple Apertures

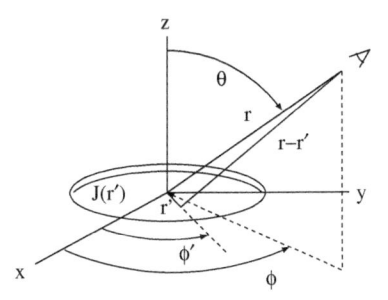

Figure 4.2 General aperture antenna analysis. The vectors $\mathbf{r}'$ and $\mathbf{J}(\mathbf{r}')$ lie in the $x - y$ plane.

$$\mathbf{A}(\mathbf{r}) = \frac{\mu_\circ}{4\pi} \int_{apt} \frac{\mathbf{J}(\mathbf{r}')e^{-jk(\mathbf{r}-\mathbf{r}')}}{|\mathbf{r}-\mathbf{r}'|} d\mathbf{r}', \qquad (4.1)$$

which is like (2.2), generalized for a continuous source distribution. Together with the Huygens source expressions above, (4.1) can be interpreted as a concrete statement of Huygens' principle. Throughout what follows, we use primed coordinates to represent source points and unprimed ones to represent observing points, as in Chapter 2. The integration in (4.1) is over the aperture plane, and the observing points are in the upper-half space. A number of simplifying approximations facilitate evaluation and interpretation of (4.1). Refer to Figure 4.2.

- We assume that the source fields vanish on the diffraction screen but have finite values everywhere within the aperture. This approximation holds for apertures that are very large compared to a wavelength ($r'_{max} \gg \lambda$).
- We seek radiation-field solutions and take $r \gg \lambda$. Furthermore, as with antenna arrays, we take $r \gg r'_{max}$ in order to simplify the formalism. This permits the substitution $|\mathbf{r} - \mathbf{r}'| \to r$ in the denominator of (4.1).
- More care must be taken with the exponential term due to its sensitivity to small changes in its argument:

$$|\mathbf{r}-\mathbf{r}'| = \left[(x-x')^2 + (y-y')^2 + (z-z')^2\right]^{1/2}$$

$$= r\left[1 - \underbrace{\frac{2xx'}{r^2} - \frac{2yy'}{r^2}}_{O(r'/r)} + \underbrace{\frac{x'^2+y'^2}{r^2}}_{O(r'/r)^2}\right]^{1/2}$$

$$\approx r\left[1 - \frac{xx'}{r^2} - \frac{yy'}{r^2} + \frac{x'^2+y'^2}{2r^2}\right].$$

Note that we retain only the leading terms in a binomial expansion in the last line.

Recall that the unprimed coordinates are observing coordinates so that terms like x/r and y/r are the direction cosines of the ray to the observer. We incorporate these into the *wavevector* $\mathbf{k}$, which has a modulus equal to the wavenumber and a direction parallel to the ray:

$$\frac{k_x}{k} = \frac{x}{r} = \sin\theta\cos\phi$$
$$\frac{k_y}{k} = \frac{y}{r} = \sin\theta\sin\phi$$
$$\frac{k_z}{k} = \frac{z}{r} = \cos\theta.$$

Altogether, the vector potential equation simplifies to

$$\mathbf{A}(\mathbf{r}) = \frac{\mu_\circ}{4\pi r} e^{-jkr} \int\int_{apt} \mathbf{J}_s(x',y',z'=0) e^{j\left(k_x x' + k_y y' - k\frac{x'^2+y'^2}{2r}\right)} dx'dy'. \qquad (4.2)$$

The quadratic term in the exponent is associated with *Fresnel diffraction* and substantially complicates the evaluation of the integral. This term is important close to the aperture in the *Fresnel diffraction* region or the *near field*, but can be neglected at large distances in the *Fraunhofer diffraction* region or the *far field*.

The demarcation point between the two regions occurs near the range of r where

$$\frac{k(x'^2+y'^2)}{2r} \sim 1 \quad \forall x',y',$$

implying

$$\frac{k(D/2)^2}{2r} \sim \frac{\pi}{4}\frac{D^2}{\lambda r} \sim 1,$$

where D is the antenna diameter, loosely interpreted. The usual rule of thumb is that the Fraunhofer diffraction region approximation is valid for $r > 2D^2/\lambda$. This distance can be quite large for large-aperture radars and also for lasers.

A few illustrative examples:

- VHF satellite ground station. Take $D = 20$ m, $\lambda = 1$ m, then $r > 2D^2/\lambda = 800$ m. Any satellite is easily within the Fraunhofer diffraction region.
- Arecibo S-band radar. This time, $D = 300$ m, $\lambda = 0.12$ m, and $r > 1800$ km. Most of the Earth's ionosphere is therefore within the Fresnel diffraction region.
- Laser + telescope. Take $D \sim 1$m, $\lambda = 0.5 \times 10^{-6}$ m (visible light). For this example, the Fresnel diffraction region extends to 4000 km!

In intuitive picture of the Fresnel and Fraunhofer diffraction regions is given by Figure 4.3. In the Fresnel diffraction region, the radiation pattern is that of a searchlight, whereas it is conical in the Fraunhofer diffraction region. The crossover point is given where $D \approx r\lambda/D$ or $r \approx D^2/\lambda$. In the Fraunhofer diffraction region, surfaces of constant range are spherical, and the phase of the emitted signal only depends on the range r. In the Fresnel diffraction region, surfaces of constant range are closer to planes, so the phase of the signal depends on range and bearing.

4.1 Simple Apertures

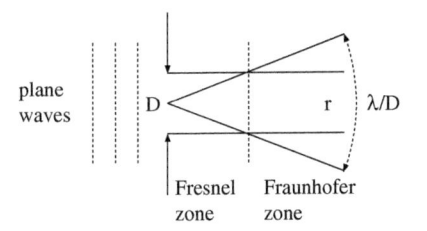

Figure 4.3 Illustration of the demarcation between the Fresnel and Fraunhofer diffraction regions.

In the Fraunhofer diffraction region, the vector potential integral simplifies to

$$\mathbf{A}(\mathbf{r}) = \frac{\mu_\circ}{4\pi r} e^{-jkr} \int\!\!\int_{apt} \mathbf{J}_s(x',y',z'=0) e^{j(k_x x' + k_y y')} dx' dy'$$
$$\propto F(k_x, k_y).$$

In the far field, the modulus of the vector potential, electric field and magnetic field are proportional, so the radiation pattern of the antenna is simply given by $|F(k_x,k_y)|^2$, which should be normalized to unity by convention. Since the wavevector components are functions of θ and ϕ, the form of the radiation pattern is as usual. Note, however, that the factor $F(k_x,k_y)$ is the 2-D *Fourier transform* of the aperture field. This is a profound and useful result, which makes the properties of Fourier transforms available for diffraction calculations.

Recall that we distinguished in Chapter 2 between the reactive fields and the radiative fields, the transition being at $r \approx \lambda$. Here, we distinguish between the near and far fields, the transition being at $r \approx D^2/\lambda$. Both terminologies exist simultaneously, and one may consider the "radiative far field," "radiative near field," "reactive far field," and "reactive near field," depending on the application. Most but not all radar work takes place in the "radiative far field." Importantly, in the near field, incident power density does not decrease with range as r^{-2}. This represents a loss of focus for nearby targets and necessitates a reformulation of the radar equation and related expressions.

4.1.1 Example – Rectangular Aperture

See Figure 4.4 for reference. The aperture excitation is given by a uniform current sheet $\mathbf{J}_s = J\hat{\mathbf{x}}$. (Note that the direction of the equivalent surface current is irrelevant for the scalar calculation.) The element factor for the aperture is

$$F(k_x, k_y) = J \int_{-L_x/2}^{L_x/2} \int_{-L_y/2}^{L_y/2} e^{j(k_x x' + k_y y')} dx' dy'$$
$$= J \int_{-L_x/2}^{L_x/2} e^{jk_x x'} dx' \int_{-L_y/2}^{L_y/2} e^{jk_y y'} dy'$$

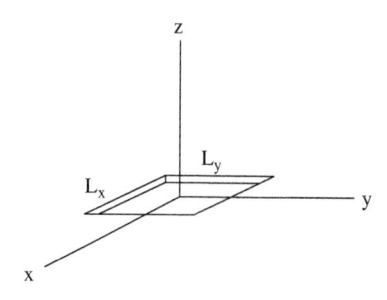

Figure 4.4 Rectangular-aperture antenna with dimensions $L_x \times L_y$.

$$= JL_xL_y \frac{\sin X}{X} \frac{\sin Y}{Y}$$
$$= JL_xL_y \operatorname{sinc} X \operatorname{sinc} Y,$$

where the new auxiliary variables have been defined as

$$X \equiv k_x L_x/2 = \frac{\pi L_x}{\lambda} \sin\theta \cos\phi$$
$$Y \equiv k_y L_y/2 = \frac{\pi L_y}{\lambda} \sin\theta \sin\phi,$$

and where the radiation pattern is then given by $|F(\theta,\phi)|^2$. The pattern has a main lobe normal to the aperture plane and numerous sidelobes outside that, the number and width depending on the aperture dimensions in wavelengths. First sidelobes of the *sinc* function are reduced $\sim$13.5 dB from the main lobe.

4.1.2 Fourier Transform Shortcuts

The properties of Fourier transforms can sometimes be exploited to simplify calculations. The most important of these properties is the convolution theorem, which states that the Fourier transform of the convolution of two functions is the product of their individual Fourier transforms. This is easily proven. The convolution of A and B is

$$A(x) \otimes B(x) \equiv \int_{-\infty}^{\infty} A(x')B(x-x')dx'.$$

Its Fourier transform is

$$\mathcal{F}(A(x) \otimes B(x)) = \int_{-\infty}^{\infty}\int_{-\infty}^{\infty} A(x')B(\underbrace{x-x'}_{x''})e^{jkx}dx'dx,$$

where we define a new variable x''. Note that $dx'' = dx$ in the integral when x' is held constant. Consequently,

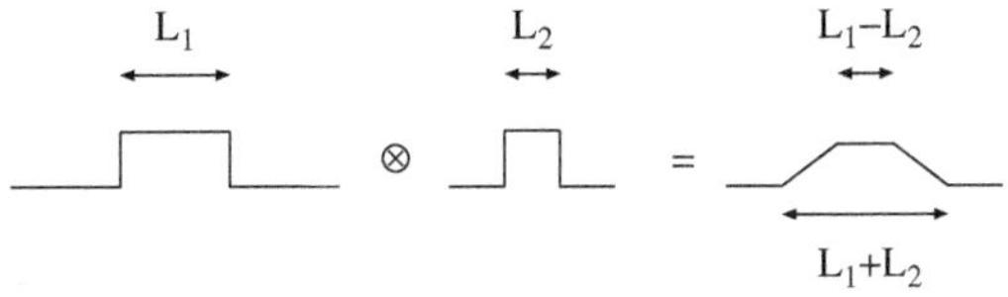

4.1 Simple Apertures

Figure 4.5 A tapered-aperture distribution is the convolution of two rectangular distributions.

$$\mathcal{F}(A(x) \otimes B(x)) = \int_{-\infty}^{\infty} \int_{-\infty}^{\infty} A(x')B(x'')e^{jk(x'+x'')}dx'dx''$$
$$= \int_{-\infty}^{\infty} A(x')e^{jkx'}dx' \int_{-\infty}^{\infty} B(x'')e^{jkx''}dx''$$
$$= \mathcal{F}(A)\mathcal{F}(B).$$

The convolution theorem is useful when the aperture distribution can be expressed as the convolution of two or more other distributions with known Fourier transforms. An example is a trapezoidal distribution, the convolution of two uniform distributions (as shown in Figure 4.5). The trapezoid is one kind of taper that can be used to shape an aperture's radiation pattern. We need do no new calculations, since the Fourier transform of a uniform (rectangular) distribution is known. Taking the product of two such functions gives the radiation pattern for the trapezoid

$$|F(\theta,\phi)|^2 \propto (\text{sinc}X_1)^2 (\text{sinc}X_2)^2$$
$$X_1 = \frac{\pi L_1}{\lambda} \sin\theta \cos\phi$$
$$X_2 = \frac{\pi L_2}{\lambda} \sin\theta \cos\phi,$$

where we have taken the taper to be in the $\hat{x}$ direction of a rectangular array, for this example. By comparison to a uniform aperture of width $L_1 + L_2$, the tapered aperture has a radiation pattern with a broader main beam, lower directivity, and lower effective area. The payoff comes in the form of reduced sidelobes, which is usually the goal of a taper. Note that this discussion has ignored the pattern in the $y-z$ plane.

An extreme example of a taper is given by an aperture distribution with a Gaussian shape. The Fourier transform of a Gaussian is another Gaussian, which obviously has no sidelobes. Since a Gaussian distribution has infinite extent, it cannot be realized in practice. A Gaussian on a pedestal has a radiation pattern with finite (but perhaps still small) sidelobes.

4.1.3 Example – Circular Aperture

Another important example is a uniformly excited circular aperture. Fourier transform properties are of little use to us here, and the calculation follows the pattern of the rectangular aperture. We again consider the Fraunhofer diffraction region. The pertinent geometry is shown in Figure 4.6. The direction of $\mathbf{J}_s$ is uniform and arbitrary:

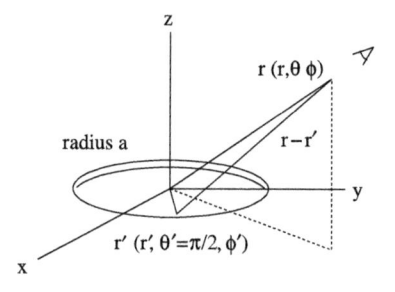

Figure 4.6 Radiation for a uniform circular-aperture antenna.

$$F(\mathbf{k}) = J \int_0^a \int_0^{2\pi} e^{j(k_x x' + k_y y')} r' dr' d\phi'.$$

As before, we use primed and unprimed coordinates to represent source and observation points, respectively. The integration is over the circular aperture with radius a. The mixed Cartesian and spherical coordinates are inconvenient. We therefore recognize that the argument of the exponential is the dot product of the source coordinate $\mathbf{r}'$ and the wavevector $\mathbf{k}$. Rewriting this term in spherical coordinates gives

$$F(\mathbf{k}) = J \int_0^a \underbrace{\int_0^{2\pi} e^{jkr' \sin\theta \cos(\phi-\phi')} d\phi'}_{2\pi J_\circ(kr' \sin\theta)} r' dr'.$$

The ϕ' integral can be performed immediately, yielding $J_\circ$, the Bessel function of the first kind of order zero. This function is superficially similar to the cosine function but has aperiodic roots and a gradual taper. The r' integral can next be performed, making use of the following identity,

$$\int x J_\circ(x) dx = x J_1(x),$$

where J_1 is the Bessel function of the first kind of order 1, which is superficially similar to the sine function. Consequently,

$$F(\mathbf{k}) = J \frac{2\pi a}{k \sin\theta} J_1(ka \sin\theta)$$
$$F(\theta) = \frac{2 J_1(ka \sin\theta)}{ka \sin\theta}, \tag{4.3}$$

where the second term has been written explicitly as a function of θ and has been normalized to unity. The radiation pattern is then given by the square of the element factor.

Evaluation of (4.3) shows it to be a function with a main lobe on axis and sidelobes (the first reduced by $\sim$17.6 dB from the main lobe). Half-power points occur where the argument $ka \sin\theta \sim 1.6$ or

4.1 Simple Apertures

$$HPBW = 2\theta_{1/2} = 2\sin^{-1}\frac{1.6}{ka}$$
$$\approx 2\frac{1.6}{\pi}\frac{\lambda}{2a}$$
$$= 1.02\frac{\lambda}{2a},$$

assuming a large aperture. For the uniformly illuminated apertures, $A_{eff} = A_{phys}$, and $D = (4\pi/\lambda^2)A_{eff} = 4\pi^2(a/\lambda)^2$. As an example, Figure 4.7 shows the radiation pattern for a circular aperture with a 20λ radius.

4.1.4 Aperture Efficiency

There are two approaches to solving problems involving aperture antennas. We are familiar with the first, which is to calculate the radiation pattern, integrate over all bearings to calculate the total radiated power and then calculate the directivity using

$$D(\theta,\phi) \equiv 4\pi\frac{|F(k_x,k_y)|^2}{\int\int |F(k_x,k_y)|^2 \sin\theta\, d\theta d\phi}.$$

From the directivity, one can calculate the beam solid angle, estimate the beamwidth, and calculate the effective area from reciprocity. This is the most complicated approach but is necessary to determine the directivity and effective area as functions of direction.

Another simpler and potentially more direct approach is to calculate the aperture efficiency

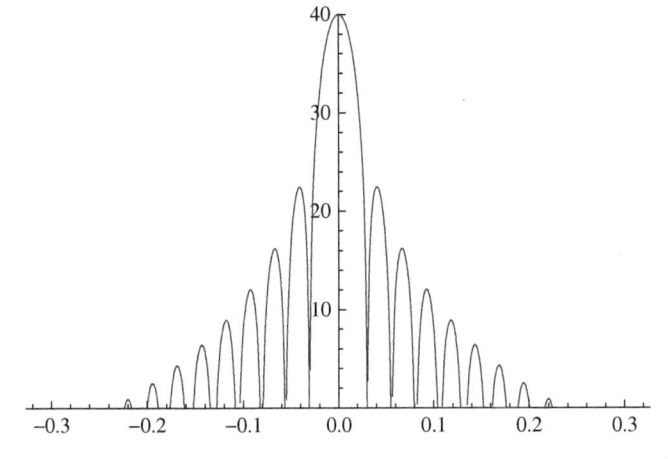

Figure 4.7 Radiation pattern for a uniform circular-aperture antenna of radius $a = 20\lambda$ versus zenith angle in radians. The plot uses a log scale that spans 40 dB of dynamic range. Note that the first sidelobes are down 17.5 dB from the main lobe. Compare this with 13.5 dB for a rectangular aperture.

$$\varepsilon_{ap} \equiv \frac{A_{eff}}{A_{phys}}$$

$$= \frac{\lambda^2}{4\pi} \frac{D}{A_{phys}}$$

and, from it, the effective area directly. Here, D represents the maximum directivity, which we take to occur in the direction normal to the antenna aperture. The aperture efficiency can be expressed entirely in terms of the aperture illumination, as is shown below, and is usually easier to calculate than $D(\theta,\phi)$.

To proceed, we will have to reintroduce the constants of proportionality relating the aperture and radiation fields heretofore neglected. Recall the definition

$$D = 4\pi \frac{P_\Omega}{P_t}.$$

All of the radiated power must necessarily pass through the aperture. Calculating the total radiated power, then, amounts to integrating the Poynting flux across the aperture area,

$$P_t = \frac{1}{2Z_\circ} \int |E_{tan}|^2 da,$$

where E_{tan} is the tangential component of the electric field in the aperture. Calculating the incident power density per unit solid angle, P_Ω, meanwhile, requires us to formalize the relationship between aperture and the far-field electric fields. We have already established that (in the direction normal to the aperture plane)

$$\mathbf{A}(r) = \frac{\mu_\circ}{4\pi r} e^{jkr} \int \mathbf{J}_s da,$$

where the integral is again over the aperture. In the far field, the electric field is given by $\mathbf{E} = -j\omega\mathbf{A}$. In the aperture, meanwhile, we have $\mathbf{J}_s = \hat{\mathbf{n}} \times \mathbf{H} = (1/Z_\circ)\hat{\mathbf{n}} \times (\hat{\mathbf{n}} \times \mathbf{E}_{tan}) = -\mathbf{E}_{tan}/Z_\circ$. Consequently, the on-axis far-field radiation and aperture fields are related by

$$\mathbf{E}(r) = j\omega \frac{e^{jkr}}{2\pi r} \frac{\mu_\circ}{Z_\circ} \int \mathbf{E}_{tan} da. \quad (4.4)$$

Note that a factor of two has been introduced in the previous equation to account for the contribution to the radiation field from the equivalent magnetic surface current, which has otherwise been neglected in our calculations. This equation is a succinct statement of the field equivalence theory and embodies the aperture method of radiation pattern calculations. The corresponding power density per unit solid angle is

$$P_\Omega = \frac{r^2}{2Z_\circ} \mathbf{E} \cdot \mathbf{E}^*$$

$$= \frac{1}{2Z_\circ^3} \left(\frac{\omega\mu_\circ}{2\pi}\right)^2 \left|\int E_{tan} da\right|^2.$$

Combining all the components and noting the definition of a spatial average over the aperture $\langle\rangle = \int da/A_{phys}$, we arrive at the simple but important formula

$$\varepsilon_{ap} = \frac{|\langle E_{\tan}\rangle|^2}{\langle|E_{tan}|^2\rangle},$$

where $E_{\tan}$ can be replaced with any measure of the aperture excitation. By definition, the aperture efficiency is at most unity, which occurs only for uniform excitation. Combining the aperture efficiency with the physical aperture size yields the effective area in the direction normal to the aperture plane. From that, one can calculate the directivity (with the reciprocity theorem) and beam solid angle, and estimate the HPBW. The calculations involved are less complicated than those involved in the previous strategy, but only yield single numbers.

A third strategy would be to know (or guess) the aperture efficiency for a number of popular aperture antennas. In practice, figures between 0.5–0.7 are typical. This strategy may be sufficiently accurate for many purposes.

4.2 Reflector Antennas

Narrow beams and high directivities accompany large apertures. One way to increase the size of an aperture is to use it to illuminate a secondary aperture or reflector. Sometimes, tertiary reflectors are also used. Large reflectors can be constructed for relatively low cost, although the manufacturing tolerances involved can be quite narrow.

A reflector antenna is illustrated in Figure 4.8. In the diagram, a primary source is situated at the focal point F. A point on the reflecting surface is labeled R. In analyzing reflector antennas, it is expedient to use three coordinate systems: Cartesian (x,y,z) with its origin at O, and cylindrical (ρ,z,ϕ) and spherical (r,θ,ϕ), with their origins at F. The vertical (z) axis is the polar axis. The plane of the aperture is the $z = z_o$ plane.

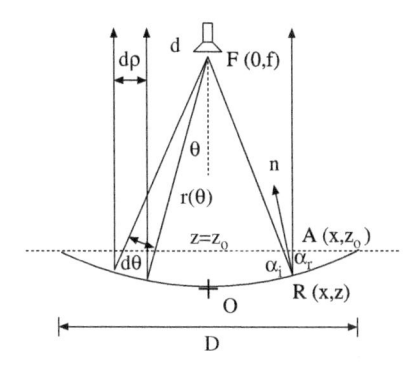

Figure 4.8 Geometry of a prime-focus parabolic reflector. F is the focal point, $z - z_o$ is the aperture plane, R is the reflection point, and A is the point where the ray intercepts the aperture plane. Taking the z-axis as an axis of revolution, the parabola becomes a paraboloid.

We can use ray theory to analyze reflection, where a ray is a line normal to the phase fronts of the electromagnetic radiation. In Figure 4.8, rays emanate radially from the focal point, change direction at the surface of the reflector so as to become collimated (aligned with the z axis), and proceed vertically upward thereafter.

Reflection implies two conditions that define the shape of the reflecting surface. The first is that the waves emitted at the focal point and reflected by the reflector should all arrive in phase at the aperture plane, where they constitute a plane wave. A plane wave maximizes the aperture efficiency (by virtue of its uniformity across the planar aperture) and will continue to propagate upward and outward according to Huygens' principle, exhibiting diffraction and forming a radiation pattern like those we have studied already.

For the phase to be uniform across the aperture plane, we require $\overline{FR} + \overline{RA} = \text{const}$, where A is the point where the ray intersects the aperture plane. If we take the base of the reflector to be at the origin of the coordinate system, this implies

$$\sqrt{x^2 + (f-z)^2} + z_o - z = \text{const} = f + z_o,$$

where $f + z_o$ is the length of the ray that travels strictly vertically. Rearranging gives $4fz = x^2$ in the plane or (generalizing) $4fz = x^2 + y^2 = \rho^2$ in a volume, where ρ is the radial distance from the axis of symmetry in cylindrical coordinates. This is the equation of a *parabolic* surface.

In terms of spherical coordinates centered on the focal point F, the equation for the surface can be written (after noting $r\cos\theta = f - z$ and $r\sin\theta = \rho$)

$$\overline{FR} = r(\theta) = \frac{2f}{1+\cos\theta}.$$

The second condition for reflection is that the angle of incidence α_i at the point R is equal to the angle of reflection α_r, both with respect to the surface unit normal vector $\hat{\mathbf{n}}$. The direction of the unit normal can be found by considering the family of surfaces defined by the equation

$$g(r,\theta) = f - \frac{1}{2}r(1+\cos\theta)$$
$$= f - r\cos^2(\theta/2).$$

By the previous equation, the $g = 0$ surface is coincident with the parabolic reflector. Other values of g specify surfaces adjacent to it. The unit normal vector can be found by calculating $\hat{\mathbf{n}} = \nabla g/|\nabla g| = -\hat{\mathbf{r}}\cos(\theta/2) + \hat{\boldsymbol{\theta}}\sin(\theta/2)$, which clearly has unit length. The angle of incidence obeys the equation $\cos\alpha_i = -\hat{\mathbf{r}}\cdot\hat{\mathbf{n}} = \cos(\theta/2)$ or $\alpha_i = \theta/2$. The angle of reflection obeys $\cos\alpha_r = \hat{\mathbf{z}}\cdot\hat{\mathbf{n}}$, where $\hat{\mathbf{z}}$ can be written $-\hat{\mathbf{r}}\cos\theta + \hat{\boldsymbol{\theta}}\sin\theta$. Consequently, it can be seen that $\alpha_r = \theta/2$ as well, so the angle of incidence is the angle of reflection. Rays emitted radially from the focal point will become collimated upon reflection.

It has already been shown that the aperture plane is a surface of constant phase for the reflected waves. What about amplitude? Suppose the aperture size of the emitter at the focal point is d while the diameter of the reflector is D. The associated HPBWs of the two are then λ/d and λ/D, respectively. Obviously, the latter can be made much narrower than for former. The gain of a reflector antenna is limited only by engineering limitations on the size and smoothness of the reflecting structure.

On the ray path $\overline{FR}$, the radiation travels mainly in the far field of the small emitter, and the power density decreases as the square of the distance from it. Since the path length of the ray $r(\theta)$ increases with θ, portions of the reflector near the periphery will receive less illumination than portions near the center. On the return ray path $\overline{RA}$, the radiation travels mainly in the near field of the large reflector, so the power density is not attenuated on that part of the path back to the aperture. Because of the outbound attenuation, however, the amplitude of the aperture excitation will automatically be tapered, even though the total distance traveled (and therefore the phase) is constant. This may be desirable from the point of view of sidelobe control. The radiation pattern of the emitter can be tailored for optimal illumination of the reflector in any case.

We can calculate the aperture distribution due to the effect discussed above with the help of a little geometry. Suppose the radiation pattern of the feed illuminating the reflector is given by $p(\theta,\phi)$, where θ and ϕ are the polar and azimuthal angles measured from the feed and where the polar axis is vertical and downward. The power incident on the part of the reflector occupying the differential solid angle $d\Omega$ is then

$$p(\theta,\phi)\sin\theta d\theta d\phi.$$

The same amount of power will be contained in the corresponding column of the reflected signal passing through the aperture. Switching to cylindrical coordinates, this implies

$$P(\rho,\phi)\rho d\rho d\phi = p(\theta,\phi)\sin\theta d\theta d\phi$$
$$P(\rho,\phi) = p(\theta,\phi)\frac{1}{r}\frac{d\theta}{d\rho},$$

where P is the power density per unit area passing upward through the aperture plane. Some manipulation shows that $d\rho/d\theta = r(\theta)$ on the parabolic surface and consequently that $P = p/r^2$:

$$P(\rho,\phi) = p(\theta,\phi)\left(\frac{1+\cos\theta}{2f}\right)^2$$
$$= p(\theta,\phi)\left(\frac{4f}{4f^2+\rho^2}\right)^2, \qquad (4.5)$$

where the second expression in cylindrical coordinates follows from solving $\rho = 2f\sin\theta/(1+\cos\theta)$ for $\cos\theta$. These results demonstrate that the aperture of a parabolic reflector is not uniformly illuminated by a hypothetical isotropic source at the feedpoint. For uniform illumination of the aperture, a specialized feed system at the focal point with a pattern able to compensate for the terms in parentheses in (4.5) would have to be constructed. Without such a feed, the aperture illumination is tapered, and the effective area, directivity, and beamwidth degraded as a result. The sidelobe level, too. will be reduced by the taper, and the objective therefore is to find a feed system that provides an acceptable tradeoff.

4.2.1 Spillover, Obstruction, and Cross-Polarization Losses

Optimal aperture efficiency is achieved for uniform illumination. The radiation pattern of the feed antenna is apt to be closest to uniform near the center of its main beam and more irregular further out. This suggests that the mean beam of the feed antenna should totally encompass the reflector. However, this implies that a considerable fraction of the radiated power will fall outside the reflector surface and be lost. The spillover efficiency refers to the fraction of the power emitted by the feed that is actually reflected. Spillover efficiency decreases as aperture efficiency increases, and an optimum tradeoff is generally sought.

The spillover efficiency can be expressed as the ratio of the power reflected to the total power emitted,

$$\varepsilon_s = \frac{\int_0^{2\pi} \int_0^{\theta_{max}} p(\theta,\phi) d\Omega}{\int_0^{2\pi} \int_0^{\pi/2} p(\theta,\phi) d\Omega},$$

where θ_{max} is the largest value of θ subtended by the reflector. Since the total power radiated by the feed is related to its peak power p_{max} and directivity D_f, this can be simplified as

$$\varepsilon_s = \frac{D_f}{4\pi} \int_0^{2\pi} \int_0^{\theta_{max}} \frac{p(\theta,\phi)}{p_{max}} d\Omega.$$

Efficiency is also affected by the degree to which the antenna feed system physically obstructs the reflected signal. Not only the feed, but also the waveguide and supporting structures block some fraction of the emitted power. This can be taken into account in the calculation of the aperture efficiency, where the aperture field is taken to be zero underneath obstructions in the geometric optics approximation. Offsetting the focal point and the feed so that it is not in the middle of the reflector's radiation pattern is one way to minimize the effect.

Still another factor influencing the efficiency of the reflector is cross polarization. The polarization of a signal changes upon reflection from a parabolic reflector, and this can mean that some of the radiated power ends up with a polarization other than the intended one and is effectively lost. In order to evaluate the effect, reflection of the radiation fields, and not just the power, must be considered.

In the limit of geometric optics, we have seen that reflection is governed by Snell's law, and the angle of incidence is the angle of reflection with respect to the unit normal

4.2 Reflector Antennas

$\hat{\mathbf{n}}$ in Figure 4.8. The boundary conditions at the surface of a conductor are such that the components of the incident and reflected electric fields tangential to the surface cancel while the normal components equate. The modulus of the field is unchanged by reflection. This can all be stated compactly as

$$\hat{\mathbf{e}}_r = 2(\hat{\mathbf{n}} \cdot \hat{\mathbf{e}}_i)\hat{\mathbf{n}} - \hat{\mathbf{e}}_i,$$

where $\hat{\mathbf{e}}_{i,r}$ are unit vectors parallel to the electric fields of the incident and reflected waves, respectively. Combining this with the result from the previous section yields an expression for the electric field over the planar aperture with an amplitude given in terms of the square root of the incident power from the feed

$$\mathbf{E}_r(\rho,\phi) = \sqrt{2Z_\circ p(\theta,\phi)} \frac{4f}{4f^2 + \rho^2} \hat{\mathbf{e}}_r,$$

with the corresponding secondary radiation field given by (4.4) generalized for all pointing directions,

$$\mathbf{E}(\mathbf{k}) = j\omega \frac{e^{ikr}}{2\pi r} \frac{\mu_\circ}{Z_\circ} \int \mathbf{E}_r e^{j(k_x x' + k_y y')} da, \quad (4.6)$$

where $da = \rho d\rho d\phi$ is the integral over the aperture plane.

An illustrative example is that of an ideal dipole antenna located at the focal point and aligned with the $\hat{\mathbf{x}}$ axis. If the dipole were oriented parallel to $\hat{\mathbf{z}}$, its electric field would be entirely in the $\hat{\boldsymbol{\theta}}$ direction. (This is stated for illustrative purposes, there being no obvious reason to orient the feed dipole this way.) In this case, it is aligned with

$$\hat{\mathbf{e}}_i = \frac{\hat{\boldsymbol{\theta}}\cos\theta\cos\phi - \hat{\boldsymbol{\phi}}\sin\phi}{\sqrt{1 - \sin^2\theta\cos^2\phi}}.$$

As shown above, the unit normal vector to the reflector surface is calculated from gradient of the equation that specifies the curve:

$$g(r,\theta) = r - \frac{2f}{1 + \cos\theta} = 0$$
$$\hat{\mathbf{n}} = -\nabla g / |\nabla g|$$
$$= \hat{\boldsymbol{\theta}}\sin(\theta/2) - \hat{\mathbf{r}}\cos(\theta/2).$$

The resulting expression for the reflected aperture field is complicated but can be simplified by converting from spherical to Cartesian coordinates. The aperture field is necessarily tangential to the aperture:

$$\hat{\mathbf{e}}_r = \frac{\hat{\mathbf{x}}(\cos^2\phi\cos\theta + \sin^2\phi) + \hat{\mathbf{y}}\cos\phi\sin\phi(\cos\theta - 1)}{\sqrt{1 - \cos^2\phi\sin^2\theta}}. \quad (4.7)$$

Note however that the denominator of this expression is identical to the element factor for an elemental dipole antenna oriented parallel to the $\hat{x}$ axis. Consequently, the integral in (4.6) involves just the numerator multiplied by the parabolic geometry factor and the appropriate constants.

From (4.7), it is evident that the cross-polarized ($\hat{y}$) component of the electric field vanishes in the E and H planes. From the symmetry in the aperture field, it is also clear that the cross-polarized component will be absent in the radiation E- and H-plane fields. A pair of crossed elemental dipoles driven 90° out of phase can be used to generate circularly polarized radiation in the principle planes. This could be very useful if the antenna was used for satellite communications. Note, however, that the polarization will be elliptical in general away from the principle planes, which could be important, particularly if the beam is broad. Similar remarks hold for phased-array antennas of crossed dipole antennas used to synthesize circular polarizations. This effect can usually be neglected for large-aperture antennas with narrow beams, however.

4.2.2 Surface Roughness

Imperfections in the reflector surface can degrade the efficiency of an antenna if those imperfections are comparable in size or large compared to the wavelength. "Efficiency" here refers to an increase in the power in the sidelobes at the expense of the power in the main lobe. A formula governing gain reduction due to surface imperfections in prime focus antennas was derived by Ruze in 1952. Ruze actually tested his law by introducing a number of dings to a perfect parabolic reflector and measuring the effects on the gain. (Photographs of his results can be found in the literature.) According to Ruze's law, the gain efficiency of a reflector is given by

$$\eta = e^{-(4\pi\sigma/\lambda)^2},$$

where σ is the RMS surface roughness. This formula is only accurate when the surface errors are numerous and random, and the effects of a few dings are difficult to estimate. Generally speaking, the larger the scale size of the dings, the more closely confined the resulting sidelobes will be to the main beam.

4.2.3 Cassegrain, Gregorian, and Spherical Reflectors

It is often desirable to add additional subreflectors to the basic prime focus reflector antenna discussed above. One advantage of multiple reflectors is that they permit the primary emitter to be located near ground level, which can be advantageous in view of weight restrictions and the need to connect the emitter to a transceiver with a potentially heavy and/or lossy feedline. Multiple reflectors also offer additional design degrees of freedom that can be used to correct for aberrations in the primary reflector. Finally, multiple reflectors facilitate the use of multiple primary emitters operating at different frequencies.

Secondary reflectors come in two classes, Cassegrain and Gregorian. A Cassegrain subreflector has a convex shape, whereas a Gregorian subreflector is concave. The two kinds of subreflectors are illustrated in Figure 4.9. Although the two systems are basically equivalent, the former is used much more widely in practice than the latter. If the primary emitter is a point source, the secondary reflectors in Cassegrain and Gregorian systems are hyperbolas and ellipses of rotation, respectively. More complicated shapes can be used for non-point-source primary emitters, however, or to correct for aberrations in the parabolic main reflector. In order to reduce obstruction loss, the direction of the main beam of the antenna can be offset from the secondary reflectors and the focal point of the primary parabolic reflector.

Finally, it is possible to construct reflector antennas using spherical rather than parabolic main reflectors. An example is the main reflector of the Arecibo Radio Observatory in Puerto Rico. A diagram of Arecibo is shown in Figure 4.10. Arecibo is not only a radio telescope but also a high-power radar operating at UHF and S-band frequencies through two different feed systems: a linefeed and a Gregorian subreflector system.

Parallel rays impinging on a spherical reflector from infinity converge not at a focal point but along a focal line. Coupling such radiation into a feed system poses a complicated engineering problem but can be accomplished with a linefeed, which is a slotted waveguide section, or with an assemblage of specialized reflectors custom designed to collect the radiation and focus it to a point, as in a conventional reflector antenna. Both approaches are implemented at Arecibo.

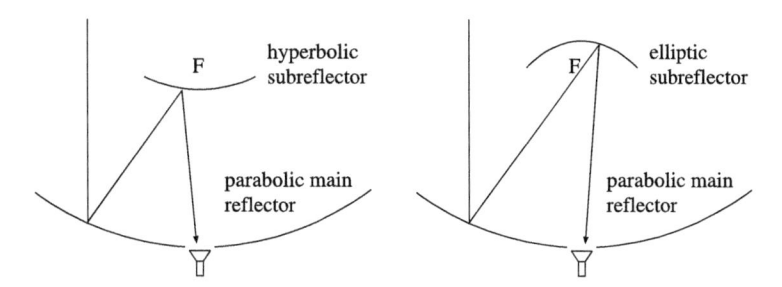

Figure 4.9 Cassegrain and Gregorian antenna systems. "F" denotes the focal point of the primary reflector.

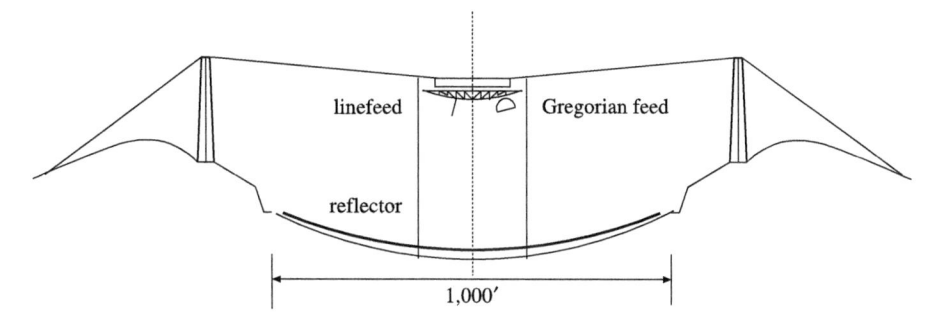

Figure 4.10 Arecibo Radio Observatory. The linefeed and Gregorian feed systems are suspended from a truss and can be mechanically steered in azimuth and zenith.

The advantage of the spherical reflector is that it allows beam steering without the necessity of moving the main reflector – something that would hardly be possible with Arecibo's 1,000' dish. Whereas a parabolic reflector has a single focal point, the focal lines of a spherical reflector are not unique but depend on the target direction. Rays from different directions focus on line segments with different locations and attitudes. The reflected signals from different targets can be coupled into either feed system by adjusting their placement and alignment mechanically. The effective area of the aperture is reduced by off-zenith steering, however, since the illuminated fraction of the main reflector is reduced. At Arecibo, only zenith angles less than about 15° can be used effectively. Even this limited steerability has greatly expanded Arecibo's scientific roles in radio and radar astronomy and aeronomy.

The total efficiency of a reflector antenna is the product of many component terms involving the aperture efficiency, the spillover efficiency, the polarization efficiency, and losses associated with the surface roughness of the reflector, obstructions, impedance mismatches, and other system imperfections. All of these parameters affect the signal power received by a reflector antenna. What ultimately matters on the receiving end of a communications or radar link is not the signal power, however, but rather the signal-to-noise power ratio, since this determines detectability. Factors influencing the performance of a receiving antenna in a noisy environment should also be considered.

In the next chapter, we will investigate sources of noise. At UHF frequencies and beyond, radiation from the warm ground itself can contributed significantly to the system noise level. Prime focus parabolic reflector antennas are particularly prone to ground noise, since the feed system spillover is pointed directly at the ground. The introduction of a second (or third) subreflector is one remedy for this situation. As shown in Figure 4.9, the feed system can be mounted in a hole cut through the main reflector. With this configuration, spillover is directed upward rather than downward, and ground noise is excluded. Furthermore, since the feed system is close to the ground, the waveguide is shortened, and the associated losses are reduced. These advantages typically come at the expense of the blockage imposed by the large secondary reflector and the reduced aperture efficiency that results. Other tradeoffs affecting the sensitivity of antennas in noisy environments are discussed in the next chapter.

4.3 Notes and Further Reading

References given in Chapter 1 covering Fourier analysis and Chapter 2 covering antenna theory are relevant to this chapter as well. A few extra references of historical interest pertaining to early work in diffraction theory are listed below. A widely used standard reference on optics is the one by Born and Wolf. Additional references pertaining to diffraction theory are provided at the conclusion of Chapter 9.

A popular reference that reviews special functions is the text by Abramowitz and Stegun. Readers interested in the history of radar at the Arecibo Radio Observatory may wish to consult the recent paper by J. D. Mathews. As noted earlier, an extensive treatment of the field-equivalence principle is given in J. A. Kong's text.

M. Abramowitz and I. A. Stegun. *Handbook of Mathematical Functions with Formulas, Graphs, and Mathematical Tables*. US Government Printing Office, Washington, DC, 1972.

M. Born and E. Wolf. *Principles of Optics*. Pergamon Press, New York, 1970.

A. Fresnel. Memoir on the diffraction of Light. In *The Wave Theory of Light: Memoirs by Huygens, Young and Fresnel*. American Book Company, Woodstock, GA, 1818.

Chr. Huygens. *Traité de la Lumiere*. Leyden, 1690.

G. Kirchhoff. Zur Theorie der Lichtstrahlen. In *Annalen der Physik*. Wiley, 1883.

J. A. Kong. *Electromagnetic Wave Theory*. EMW Publishing, Cambridge, Massachusetts, 2008.

J. D. Mathews. A short history of geophysical radar at Arecibo Observatory. *Hist. Geo-Space Sci.*, 4:19–33, 2013.

J. Ruze. The effect of aperture errors on the antenna radiation pattern. *Nuovo Cimento Suppl.*, 9(3):364–380, 1952.

A. Sommerfeld. Lectures on theoretical physics, v. 4. In *Optics*, pages 198–201, 211–212. Academic Press, New York, 1954.

W. L. Stutzman and G. A. Thiele. *Antenna Theory and Design*, 3rd ed. John Wiley & Sons, New York, 2012.

Exercises

4.1 Consider a circular aperture with diameter $D = 133\lambda$ with an excitation given by

$$E(r) \propto \cos^n(\pi r/D),$$

such that the excitation drops to zero at the aperture boundary.

a) Find the aperture efficiency for $n = 0, 1, 2$.
b) Find the directivity (in dB) for each case.
c) Estimate the HPBW for each case.

Try to work this problem using the simplest means possible.

4.2 We saw that the angular spectrum of an aperture antenna can be expressed as

$$|F(k_x, k_y)|^2 \propto \left| \int \mathbf{E}(x', y') e^{j\mathbf{k} \cdot \mathbf{r}'} ds' \right|^2, \qquad (4.8)$$

where the integration is over the aperture and where $\mathbf{r}'$ is a vector in the plane of the aperture. For a uniform phase aperture, the main beam will be in the direction normal to the aperture ($\mathbf{k} \cdot \mathbf{r}' = 0$), and the power density in the direction of the maximum can then be written as

$$p_{\max} = \frac{k^2}{32\pi^2 Z_\circ} \left| \int \mathbf{E}(x', y') ds' \right|^2, \qquad (4.9)$$

where the appropriate constants have been filled in for the first time. Finding the *total* radiated power by integrating (1) over all angles would be quite difficult in practice. However, since we know all the power radiated had to pass through the aperture, we can cleverly calculate the total radiated power by integrating the usual expression for the power density of an electromagnetic wave over the aperture,

$$p_{\text{total}} = \frac{1}{2Z_\circ} \int |\mathbf{E}(x',y')|^2 ds'. \tag{4.10}$$

Combining (2) and (3), we can calculate the directivity of the aperture radiator:

$$D \equiv 4\pi \frac{p_{\text{max}}}{p_{\text{total}}}$$

$$= \frac{4\pi}{\lambda^2} \frac{|\int \mathbf{E}(x',y')ds'|^2}{\int |\mathbf{E}(x',y')|^2 ds'}$$

$$= \frac{4\pi}{\lambda^2} A_{\text{phys}} \frac{|\langle \mathbf{E} \rangle|^2}{\langle |\mathbf{E}|^2 \rangle}$$

$$= \frac{4\pi}{\lambda^2} A_{\text{eff}}. \tag{4.11}$$

The above expresses the reciprocity theorem and the definition of the effective area of an aperture in terms of its physical area. The aperture efficiency is the ratio of the effective to the physical area.

Now the question: using the Schwartz inequality given below, prove that an aperture with uniform amplitude excitation has the largest effective aperture and hence the highest directivity.

$$\left| \int AB\, ds \right|^2 \leq \int A^2 ds \int B^2 ds.$$

Hint: let $A = 1$ and B be the aperture field.

4.3 The above calculations should demonstrate that the aperture efficiency for a uniformly illuminated aperture is unity. Now, show this explicitly. Consider a large square aperture in the $x - y$ plane (at $z = 0$) measuring 30 wavelengths on a side containing a uniform tangential field.

a) Write down the spatial Fourier transform expressing the radiation field.
b) Calculate the power pattern in the $x - z$ plane ($\phi = 0$) and plot it. The plot should make clear the HPBW and the level of the first sidelobe at least.
c) Now integrate the entire pattern to get the total radiated power, and from that, the directivity. Be careful with your limits of integration. Find the effective area from the directivity and compare it with the physical area. If they don't agree to within a few percent, you've done something wrong.

Exercises

4.4 Tapered aperture distributions produce beams with lower sidelobes than distributions with sharp edges. Suppose we have a circular aperture of diameter $D = 50\lambda$ where the aperture distribution depends only on the distance r' from the center. Neglecting the Fresnel term, we have seen that the radiation field F is then related to the aperture distribution E through the integral

$$F(\theta) \propto \int E(r') J_\circ(kr' \sin\theta) r' dr',$$

where the primed variables refer to the position in the aperture plane, which we take to be the $x - y$ plane here, and where the radiated power density is proportional to $|F(\theta)|^2$.

a) Verify numerically the known result that, for a uniform aperture distribution, the half-power beamwidth (HPBW) is $1.02\lambda/D$ radians and the first sidelobe level is 17.6 dB below the main beam. Make a properly scaled plot to show this.

b) Now, suppose the aperture field is uniform out to $r' = D/4$ and then tapers linearly with r' to zero at $r' = D/2$. Repeat part a), finding the new HPBW and the first sidelobe level, and plot the radiation pattern.

c) Calculate the aperture efficiency and the effective area for both cases by the easiest means you know.

4.5 Find the effect of a small quadratic phase error on the Fraunhofer zone radiation pattern of a rectangular aperture. Take the excitation to be

$$\mathbf{J}_s = J_s \hat{\mathbf{x}} e^{j\alpha x^2}, \quad |x| \leq L_x/2, \, |y| \leq L_y/2,$$

where J_s is a constant and where $\alpha(L_x/2)^2 \ll 1$.

a) First, calculate the radiation pattern analytically (pencil-and-paper style), making use of the following Fourier transform pairs,

$$\mathcal{F} f(x) = F(k_x)$$
$$\mathcal{F} jx f(x) = \frac{d}{dk_x} F(k_x)$$
$$\mathcal{F} x^2 f(x) = -\frac{d^2}{dk_x^2} F(k_x),$$

and the approximation $e^{j\alpha x^2} \sim 1 + j\alpha x^2$. Your answer should look like the radiation pattern of a uniform aperture, only multiplied by another function. Plot the pattern, and make clear what has changed.

b) Calculate the half-power beamwidths in the $x - z$ and $y - z$ planes and use this to estimate the directivity. Take $L_x = L_y = 4\lambda$ and $\alpha(L_x/2)^2 = 1.0$.

c) Next, calculate the directivity exactly and compare with (b). Give some thought to the easiest way to do this. Comment on the results, and explain any discrepancy.

4.6 Investigate the power pattern of an antenna in the Fresnel zone. Consider a 1-D aperture that is uniformly illuminated in the $\hat{\mathbf{x}}$ direction. Neglect the $\hat{\mathbf{y}}$ dependence in this problem (no integrals in the $\hat{\mathbf{y}}$ direction needed). Take $L_x = 50$ m and $\lambda = 1$ m. Calculate the distance where the transition from the Fresnel zone to the Fraunhofer zone occurs, and call this $r_\circ$. Now, using the full expression for the aperture integral, *including the quadratic Fresnel zone term*, calculate the power pattern $|F(k_x)|^2$ at distances of $0.1r_\circ$ and $10r_\circ$. Plot your results. In the Fraunhofer zone, the pattern should agree with the far field $(\sin X/X)^2$ pattern. In the Fresnel zone, meanwhile, the pattern should start to resemble a searchlight pattern with steep walls.

5
Noise

At this point, we are able to evaluate the performance of communication and radar links and estimate the amount of power received. What is important for detectability, however, is not the received signal power (which can always be improved through amplification), but the signal-to-noise ratio, the SNR (which cannot). By the end of the chapter, we will have investigated strategies for optimizing the SNR. Such strategies constitute a significant fraction of the radar art. First, we must identify and learn about natural and instrumental noise sources. We confine the analysis to noise due to thermally agitated charge carriers. Thermal noise is usually the dominant noise source in radio and radar, and is common to all electrical systems. To understand thermal noise, we will have to take a brief detour outside the realm of classical physics.

5.1 Blackbody Radiation

Consider electromagnetic radiation existing within the cavity of a conductor. Classical physics tells us that the boundary conditions at the walls of the cavity will force the radiation to manifest as an infinite set of discrete modes, each with a characteristic frequency. Modern physics showed that each of these modes can be associated with particles, photons, with energy hf, where $h = 6.626 \times 10^{-24}$ is Planck's constant and f is the frequency of the mode.

The number of photons in each mode fluctuates continuously. Photons obey Bose–Einstein statistics, and so the average number of photons per mode is given by

$$n = \frac{1}{e^{hf/KT} - 1},$$

where $K = 1.38 \times 10^{-23}$ J/K is Boltzmann's constant and T is the Kelvin temperature. The average energy per mode is therefore $E = nhf$.

To simplify the problem, we can take the cavity to be rectangular with dimensions $l_x \times l_y \times l_z$. The frequency of a particular mode can easily be shown to be

$$f_{ijk} = \frac{c}{2}\left[(i/l_x)^2 + (j/l_y)^2 + (k/l_z)^2\right]^{1/2},$$

where the mode numbers i, j, k are positive integers. In fact, there are two degenerate modes for every set of mode numbers – the TE and TM modes. The total number of modes with frequency less than f can therefore readily be found to be

$$N(f) = \frac{8\pi f^3}{3c^3}V,$$

where $V = l_x l_y l_z$ is the volume of the cavity. The number of modes per unit volume with frequencies between f and $f+df$ is then $V^{-1}(dN(f)/df)df$. Consequently, the energy density in this band of frequencies is

$$dU(f) = \frac{8\pi f^2}{c^3} \frac{hf}{e^{hf/KT} - 1} df,$$

which has units of J/m^3. Integrating this expression over all frequencies gives the total energy density $U = (8K^4\pi^5/15h^3c^3)T^4$.

The connection between the foregoing analysis and thermal noise is the following. Imagine that a small hole is drilled into the cavity. In thermodynamic equilibrium, the power emitted through the hole must equal the power that enters through it. The hole constitutes an ideal black body, an idealized source and sink of radiation. The energy carried by photons exiting through a hole of differential size dA in the time interval dt through the differential solid angle $d\Omega$ will be

$$dE = dU(f)(cdt)(dA\cos\theta)\frac{d\Omega}{4\pi}, \tag{5.1}$$

where θ is the direction of exit relative to the outward unit normal. The corresponding radiated power is dE/dt, and the total power radiated into all solid angles is

$$dP = cdU(f)\frac{dA}{4\pi}\int \cos\theta\, d\Omega$$
$$= cdU(f)\frac{dA}{4\pi}\int_0^{2\pi}\int_0^{\pi/2} \cos\theta\sin\theta\, d\theta d\phi$$
$$= \frac{c}{4}dU(f)dA$$
$$= \frac{2\pi f^2}{c^2}\frac{hf}{e^{hf/KT}-1}dfdA,$$

which is Planck's law for blackbody radiation. In predicts the power emitted from a blackbody surface in a band of frequencies on the basis of nothing more than the body's temperature. This is the foundation of the thermal noise problem.

In view of the fact that $hf/KT \ll 1$ in most applications related to radio and radar, Planck's law can be simplified to the form of the Rayleigh–Jeans law:

$$dP \approx \frac{2\pi f^2}{c^2} kT \, df \, dA.$$

Integrating the Rayleigh–Jeans law over all frequencies gives the Stephan–Boltzmann law, $dP = \sigma_B T^4 dA$, where $\sigma_B = 5.67 \times 10^{-8}$ W/m^2/K^4 is the Stephan–Boltzmann constant. Although they are idealizations, the blackbody radiation model and the quantities derived from it are widely used to analyze emissions from actual bodies, from circuit elements to regions of the celestial sphere.

5.2 Nyquist Noise Theorem

The connection between blackbody radiation and noise in passive linear circuit elements was made in a 1928 Physical Review paper by Harry Nyquist still renowned for its incisiveness and simplicity. The Nyquist noise theorem accounts for noise emitted from a resistor arising from the thermal agitation of its electrons. The electron motion is random and causes no average current to flow, but current does flow in an RMS sense, and associated with this is the emission of power. In thermodynamic equilibrium, an equal amount of power must be absorbed by the resistor from the environment. The temperature of the resistor (and the environment) will vary until equilibrium can be achieved.

An isolated resistor can be modeled as a 1-D resonant cavity of length l terminated at each end in a short circuit. The cavity will support wave modes for which its length is an integer number of half wavelengths. The number of modes per unit length per unit bandwidth is consequently $2/c$, and, by analogy to the cavity resonator analyzed above, the average energy density in the frequency interval between f and df will be

$$dU(f) = \frac{2}{c} \frac{hf}{e^{hf/KT} - 1} df$$
$$\approx \frac{2}{c} KT \, df,$$

where the second line reflects the limit $hf/KT \ll 1$ suitable for most radio and radar applications once more. In such applications, the channel will have a passband occupying a well-defined bandwidth Δf. The energy density per unit length of the resistor falling into this passband will be $U = 2KT\Delta f/c$, which has units of J/m.

The rate of power flow at any point along the resistor is given by the product of the energy density and the speed of propagation, c. On average, an equal amount of power will flow in either direction at every point on the resistor. That amount of power will be

$$P = KT\Delta f,$$

which is the key result of the analysis. It gives the noise power in a passband Δf associated with the thermal agitation of electrons in a passive device at temperature T.

Now, if one of the resistor shorts were replaced with a matched termination, this power would flow out through the termination. If we model the noisy resistor R as an ideal resistor R in series with a voltage source, the RMS value of the voltage required to deliver this power to the matched load (also R) would be

$$v_{RMS} = \sqrt{4KTR\Delta f} \longrightarrow P = \langle v_{RMS}^2 \rangle / 4R. \tag{5.2}$$

This is Nyquist's noise theorem. The noise voltage across the terminals of the resistor will be $N(0, v_{RMS})$. The noise is broadband and often referred to as "white noise." Note that the power outflow must be matched by an equal inflow from the environment if thermodynamic equilibrium is to be preserved.

5.2.1 Noise in Passive Linear Networks

Nyquist's noise theorem applies not only to resistors but to any passive linear network. The principle is illustrated in Figure 5.1. Consider a resistor R at temperature T connected by means of a lossless bandpass filter centered at frequency f with bandwidth $B = \Delta f$ to a passive linear network with impedance $Z_l = R_l + j\chi_l$ as shown. According to Ohm's law, the power delivered by the resistor to the network will be $P = I_{RMS}^2 R_l$, or

$$P = \left| \frac{v_{RMS}}{R + Z_l} \right|^2 R_l$$
$$= \frac{4KTR\Delta f}{|R + Z_l|^2} R_l.$$

The resistor and the network must eventually achieve thermodynamic equilibrium, at which point their temperatures will be the same, and the power delivered from the network to the resistor will match the power delivered from the resistor to the network found above. This is the principle of detailed balance. Define v' as the Thévenin equivalent noise voltage of the network. Then the power delivered from the network to the resistor must be

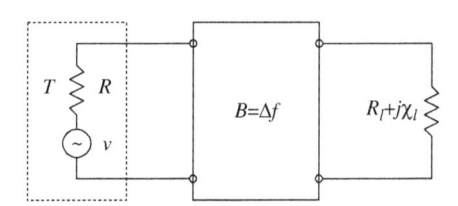

Figure 5.1 Illustration of Nyquist's theorem. A resistor R is connected to a passive linear network ($Z_l = R_l + j\chi_l$) via a bandpass filter with bandwidth $B = \Delta f$ centered on frequency f.

5.2 Nyquist Noise Theorem

$$P' = \left|\frac{v'_{RMS}}{R+Z_l}\right|^2 R = P.$$

Solving this equation for v' yields

$$v'_{RMS} = \sqrt{4KTR_l\Delta f},$$

which is Nyquist's noise theorem generalized for any passive linear network with Thévenin equivalent resistance R_l.

In the event that the load is matched to the source, $R_l = R$, $\chi_l \to 0$, the power delivered by the noise source over the bandwidth B is just $P = KTB$ (W). This is the so-called available noise power under matched-load conditions. This result is also referred to as Nyquist's noise theorem and will be used throughout the rest of the chapter. (In the event of an impedance mismatch, an extra factor of $(1-|\Gamma|^2)$ should be included, where Γ is the reflection coefficient.) For example, taking $T = 300$ K (room temperature) and considering a bandwidth $\Delta f = B = 1$ MHz gives a noise power of 4×10^{-15} W. This may seem like a very small amount of power, but it could be significant (even dominant) compared to the signal power received directly from the terminals of an antenna, depending on the application.

In the event that more than one thermal noise source are involved, we can expect the RMS noise voltages of the sources to be statistically uncorrelated. Consequently, the noise powers rather than the noise voltages will add. The weighting of the addition can be summarized as

$$P = \frac{\langle v_1^2\rangle + \langle v_2^2\rangle}{4(R_1+R_2)}$$
$$= \frac{KB(T_1R_1 + T_2R_2)}{R_1+R_2},$$

which gives the power delivered to a matched load. Note especially that the equivalent temperature weighting is proportional to the resistance.

5.2.2 Reciprocity Theorem, Revisited

The foregoing discussion provides the basis for an alternative proof of the reciprocity theorem. Outlining the proof will also provide insights for calculating the noise power delivered by an antenna. We will see that this noise power is not mainly due to the thermal agitation of charge carriers in the antenna itself but rather to distant radiation sources in the environment.

Figure 5.2 shows an antenna connected to a matched resistor R_L at temperature T_L situated within a blackbody cavity at uniform temperature T. The resistor will deliver power to the antenna, which will then be radiated into the cavity. At the same time, the antenna will

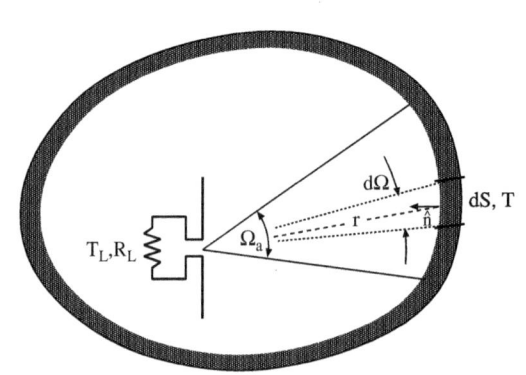

Figure 5.2 Diagram illustrating the reciprocity theorem and the concept of antenna noise temperature. The surface area element dS on the cavity is at a distance r from the antenna and has an inward unit normal vector $\hat{\mathbf{n}}$.

receive power from the cavity and deliver it to the resistor. In thermodynamic equilibrium, $T_L = T$, and the two powers will be equal.

Consider first the power received by the antenna from the wall of the blackbody cavity. Recall that (5.1) gives the energy density of photons emitted from an area element dA in the time interval dt into a differential solid angle $d\Omega$. According to this equation, the differential power received by the antenna will be

$$\begin{aligned} dP &= \frac{c}{2} dU (\hat{\mathbf{r}} \cdot \hat{\mathbf{n}} dS) \frac{A_{\text{eff}}}{4\pi r^2} \\ &= \frac{4\pi f^2}{c^2} KT df (\hat{\mathbf{r}} \cdot \hat{\mathbf{n}} dS) \frac{A_{\text{eff}}}{4\pi r^2} \\ &= \frac{4\pi f^2}{c^2} KT df A_{\text{eff}} \frac{d\Omega}{4\pi}, \end{aligned} \tag{5.3}$$

where we have made use of the fact that $\cos\theta = \hat{\mathbf{r}} \cdot \hat{\mathbf{n}}$ and where the solid angle subtended by the antenna from the surface-area element is A_{eff}/r^2. The factor of one-half reflects the fact that the antenna can only intercept the power in one polarization, regardless of the type of antenna, whereas (5.1) accounts for the power in both polarizations. Finally, $d\Omega = \hat{\mathbf{r}} \cdot \hat{\mathbf{n}} dS/r^2$ is the solid angle subtended by the surface-area element from the feedpoint of the antenna.

We can now determine the total power received by the antenna from its environment by integrating over all the sources within the antenna beam solid angle Ω_a

$$\begin{aligned} P &= KT df \frac{A_{\text{eff}}}{\lambda^2} \iint_{\Omega_a} d\Omega \\ &= KT df \frac{A_{\text{eff}}}{\lambda^2} \Omega_a, \end{aligned} \tag{5.4}$$

which gives the power delivered to the load resistor R_L.

In thermodynamic equilibrium, with $T_L = T$, this power must be balanced by the power delivered by the load resistor to the antenna, $KT df$. With the recognition that $\Omega_a \equiv 4\pi/D$, the balance equation can be expressed as

$$A_{\text{eff}} = \frac{\lambda^2}{4\pi} D,$$

which restates the reciprocity theorem. The result can be generalized and made to relate $A_{\text{eff}}(\theta, \phi)$ and $D(\theta, \phi)$ by restricting the domain of integration in (5.4).

5.3 Antenna and System Noise Temperature

The preceding analysis considered the balance between power received by an antenna from thermal noise sources around it and power emitted by the antenna in thermodynamic equilibrium. We do not demand thermodynamic equilibrium and can now consider the noise power received by an antenna and delivered to a load. In this case, the load is a receiver and the surrounding noise sources are warm bodies in the environment. These bodies do not have a common temperature as a rule, meaning that the noise should vary with antenna pointing. The range to the sources is immaterial; what matters is their temperatures and the solid angles they occupy.

Combination of (5.3) and the reciprocity theorem gives an expression for the differential noise power delivered by an antenna to a matched receiver per unit solid angle occupied by a noise source:

$$dP = KT(\theta, \phi) B G(\theta, \phi) \frac{d\Omega}{4\pi},$$

where B is the receiver bandwidth and where the temperature of the noise source is allowed to vary with bearing. The total received noise power is given by the integral over all solid angles

$$P = KB \frac{1}{4\pi} \oint T(\theta, \phi) G(\theta, \phi) \, d\Omega. \qquad (5.5)$$

It is conventional to define an equivalent antenna noise temperature T_{ant} as the temperature of a hypothetical noise source that would give rise to the same noise power, i.e., $KBT_{\text{ant}} = P$. Employing this definition and rewriting (5.5) with the help of the properties of the antenna gain, we get an experimental definition of the antenna noise temperature:

$$T_{\text{ant}} = \frac{\oint T(\theta, \phi) G(\theta, \phi) \, d\Omega}{\oint G(\theta, \phi) \, d\Omega}.$$

The antenna noise temperature therefore reflects not the physical temperature of the antenna but rather the temperature of the thermal noise sources in the field of view. The weighting is determined by the antenna gain in the direction of the various noise sources. For steerable antennas, different values of T_{ant} will be associated with different pointing positions and also different times of day. It is common practice to measure an antenna noise temperature map early in the operation of a new radio telescope or radar.

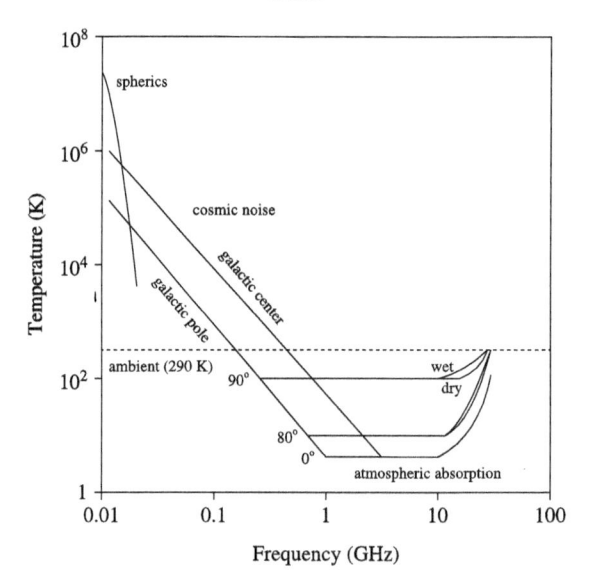

Figure 5.3 Sketch showing antenna noise temperature versus frequency.

The preceding analysis assumed matched-impedance conditions between the antenna and the receiver and also neglected the Ohmic resistance of the antenna by comparing the radiation resistance. These are good assumptions in most radar applications. If Ohmic resistance is not negligible, as might be the case for electrically short antennas, a series combination of noise sources would have to be considered.

A sketch depicting antenna noise temperature versus frequency observed on Earth is shown in Figure 5.3. At frequencies below about 15 MHz, antenna noise is dominated by radio emissions from lightning, called "atmospherics" (or "spherics," for short). While lightning might not be present nearby, lightning strikes are always occurring somewhere on the planet. At HF frequencies, emissions from lightning can propagate from the source to the receiver by reflecting off the ionosphere. The equivalent noise temperature for these emissions is of the order of 10^6–10^8 K.

Between about 15 MHz and 1 GHz, antenna noise is dominated by cosmic radio noise or "sky noise." This is made up of emissions from radio stars in our galaxy. The sky noise temperature decreases with frequency at a rate of about $f^{-2.4}$. Cosmic noise at frequencies below 10–15 MHz cannot penetrate the ionosphere. A typical cosmic sky noise temperature at VHF (UHF) frequencies is about 1000 K (100 K). The actual temperature depends on frequency and also on direction; cosmic noise is strongest in the direction of the Milky Way's center and weakest looking out of the ecliptic plane.

Cosmic noise continues to decrease with frequency until reaching the 2.7 K background noise temperature at about 10 GHz. However, antenna noise observed on the ground starts increasing above about 1 GHz due to atmospheric absorption noise (see next section). This is radio emission mainly from water vapor in the Earth's atmosphere. The antenna noise temperature due to atmospheric absorption varies with zenith angle; rays with small

5.3 Antenna and System Noise Temperature

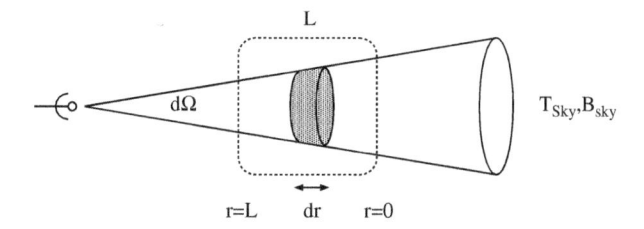

Figure 5.4 Diagram illustrating atmospheric absorption noise.

zenith angles penetrate the atmosphere quickly and so have smaller noise temperatures than rays with low elevations, which must traverse a thick atmospheric layer. Atmospheric absorption noise temperatures can be as great as the ambient temperature, about 200 K.

The *radio window* refers to the band of frequencies where the terrestrial noise temperature has a minimum. It is centered in the vicinity of 2 GHz, which is consequently the band where communications links with satellites are concentrated. The noise temperature here is between 10 and 100 K, depending on time and season, direction, and atmospheric conditions (relative humidity).

5.3.1 Atmospheric Absorption Noise and Radiative Transport

Atmospheric water vapor absorbs radiation from galactic radio sources and also from the sun. While the photosphere (the visible part of the sun) has a temperature of about 6000 K, the corona (the hot plasma surrounding the sun) has a temperature of $1-2 \times 10^6$ K and is a source of strong background RF radiation. Atmospheric water vapor blocks some of this radiation from reaching the ground but also emits additional noise as a result. The receiver consequently sees a combination of sky noise and atmospheric absorption noise.

What is the connection between atmospheric absorption and noise? To maintain thermodynamic equilibrium, isolated bodies that absorb radiant energy must also radiate. This is the same principle of detailed balance that lies at the root of Nyquist's theorem and the concept of antenna noise temperature. A body absorbing energy increases in temperature until the energy it radiates matches the energy it absorbs. The terminal temperature may be calculated using Planck's law. We once again do not demand that thermodynamic equilibrium actually exist and can use the radiative transport equations to evaluate noise from atmospheric constituents at any temperature.

We can quantify the effects of atmospheric absorption with the help of Figure 5.4, which depicts a region of the sky to the right being observed through a solid angle $d\Omega$ by an antenna on the left. In between is a cloud of absorbing (and therefore emitting) material of width L. The problem is usually described in terms of the noise brightness B which is the noise power density per solid angle, $dB = dP/d\Omega$.

Suppose the brightness in the absence of the cloud is given by B_{sky}. An equation for the differential change in brightness due to a cloud layer of width dr is

$$dB = -B\alpha\rho dr + dB_{cloud}, \tag{5.6}$$

where α is the absorption cross-section of a cloud molecule and ρ is the volume density of the molecules. The first term on the right side of (5.6) represents the absorption of noise in the cloud layer, and the second term represents the isotropic emission of noise from the cloud layer due to its thermal energy. Note that the absorption cross-section for water vapor molecules and droplets depends strongly on drop size and differs for rain, snow, fog, and humidity. It is also a strong function of frequency. A more detailed treatment of absorption is given in Chapter 11.

Concentrating on the second term for the moment, we can write the power in watts emitted by the cloud in a volume dv as $w\rho dv$, w being the power emitted by each molecule. If the radiation is isotropic, then the associated noise power density back at the receiver a distance r away and the associated brightness are

$$dP_{cloud} = \frac{w\rho dv}{4\pi r^2}; \quad dv = r^2 dr d\Omega$$

$$dB_{cloud} = dP/d\Omega = \frac{w\rho}{4\pi} dr.$$

Altogether, the equation for the variation of the brightness from the distant side of the cloud to the near side is given by

$$\frac{dB}{dr} + B\alpha\rho = \frac{w\rho}{4\pi}. \tag{5.7}$$

This is an elementary ordinary first-order differential equation. Before solving it, we can use it to calculate the brightness level at which radiative equilibrium is achieved within the cloud. Setting $dB/dr = 0$ yields $B_{cloud} = w/4\pi\alpha$. This is the brightness level at which absorption and radiation balance. Greater (lesser) *in situ* brightness implies net heating (cooling) within the cloud.

Assuming that all the coefficients are constant, the general solution to the full differential equation is

$$B(r) = B_\circ e^{-\gamma r} + B_{cloud},$$

where $\gamma = \alpha\rho$ and $B_\circ$ is an undetermined coefficient. Note that r is measured from the far (right) side of the cloud here. The particular solution comes from enforcing the boundary condition that $B(r=0) = B_{sky}$. The result is

$$B = B_{sky} e^{-\gamma r} + B_{cloud}\left(1 - e^{-\gamma r}\right).$$

What is of interest is the brightness at the near (left) side of the cloud, since this is what will be detected by the receiver. Defining the *optical depth* of the cloud to be $\tau \equiv \gamma L$ gives

$$B(r=L) = B_{sky} e^{-\tau} + B_{cloud}\left(1 - e^{-\tau}\right).$$

It is noteworthy that, at a given frequency, the same coefficient ($\gamma = \alpha\rho$) governs the emission and absorption of noise in a medium. This is a consequence of detailed balance and confirms the proposition that good absorbers of radiation are also good radiators. That the solar corona is a strong absorber/emitter at radio frequencies but not in the visible spectrum is a vivid example of how these properties can depend strongly on frequency, however.

Finally, by applying Nyquist's theorem, we can quantify the noise equally well by an equivalent noise temperatures:

$$T_{ant} = T_{sky}e^{-\tau} + T_{cloud}\left(1 - e^{-\tau}\right). \qquad (5.8)$$

Altogether, atmospheric absorption is quantified by the optical depth of the absorbing medium. Large (small) values of τ denote an opaque (transparent) atmosphere. To a ground-based observer, atmospheric absorption decreases the direct radiance of the sun. However, the observer will also see RF radiation from the atmosphere itself, which is being heated by absorption and which must radiate in pursuit of thermal equilibrium. The antenna noise temperature consequently equals the sky noise temperature for a perfectly transparent atmosphere, but the atmospheric temperature (called T_{cloud} here) for a perfectly opaque one. Note that whereas T_{sky} can depend strongly on frequency, T_{cloud} is the actual atmospheric temperature at cloud level (~200–300 K) and does not vary greatly with frequency, although the optical depth does.

The idea presented above can be applied to other situations. For example, radio receivers on spacecraft sometimes stare down at the Earth rather than up into space. In this case, the background noise source is the Earth itself, which radiates with the noise temperature of the ground. Equation (5.8) still applies, but with T_{sky} replaced by T_{ground}. Not only can atmospheric water vapor and other constituents be observed from variations in the antenna noise temperature, so can other features that effect emission and absorption. These include ground cover, forest canopies and features in the terrain.

5.3.2 Feedline Absorption

We can extend the ideas from the previous section to a treatment of non-ideal, absorbing transmission lines. These too must radiate (emit noise) if thermodynamic equilibrium is to be maintained, and the noise they contribute can be modeled according to (5.8).

As shown in Figure 5.5, there are actually three sources that contribute to the output noise of a receiver. These are the antenna (with contributions from spherics, cosmic sky noise, and/or atmospheric absorption), the transmission line, and the receiver itself. The last of these will be analyzed in the next section of the chapter. For now, it is sufficient to

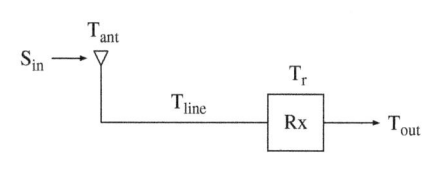

Figure 5.5 Illustration of the noise budget in a receiving system.

say that each of these noise sources can be characterized by an equivalent noise temperature and that the receiver noise temperature T_r is additive. We have already discussed T_{ant}. The noise temperature for the transmission line T_{line} (which we assume to be the ambient temperature, ~ 300 K) can be incorporated just as in (5.8):

$$T_{out} = T_{ant} e^{-\gamma L} + T_{line}(1 - e^{-\gamma L}) + T_r.$$

Here, γ is the transmission line attenuation constant per unit length, L is the line length and γL is the equivalent optical depth for the line. A common shorthand is to write the exponential term as a single factor ε, representing the fraction of the input power that exits the line, i.e.,

$$T_{out} = T_{ant}\varepsilon + T_{line}(1-\varepsilon) + T_r.$$

Referring back to Figure 5.5, we are finally in position to write the signal-to-noise ratio for the receiving system, including all the relevant sources. Take the receiver to have a power gain given by G_{rx}, and let B represent the bandwidth in Hz (not the brightness). Then

$$\begin{aligned}\left(\frac{S}{N}\right)_{out} &= \frac{S_{in}\varepsilon G_{rx}}{KB(T_{ant}\varepsilon + T_{line}(1-\varepsilon) + T_r)G_{rx}} \\ &= \frac{S_{in}}{KB(T_{ant} + T_{line}(1-\varepsilon)/\varepsilon + T_r/\varepsilon)} \\ &= \frac{S_{in}}{KBT_{sys}} \\ T_{sys} &= T_{ant} + T_{line}(1-\varepsilon)/\varepsilon + T_r/\varepsilon,\end{aligned}$$

where the concept of a system noise temperature T_{sys} has been introduced to characterize the noise level of the entire receiving system. We can think of T_{sys} as the equivalent temperature of a single, hypothetical source responsible for all the system noise. The challenge then becomes one of minimizing T_{sys} (cost effectively). Note that these results do not depend on the receiver gain, which cannot be used to salvage poor SNR.

The antenna noise temperature is a function of frequency, local time, and direction. To the extent it can be minimized in a radar design, it should be, but there may be little design latitude to work with here, as these fundamental radar design elements are usually forced by the application at hand. Transmission line noise can be minimized by using short cable runs and high-quality, low-loss cabling. It can also be mitigated by placing the first amplifier stage (the front end) directly at the antenna terminals and bypassing the transmission line (see below). Finally, receiver noise is minimized by investing in low-temperature receivers. How aggressively one works to minimize line and receiver noise depends on the relative strength of the antenna noise.

Some examples illustrate the point. Consider using a 30 m run of RG-58 transmission line for reception at 100 MHz. The attenuation coefficient for RG-58 at this frequency is 5 dB/100'. The corresponding value of ε is then about -5 dB, or 0.32. The fraction

$(1-\varepsilon)/\varepsilon$ is close to 2. Given that the sky noise temperature is about 1000 K and assuming a receiver noise temperature of $\sim$100 K (such VHF receivers are common and inexpensive), the system noise temperature will be about twice the sky noise temperature using this transmission line. This may be acceptable, depending on the application. (Note, however, that 5 dB losses would probably never be acceptable for transmission, but that isn't a noise-related issue.)

Coaxial cable is quite lossy at UHF frequencies and above. Consider next the example of using RG-17 (heavy-duty coax cable) to feed the 430 MHz radar receiver at Arecibo. The cable run from the antenna feed to the receiver building would have to be 600' long. Given a loss coefficient of 2 dB/100' at this frequency, we find $\varepsilon = -12$ dB $\sim$ 0.063. The ratio $(1-\varepsilon)/\varepsilon$ is then close to 15! The sky noise temperature at this frequency is meanwhile only about 100 k. A receiver with a lower T_r than this will have to be purchased, but the full benefits of the low-noise receiver cannot be enjoyed without ε much closer to unity. Transmission line noise will dominate in any case in this example. Coaxial cable is obviously unsuitable for this application. The situation is even more grave at microwave frequencies, which is why waveguide is preferred over coaxial cable.

5.3.3 Cascades

Complicating matters, a complete receiver is usually composed of several stages. Each stage has a gain element but may also include mixers, filters, etc. A three-stage receiver is illustrated in Figure 5.6. Each stage supplies a certain amount of gain, typically about 20 dB, but also contributes to the output noise. We can regard all the noise-generating components of a receiver stage as a single component with a noise temperature T_i to which the gain of the stage G_i is applied. In the example shown, assuming that the input noise power is characterized by a temperature T_{in}, the output noise power would be

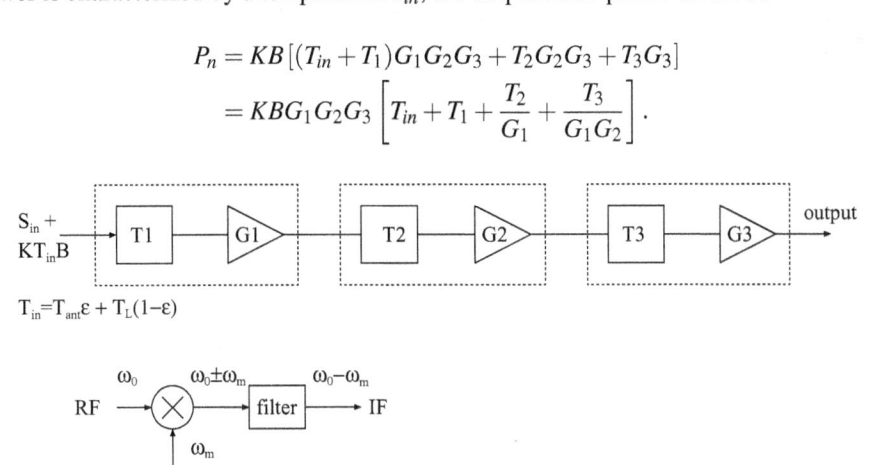

$$P_n = KB[(T_{in}+T_1)G_1G_2G_3 + T_2G_2G_3 + T_3G_3]$$
$$= KBG_1G_2G_3\left[T_{in}+T_1+\frac{T_2}{G_1}+\frac{T_3}{G_1G_2}\right].$$

$T_{in} = T_{amt}\varepsilon + T_L(1-\varepsilon)$

Figure 5.6 Cascaded receiver chain, possibly including mixer stages.

Clearly, the noise contribution from the first of the receiver stages, called the *front end*, dominates all the other receiver contributions. For this reason, designers concentrate on the noise performance of the front end of a receiver, investing resources in a low front-end noise temperature as necessary.

The formulas given earlier leading the derivation of the system noise temperature can be generalized to include contributions from the receiver cascade. In practice, it is usually acceptable to regard the front end temperature as the receiver noise temperature and use the earlier results as written. An important thing to consider, however, is the optimal placement of the front end. It is often practical and always advantageous to dislocate the front end from the rest of the receiver and place it ahead of the transmission line, as close to the antenna as possible. This is a simple and cost-effective means of mitigating the effects of the transmission line on noise performance.

5.3.4 Receiver Noise Temperature, Noise Figure

Receivers are made of active as well as passive devices; receiver noise temperature is consequently not governed by Nyquist's theorem, and the output noise of a receiver stage (e.g., an amplifier) is generally something that has to be measured. The measurement is made under "standard" conditions, which are defined by the IEEE as being at room temperature ($T_\circ = 290$ K) with matched impedances at the input and output. The test is illustrated in Figure 5.7, which shows a noise source connected to an amplifier with power gain G. Under standard conditions, $T = T_\circ$ and $Z_s = Z_{in} = R$.

We seek a figure of merit for the amplifier that characterizes the noise it produces. One such figure of merit is the noise figure F, which is the ratio of the signal-to-noise ratios at the input and output. In an amplifier, the signal power grows by a factor of the gain, but the noise power grows by at least this much, meaning that the noise figure is always greater than unity. In terms of the parameters in Figure 5.7, the noise figure can be expressed as

$$F = \frac{P_{Si}/KT_\circ B}{GP_{Si}/(GKT_\circ B + P_{Na})}$$
$$= 1 + \frac{P_{Na}}{GKT_\circ B},$$

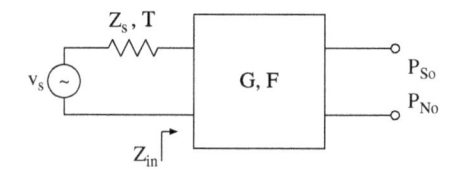

Figure 5.7 An active receiver with power gain G and noise figure F driven by a source v_s with series impedance Z_s at temperature T. The signal power at the input is P_{Si}. The output signal and noise power are P_{So} and P_{No}, respectively. The input impedance is $Z_\circ$.

where P_{Na} is the noise power produced within the amplifier itself. With some rearranging, we can find that $P_{Na} = G(F-1)KT_\circ B$. This is the noise power that would be produced by the input resistor if it were raised to a temperature $T_f = (F-1)T_\circ$. This temperature is called the amplifier noise temperature. It typically does not correspond to the actual temperature of the device, although the two may be related.

Noise temperature is one figure of merit of a receiver, and noise figure is another. The two metrics are really equivalent; manufacturers tend to quote the former when the noise temperature is less than ambient (300 K) and the latter otherwise. Noise figures are often quoted in dB. For example, a receiver with a very respectable noise temperature of 30 K has a noise figure of about 0.4 dB. Such a receiver would probably be cryogenically cooled.

Under nonstandard conditions, the actual noise performance of the amplifier will deviate from what is described above. Referring back to Figure 5.7, consider what happens when the input noise source temperature, source impedance and input impedance are allowed to assume arbitrary values. In this case, the input noise power must be written as

$$P_{Ni} = KTB \frac{4R_s}{|Z_s + Z_{in}|^2} R_{in} \equiv MKTB,$$

where Nyquist's theorem has been invoked, KTB is the available noise power and M is therefore the impedance mismatch factor. The output noise power is consequently

$$P_{No} = GMKTB + P_{Na} = GKTB(M + (F-1)T_\circ/T).$$

The signal input power is

$$P_{Si} = \left|\frac{v_s}{Z_s + Z_{in}}\right|^2 R_{in} = \frac{|v_s|^2}{4R_s} M$$

and the output signal power is just G times this. The output signal-to-noise ratio can then be expressed as

$$\frac{P_{So}}{P_{No}} = \frac{|v_s|^2}{4R_s KTB} \frac{M}{M + (f-1)T_\circ/T}.$$

Finally, we can calculate the ratio of the input to output signal-to-noise ratios. Note that this is not the noise figure, which is defined for standard conditions, but is instead the effective noise figure F':

$$F' = 1 + \frac{(F-1)T_\circ/T}{M}.$$

Under matched conditions, this reverts to the usual noise figure as it must. Improper impedance matching introduces a value of M less than unity and implies degraded noise performance. A value of T that is greater than $T_\circ$, meanwhile, implies strong input noise, making the effects of amplifier noise less significant and reducing the effective noise figure.

5.3.5 Receiver Bandwidth

Noise power is proportional to bandwidth, and in the context of radar receivers it is the receiver bandwidth that is in question. Decreasing the bandwidth decreases the noise at the cost of possibly decreasing the information available from the received signal. In radar applications, the bandwidth of the signals is proportional to the reciprocal of the radar pulse width. A 1μs pulse has a bandwidth of about 1 MHz. Narrower radar pulses permit measurements of increasingly fine details in the target at the expense of increased noise power.

5.3.6 Bandwidth and Channel Capacity

A central question in information theory concerns the amount of information that can be carried on a noisy channel with finite bandwidth. We have established that bandwidth comes at the expense of noise. We might also expect both low noise and broad bandwidth to be conducive to the information-carrying capability of a channel. Understanding the tradeoffs will be essential to designing a communication link.

Suppose that information on a channel is being carried by a continuous, time-varying signal $V_s(t)$ in which messages can be encoded. The signal will necessarily be accompanied by noise, $V_n(t)$. If the signal and noise are uncorrelated, then the total power in the channel will be the sum of the signal and noise power, $P = S + N$. The corresponding typical RMS value of V will obey the rule $V_t^2 = V_s^2 + V_n^2$. As a practical matter, values of V will be limited to an operating band with a peak value that we will call ηV_t.

Consider next dividing the operating band into 2^b sub-bands of equal width. Each sub-band will cover a range of values $\delta V = 2\eta V_t / 2^b$ wide. The sub-band occupied by V at any given time can be labeled with a b-bit binary number. Now, if we wished to encode information into this number, b should not be made so large that $\delta V < 2\eta V_n$, since in that case the number would be randomized by the noise. As a result, the maximum number of bits of information is limited by

$$2^b = \frac{V_t}{V_n} = \sqrt{\frac{V_t^2}{V_n^2}} = \sqrt{\frac{V_n^2}{V_n^2} + \frac{V_s^2}{V_n^2}}$$
$$= \sqrt{1 + S/N}$$
$$b = \log_2\left[(1 + S/N)^{1/2}\right].$$

Now, if N such measurements are made in the time interval T, the bit-rate for information exchange C will be

$$C = \frac{N}{T} \log_2\left[(1 + S/N)^{1/2}\right].$$

Finally, the sampling theorem limits the sampling rate N/T of a channel to twice the channel bandwidth, $2B$. Making the substitution yields an expression for the channel capacity,

$$C = B\log_2(1+S/N).$$

This is the Shannon–Hartley theorem, which gives the theoretical maximum capacity of a channel to pass information. The formula indicates that increasing the SNR at the expense of bandwidth is conducive to higher channel capacity in the low-SNR limit but not in the high-SNR limit.

5.4 The Matched Filter

Perhaps the most critical component in a receiver is the filter, which determines which parts of the received signal and noise are passed along for processing. Most often (but not always), filters are designed to maximize the signal-to-noise ratio at their output. The matched-filter theorem discussed below provides a recipe for accomplishing this. "Matched," in this case, refers not to an impedance or polarization match (both of which are also important), but to an optimal match between the filter characteristics and the signals likely to be encountered.

We consider a linear filter with input $x(t) = f(t) + n(t)$ and output $y(t) = y_f(t) + y_n(t)$, where f and n refer to the signal and noise components, respectively. The objective is to optimize the filter spectrum $H(\omega)$ or, equivalently, the impulse response function $h(t)$ to maximize the output signal-to-noise power ratio at some time $t_\circ$. Note that $H(\omega)$ and $h(t)$ are a Fourier transform pair, $H(\omega) = \int f(t)\exp(-j\omega t)dt = \mathcal{F}(f(t))$. The output signal power is just the square modulus of $y_f(t)$. Noise is a random variable, and the noise power is the expectation of the square modulus of $y_n(t)$. Therefore, we maximize the ratio R:

$$R = \frac{|y_f(t_\circ)|^2}{E(|y_n(t_\circ)|^2)}.$$

While we do not have access to the expectation of the noise power, we can assume the noise to be statistically stationary and approximate the expectation with a time average.

Let $F(\omega)$ and $N(\omega)$ be the frequency-space representations of the input signal and noise, respectively. We can define the signal and noise as being nonzero over only a finite span of time T to avoid problems associated with the definitions of $F(\omega)$ and $N(\omega)$, which involve integrals over all times (see Chapter 1). The units of these two quantities are the units of $y(t)$ per Hz. Still in the frequency domain, the output signal and noise are the products $F(\omega)H(\omega)$ and $N(\omega)H(\omega)$. In the time domain, the output noise is then

$$y_n(t) = \mathcal{F}^{-1}(N(\omega)H(\omega))$$
$$= \frac{1}{2\pi}\int N(\omega)H(\omega)e^{j\omega t}d\omega.$$

At any instant in time, the output noise power is

$$y_n(t)y_n^*(t) = \left(\frac{1}{2\pi}\right)^2 \int N(\omega)H(\omega)e^{j\omega t}d\omega \int N^*(\omega')H^*(\omega')e^{-j\omega't}d\omega'.$$

The time average of the power over the interval T is

$$\langle y_n(t)y_n^*(t)\rangle = \frac{1}{T}\left(\frac{1}{2\pi}\right)^2 \int\int\int N(\omega)N^*(\omega')H(\omega)H^*(\omega')\underbrace{e^{j(w-w')t}dt}_{2\pi\delta(\omega-\omega')}d\omega d\omega'$$

$$= \frac{1}{T}\frac{1}{2\pi}\int |N(\omega)|^2|H(\omega)|^2 d\omega$$

$$= \frac{1}{2\pi}\int S_n(\omega)|H(\omega)|^2 d\omega,$$

where $S_n(\omega)$ is the power spectrum of the noise with units of power per Hz. The noise usually has a nearly flat spectrum over the bandwidth of interest (white noise), and so this is just a constant $S_\circ$.

Incorporation of these results in the signal-to-noise ratio expression gives

$$R|_{t_\circ} = \frac{|\frac{1}{2\pi}\int F(\omega)H(\omega)e^{j\omega t_\circ}|^2}{\frac{S_\circ}{2\pi}\int |H(\omega)|^2 d\omega}. \tag{5.9}$$

We are not interested in evaluating (5.9) but rather on conditioning its upper limit. This can be done with the help of the *Schwartz inequality*,

$$\left|\int_a^b f(x)g(x)dx\right|^2 \leq \int_a^b |f(x)|^2 dx \int_a^b |g(x)|^2 dx. \tag{5.10}$$

Identifying $f = F$, and $g = He^{j\omega t_\circ}$ such that $|g|^2 = |H|^2$ reveals the relevance of the Schwartz inequality to the signal-to-noise ratio expression in (5.9). The maximum signal-to-noise ratio is consequently achieved when the equality above is satisfied. This obviously occurs when $f^* = g$, or in our case when

$$H(\omega)e^{j\omega t_\circ} = F^*(\omega),$$

which is the matched-filter theorem. The meaning is clearer in the time domain:

5.4 The Matched Filter

$$H(\omega) = AF^*(\omega)e^{-j\omega t_\circ}$$

$$\begin{aligned}h(t) &= \frac{1}{2\pi}\int AF^*(\omega)e^{-j\omega t_\circ}e^{j\omega t}d\omega \\ &= \frac{1}{2\pi}\int AF^*(\omega)e^{j\omega(t-t_\circ)}d\omega \\ &= \left[\frac{1}{2\pi}\int AF(\omega)e^{j\omega(t_\circ-t)}d\omega\right]^* \\ &= Af^*(t_\circ - t),\end{aligned}$$

which is to say that the impulse response function of the optimal filter is proportional to the complex conjugate of the time reverse of the anticipated signal.

Figure 5.8 shows an anticipated signal $f(t)$ (the reflection of a transmitted radar pulse, for example) and the optimal filter impulse response function for detection at time $t_\circ$, which is set here to coincide with the trailing edge of $f(t)$. Noise has been neglected for the moment. By selecting $t_\circ$ this way, the filter impulse response function will always "turn on" at $t = 0$, which will be convenient for implementation. The filter output is the convolution of the input and the impulse response function, a kind of integral over past history,

$$y_f(t) = \int_{-\infty}^{t} f(t')h(t-t')dt',$$

where the impulse response function determines how much of the past to incorporate. Using a matched filter,

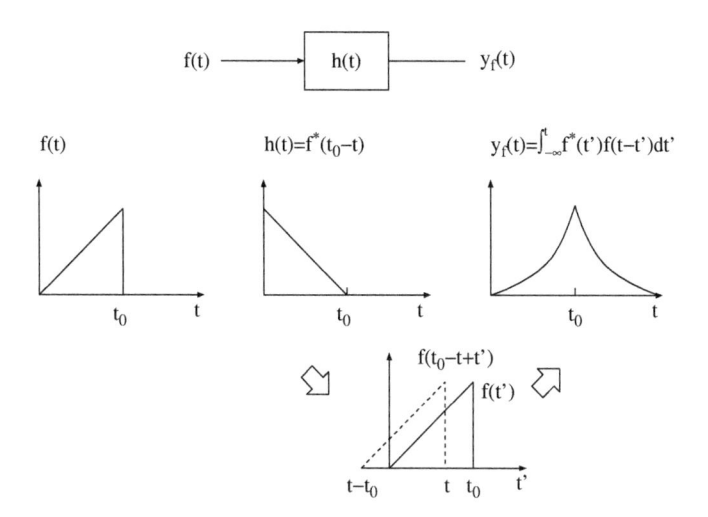

Figure 5.8 Matched filter detection of an anticipated signal. The filter is optimized for sampling at time $t_\circ$, which is set here to coincide with the trailing end of $f(t)$.

$$h(t-t') = f^*(t_\circ - (t-t'))$$
$$y_f(t) = \int f(t')f^*(t_\circ - t + t')dt'.$$

This integral is the inner product of a function with itself or its *autocorrelation* function. The convolution operator is therefore made to be the correlation of the input (signal plus noise) with a delayed copy of the anticipated signal. The former (latter) will be well (poorly) correlated, and the signal-to-noise ratio will be maximized at the time $t = t_\circ$. Note that $y_f(t_\circ)$ is related to the total energy in the $f(t)$ waveform.

The impulse response function is so named because it gives the response of the filter to an impulse

$$y_\delta(t) = \int \delta(t')h(t-t')dt'$$
$$= h(t).$$

In the laboratory, this can be measured either by applying an impulse at the input and monitoring the output on an oscilloscope or by applying white noise at the input and viewing the output with a spectrum analyzer. Analog filters can be designed to replicate many desirable functions approximately. Digital filters can be much more precise.

5.5 Two Simple Radar Experiments

All of this material can be placed in context with the consideration of a some simple radar experiments. The first is illustrated in Figure 5.9, which depicts the transmission and reception of a square radar pulse (width τ) which is scattered or reflected from a mirror target so

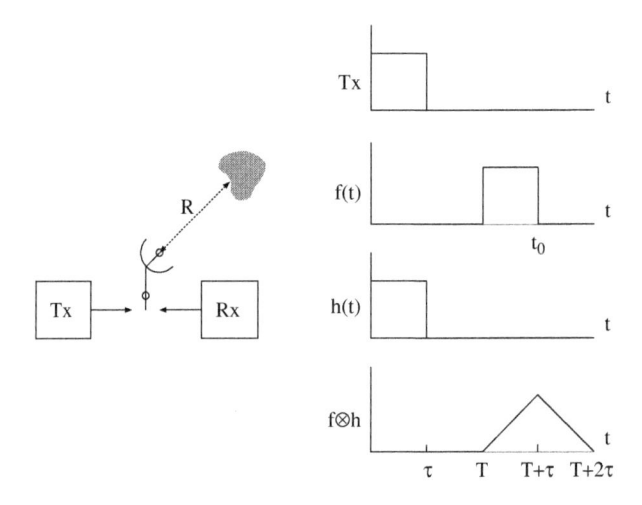

Figure 5.9 Radar experiment using a square pulse.

that the received signal is a time-delayed copy of the transmitted pulse. The delay time is $T = 2R/c$, the travel time out and back.

Figure 5.9 also shows the impulse response function $h(t) = f(t_\circ - t)$ (we take f to be real here). This is just the time reverse of the original pulse, which remains rectangular. By taking $t_\circ = T + \tau$, the time corresponding to the trailing edge of $f(t)$, we demand the matched filter impulse response function to turn on at $t = 0$. In fact, the reasoning goes the other way around: we will always by convention design $h(t)$ so that it turns on at $t = 0$, meaning that the filter output will always be a maximum at the trailing edge of $f(t)$, no matter what the shape or delay of that function. This will ultimately allow us to estimate the travel time T of an echo from the peak in the filter output.

Finally, the figure shows the convolution of the received signal with the impulse response function. This is the filter output, essentially the correlation of the transmitted waveform with a delayed replica of itself. It spans the time interval from T to $T + 2\tau$ and has a half-width τ. The output is greatest at $t = t_\circ = T + \tau$, as it was designed to be. The shape is triangular, but the output power (which is proportional to the square of the output voltage) is more sharply peaked. Note that the apparent range of the target is $R + c\tau/2$ here and that the extra delay τ has to be considered when assigning radar range to an echo. Of course, we do not generally know T a priori but rather seek to determine it from peaks in the output. The filter is usually sampled continuously, or at least over a broad span of times (the sample raster), and potential targets are identified from peaks in the output.

The radar pulse width determines the bandwidth of the matched filter, which determines the noise level and ultimately limits detectability. The pulse width also determines the range resolution of the experiment, or the ability to discern echoes closely spaced in range. In this example, the range resolution is roughly the half width of the filter, or roughly the pulse length. Echoes spaced more closely than $\delta r = c\tau/2$ would be difficult to distinguish. Finally, the pulse width determines (or is limited by) the duty cycle of the pulsed radar. The fraction of the time that high-power transmitters can be in the "on" state varies from about 2–20 percent, depending on their design. Compared to vacuum tube transmitter amplifiers, solid-state amplifiers tend to operate at lower peak power levels but with higher maximum duty cycles.

The same reasoning can be applied to the analysis of a slightly more complicated radar experiment. Figure 5.10 shows an experiment in which the transmitted waveform is phase modulated (0° or 180°) so as to form the numerical code 1, 1, −1. The bit width is τ, and the pulse width is 3τ. Such binary coded sequences are readily generated using simple RF switching networks or through digital synthesis. The corresponding impulse response function $h(t)$ for matched filtering is also shown. Once again, the convention is to design the matched filter to have an impulse response function that starts at $t = 0$. This makes the time for maximum filter output $t_\circ = T + 3\tau$ coincide with the trailing edge of the input $f(t)$, no matter what the shape or range delay of that function.

142   Noise

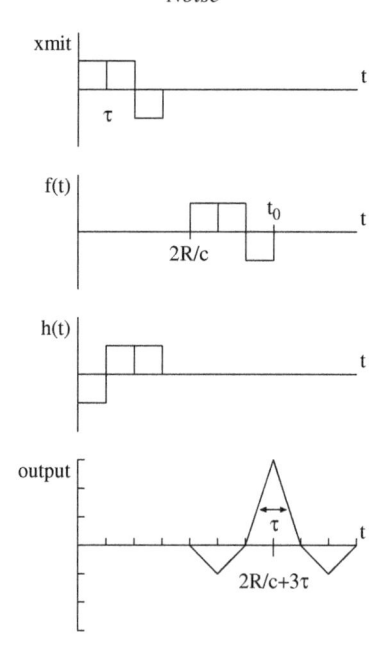

Figure 5.10 Radar experiment using a 3-bit coded square pulse.

The convolution of the received signal and the impulse response function gives the filter output for an echo from a single mirror target. Compared to the uncoded pulse example where the pulse width was just τ, we find that the amplitude of the filter output has grown by a factor of 3. The half width of the main response, however, has not grown, and remains roughly τ. Some range sidelobes, which will be detrimental to detection, have also appeared, however. While there is no noise explicitly in this problem, we can show that the noise level for the coded pulse experiment would be greater than that for the uncoded pulse by a factor of $\sqrt{3}$. Consequently, the signal-to-noise power ratio grew by a factor of 3. Another way to think about signal-to-noise improvement in pulse coding is this: since the coded pulse contains three times the energy of an uncoded pulse of length τ but has roughly the same bandwidth, we should expect a signal-to-noise ratio improvement of a factor of 3.

By using the same filter to process all possible echo signals (one with an $h(t)$ that begins at $t = 0$), we get a maximum response at a delay time related to the delay of $f(t)$ and the range of the target. Note, however, that the maximum filter output occurs at a delay time $t = 2R/c + 3\tau$ in this example. As in the previous example, the length of the pulse must be taken into account when associating the filtered echo delay with the range of the target.

Pulse coding provides a means of increasing the duty cycle of a radar experiment and the signal-to-noise ratio without degrading range resolution. A downside of pulse coding is the introduction of *clutter* or echoes from ranges other than the "correct" one. We will

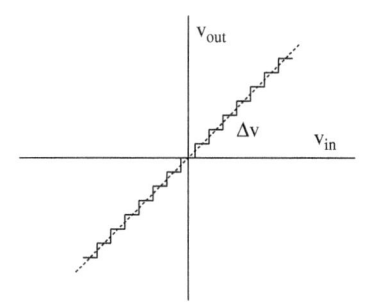

Figure 5.11 Quantization noise.

study and evaluate various pulse coding schemes in Chapter 8. What we have seen already hints about how radar waveforms can be optimized to achieve different objectives.

5.6 Quantization Noise

Another kind of noise unrelated to nature and Nyquist's theorem usually turns up in radar experiments. This is noise associated with the quantization (digitization) of an analog signal. As depicted in Figure 5.11, quantization causes small, pseudo-random discrepancies between the analog and digital representations of a signal, and these discrepancies constitute noise. Quantization noise is often but not always negligible and depends on the quantization resolution, Δv.

The largest error associated with quantization is $\pm \Delta v/2$. Let us treat the instantaneous error as a zero-mean random variable δv uniformly distributed on the interval $[\pm \Delta v/2]$. The associated noise power is proportional to the noise variance,

$$\langle \delta v^2 \rangle = \frac{1}{\Delta v} \int_{-\Delta v/2}^{\Delta v/2} \delta v^2 d\delta v$$
$$= \frac{1}{12} \left(\Delta v^2 \right).$$

Is this significant compared to background noise power? Suppose we use an 8-bit digitizer (256 levels) and set the receiver gain so that the background noise level has an RMS voltage level of 10 Δv. In that case, the ratio of quantization noise to background noise is given by

$$\frac{N_q}{N} = \frac{1/12}{10^2} = -31\,\text{dB}.$$

Quantization noise is therefore not very significant. Next, consider what happens if the receiver gain is set so that the background noise toggles just one bit of the digitizer:

$$\frac{N_q}{N} = \frac{1/12}{1} = -11\,\text{dB}.$$

This is still small, but perhaps not negligible. The real danger occurs when the background noise fails to toggle reliably even one bit of the digitizer. In that case, quantization noise may dominate, leading to significant degradation of the overall signal-to-noise ratio. It is best to set the receiver gain so that at least a few bits are continuously toggled by background noise. It may be difficult to do this while simultaneously avoiding saturation by strong signals. If so, a digitizer with finer precision (more bits) may be necessary. Even a saturated signal contains a surprising amount of information, however; one-bit digitization was once used extensively, before high-speed wide-word digitizers became available.

5.7 Notes and Further Reading

The fundamentals of blackbody radiation are thoroughly discussed in Kraus' book on radio astronomy, which also gives a description about sources of antenna noise at different frequencies. Channel capacity and other elements of information theory are covered in the text by Cover and Thomas. The treatment of the matched filter given here follows that given by Papoulis. A practical source of information regarding noise and interference in radio is "Reference Data for Radio Engineers." More practical information on electronics used in RF circuits can be found in the texts by Coleman and Hagan.

S. Chandrasekhar. *Radiative Transfer*. Oxford University Press, Oxford, UK, 1950.

C. Coleman. *An Introduction to Radio Frequency Engineering*. Cambridge University Press, 2004. Cambridge, UK, 2004.

J. B. Hagan. *Radio-Frequency Electronics*. Cambridge University Press, Cambridge, UK, 1996.

J. D. Kraus. *Radio Astronomy*, 2nd ed. Cygnus-Quasar Books, Powell, OH, 1966.

B. P. Lathi. *Modern Digital and Analog Communication Systems*. Holt, Rinehart and Winston, New York, 1983.

W. M. Middleton and M. E. Valkenburg, eds. *Reference Data for Radio Engineers*, 9th ed. Butterworth-Heinemann, Woburn, MA, 2002.

H. Nyquist. Thermal agitation of electric charge in conductors. *Phys. Rev.*, 32:110–113, 1928.

A. Papoulis. *Probability, Random Variables, and Stochastic Processes*. McGraw-Hill, 1984.

C. E. Shannon and W. Weaver. *The Mathematical Theory of Communication*. University of Illinois Press, Urbana, 1949.

Exercises

5.1 A receiver operates at a frequency of 400 MHz where the cosmic noise temperature is 50 K. There is a 50 m length of coaxial cable between the antenna and receiver made of RG-8, a common, moderately lossy cable with an attenuation rate of 5 dB/100 feet at 400 MHz. The cable temperature is 300 K. The receiver has three stages, each with 20 dB gain. The noise temperatures of the three stages are 40 K (front end), 200 K and 200 K.

a) Calculate the total system noise temperature, T_{sys}.
b) Repeat a), only replace the cable with RG-17, a heavy-duty cable with an attenuation rate of 2.3 dB/100 feet at this frequency.
c) Repeat a) again (using RG-8), only this time move the receiver front end up to the antenna, with the transmission line then connecting to the other two stages.

In practice, the losses involved with coax cable are prohibitively high at UHF frequencies and above, motivating designers to use waveguide instead, despite the much higher cost. At VHF frequencies and below, coaxial cable losses are more modest, and some new low-loss cable types have recently become available.

5.2 Suppose a radar transmits a pulse that is equal to $+1$ from $t=0$ to $t=\tau$ and equal to -1 from $t=\tau$ to $t=3\tau$.

a) Sketch the signal that will be received from a point target scatterer at a range R from the radar.
b) Sketch the filter response function of a matched filter for this situation.
c) Calculate and sketch the output from such a filter for the signal described in a).

5.3 A police radar has the following characteristics:

- transmit power: 100 MW
- antenna: 20 cm diameter parabola with $\varepsilon_a = 0.6$
- frequency: 10.55 GHz
- system noise temp: 1000 K
- bandwidth: 1 kHz

What is the maximum distance at which a car with a radar cross-section of 0.2 m^2 can be detected with a signal-to-noise ratio of 10?

Now, suppose you use a cheap radar detector to evade the radar. The gain of your antenna is 6 dB, the bandwidth of your receiver is 10 kHz, and the system noise temperature is 5000 K. At what distance will the radar produce a signal-to-noise ratio in the receiver of 10? What factor(s) give you the advantage over the police officer? In practice, what might limit the distance in the second case?

5.4 Prove the Schwarz inequality:

$$\left| \int_a^b \psi_1(x)\psi_2^*(x)dx \right|^2 \leq \int_a^b |\psi_1(x)|^2 dx \int_a^b |\psi_2(x)|^2 dx,$$

where ψ_1 and ψ_2 are complex integrable functions on $[a,b]$. There are any number of ways to do this. One way is to consider the complex function $\phi = \psi_1 + \lambda \psi_2$, where λ is a complex constant, a free parameter. Obviously, $\int_a^b |\phi(x)|^2 dx \geq 0$. Substitute ψ_1 and ψ_2 into this inequality and prove the Schwarz inequality by making the correct choice of λ. This should not be a long derivation.

5.5 Suppose you want to receive signals from the Mars rovers using NASA's deep space network (DSN) antennas, which are 34 m diameter parabolic reflectors. (NASA also has 70 m antennas, but suppose those are unavailable.) The signals of interest are X-band, 8.4 GHz, where the worst-case antenna noise temperature on Earth is about 100 K. The high-gain antenna on the rover is a 0.28 m diameter circular phased array, which we will regard as a circular aperture antenna for this problem. We can estimate that both it and the DNS have aperture efficiencies of 70 percent. The high-gain X-band transmitter on the rover emits 30 W average power. Mars, by the way, has an average distance to Earth of 225 million km.

Suppose we desire a communication link with a 10 dB signal-to-noise ratio. What will be the maximum bandwidth of the link (Hz) and the corresponding maximum channel capacity (bits/s)? How does this compare with your internet connection at home?

5.6 SETI (Search for ExtraTerrestrial Intelligence) researchers have claimed that it can cost less than a penny per word to communicate with alien civilizations. Test this claim by calculating the actual cost (cents per word) of sending intelligible data to the ETs. Assume that there is an alien civilization on a planet orbiting the star Tau Ceti, 11 light years (1.04×10^{14} km) from Earth. We will communicate with them using the Arecibo S-band radar, of which they just happen to have a duplicate for reception. The radio frequency in this case falls within the microwave window, where the sky noise temperature falls to just 10 K.

For this problem, take the gain of the radars to be 73 dB, the transmitter power to be 1 MW, and the receiver noise temperature to be 38 K. The S-band radar wavelength is 0.12 m. Let a word in this case be 5 8-bit ASCII characters. For viable communications, the aliens must receive our signal with a signal-to-noise ratio of 10. The cost of electricity here on Earth is about $0.15 per kW-hour.

6
Scattering

The objective of radar is to derive meaningful information from radio signals scattered by distant targets. What is the nature of the scattering, and how do we quantify it? The scattering cross-section is a parameter that describes the power scattered by an object exposed to electromagnetic radiation. For some objects, it can be calculated from first principles (electrons, raindrops). For others (aircraft, ships), it usually must be measured. Sometimes it is simply related to the physical size of the object. Sometimes it is not. The scattering cross-section also generally depends on the radar frequency, polarization and orientation with respect to the target. Some definitions help clarify the distinctions.

Broadly defined, scattering is the redirection of radiation away from the original direction of propagation. Reflection, refraction and diffraction all accomplish this, but often involve environmental effects and are sufficiently distinct and important in their own right to deserve independent treatments. In this text, these phenomena are grouped together under the heading "propagation." "Scattering" is reserved for wave interaction with lumped materials. Scattering is elastic (inelastic) if the incident and scattered wavelengths are the same (different). When a Doppler shift is imparted to the scattered wave, the scattering is termed "quasi-elastic." This chapter is concerned with elastic and quasi-elastic scatter only. Inelastic scattering is mainly associated with optical phenomena, the phenomenon of stimulated electromagnetic emission (SEE) of radio signals being one important counterexample.

Referring to the schematic diagram in Figure 6.1, we can express the power density incident on the target in terms of the transmitted power P_{tx}, the range to the target r, and the antenna gain G as

$$P_{inc} = \frac{P_t G}{4\pi r^2} (W/m).$$

Define the *total scattering cross-section* σ_t as the total power scattered per incident power density, or

$$P_{sc} = P_{inc}\sigma_t.$$

Scattering is a kind of radiation, and the scattering can no more be isotropic than the initial radiation. Therefore, define the scattering directivity D_s as the ratio of the scattered power density back at the receiver to the power density that would exist if the scattering were isotropic. Then the received power will be

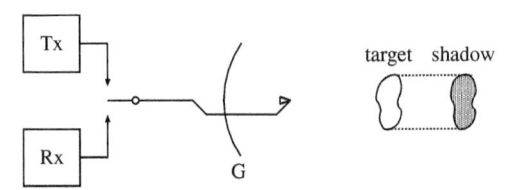

Figure 6.1 Radar illuminating a target.

$$P_{rx} = \frac{D_s P_{sc}}{4\pi r^2} A_{eff}$$
$$= \frac{P_{tx} G}{4\pi r^2} \sigma_t D_s \frac{A_{eff}}{4\pi r^2},$$

where the antenna gain and effective area are related by the reciprocity theorem if the same antenna is used to transmit and receive, as in Figure 6.1. The total scattering cross-section and scattering directivity are often combined into a single parameter, the *radar-scattering cross-section*, $\sigma = \sigma_t D_s$. This is defined as the total scattered power per incident power density that, if scattered isotropically, would give rise to the observed power back at the receiver. This is a useful engineering definition.

Finally, assuming no impedance or polarization mismatches, the entire power budget can be represented by the radar equation

$$\frac{P_{rx}}{P_{tx}} = \frac{G^2 \lambda^2 \sigma}{(4\pi)^3 r^4},$$

which applies to scatter from lone, "hard" targets. Note that the ratio above does not actually scale as λ^2; both the gain and the radar cross-section include additional wavelength dependence.

Some confusion surrounds the definitions of σ_t and σ, and it is worthwhile to reemphasize their differences. The former is the effective aperture of the scatterer that collects incident power and scatters it away from the original direction. This concept is useful for studying signal attenuation along its propagation path, for example. The latter is the effective area which, if scattering power isotropically, would result in the power actually observed back at the radar. This is usually what concerns radar specialists. A third definition sometimes used is the geometric or physical scattering cross-section, $A_g = A_p \sim \pi a^2$, where a is something like the scatterer dimension. This is the shadow area cast by the target. We can also define the backscatter efficiency as the ratio of the radar-scattering cross-section to the physical cross-section, $\varepsilon_s \equiv \sigma/A_g$.

6.1 Rayleigh Scatter

Based on the size of a target relative to the wavelength, scattering falls into one of three regimes: the *Rayleigh* regime ($a \ll \lambda$), the *optical* regime ($a \gg \lambda$) or the *resonance* regime ($a \sim \lambda$). In the Rayleigh regime, the scatterer is small compared to a wavelength, and at any given time the electric field supplied by the radiation source is essentially uniform spatially throughout the scatterer volume. What happens next is illustrated in Figure 6.2. While a spherical target is considered to simplify the calculations, the shape of the target does not greatly affect the backscatter in the Rayleigh regime.

When exposed to a uniform background electric field, charges within the target (a spherical dielectric or conductor) rearrange themselves so as to minimize their configurational energy. If the target is uniform in composition, electric dipoles will be induced within and the polarization (the dipole moment per unit volume) inside will also be uniform. At the boundaries of the target, the polarization will be discontinuous, and equivalent surface charge density will appear, as shown. The final configuration is a single electric dipole. As indicated in the figure, the electric field within the target remains uniform but is smaller than the background field. Outside, the electric field is the background field plus the field of an ideal electric dipole. As the phase of the wave changes over time with angular frequency ω, so does the polarity of the induced electric dipole. Associated with the constant rearrangement of charge is an elemental current element with an amplitude equal to the time derivative of the dipole moment (see derivation in Appendix B)

$$Idl = j\omega 4\pi a^3 \frac{\varepsilon - \varepsilon_\circ}{\varepsilon + 2\varepsilon_\circ} \varepsilon_\circ E_\circ,$$

where ε is the permittivity of the dielectric sphere. What has therefore been created is an ideal electric dipole antenna. We already have expressions for calculating the incident power density associated with $\mathbf{E}_\circ$ and the total power radiated associated with Idl. The ratio gives the total scattering cross-section σ_t for the target:

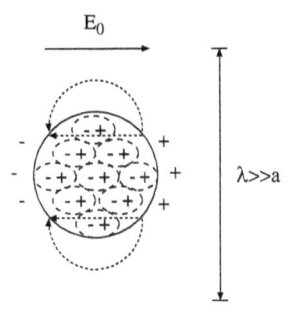

Figure 6.2 Rayleigh scatter from a spherical target of radius *a*. The illuminating signal propagates vertically and is polarized in the plane of the figure.

$$\sigma_t = \frac{128}{3}(\pi a^2)\pi^4\left(\frac{\varepsilon-\varepsilon_\circ}{\varepsilon+2\varepsilon_\circ}\right)^2\left(\frac{a}{\lambda}\right)^4.$$

In the event the scatterer is a perfect conductor rather than a dielectric, the term involving the ratio of the permittivities should be replaced by unity. Finally, the radar-scattering cross-section can be calculated by multiplying σ_t by the directivity, which we know to be 1.5 for backscatter from an elemental dipole antenna. Taking the ratio of the radar-scattering cross-section to the physical area of the target gives the backscatter efficiency

$$\varepsilon_s \propto \left(\frac{2\pi a}{\lambda}\right)^4,$$
$$= C_\lambda^4,$$

which gives the Rayleigh scattering result – that the backscatter efficiency is proportional to the target circumference (in wavelengths) to the fourth power. This result explains, among other things, why the sun appears red and the sky blue to a terrestrial observer.

If the permittivity of the dielectric is complex, power will be absorbed within it as well as scattered from it. Ohm's law states that the average power dissipated in the dielectric is the real part of $I(dlE_\circ)/2$. Dividing by the average incident power density of the wave illuminating the dielectric yields the absorption cross-section

$$\sigma_a = \frac{8\pi^2}{\lambda}a^3\Im\left(-\frac{\varepsilon-\varepsilon_\circ}{\varepsilon+2\varepsilon_\circ}\right),$$

which applies in the Rayleigh long-wavelength limit. Together, scattering and absorption contribute to extinction or the removal of power from the incident wave.

6.1.1 Polarization and Scattering Matrices

Scattering or S-matrices (or Jones matrices) capture the relationship between the incident and scattered electric field components perpendicular and parallel to the scattering plane in the far field. Here, the scattering plane is the plane containing the source, target and observation point, which need not be collocated with the source in general. The most common definition is given for the case of scattering from an incident plane wave propagating in the $+z$ direction:

$$\begin{pmatrix} E_\| \\ E_\perp \end{pmatrix}_{scat} = \frac{\exp(-jk(r-z))}{jkr}\begin{pmatrix} S_2 & S_3 \\ S_4 & S_1 \end{pmatrix}\begin{pmatrix} E_\| \\ E_\perp \end{pmatrix}_{inc}, \qquad (6.1)$$

where r is the distance traveled after scattering, the $\exp(ikz)$ term relates the phase of the incident plane wave at the target to its phase at the source and the $\exp(-ikr)/ikr$ term relates the amplitude and phase of the observed spherical wave to that of the scattered wave. The S-parameters in the matrix are complex in general. Off-diagonal or depolarizing terms indicate cross-polarization. These can be neglected in the cases of Rayleigh and Mie scattering. The relationship between the incident and scattered wave intensity is then

$$I_s(\theta,\phi) = \frac{I_i}{k^2 r^2}\left(|S_1(\theta)|^2 \sin^2\phi + |S_2(\theta)|^2 \cos^2\phi\right), \qquad (6.2)$$

where ϕ specifies the polarization angle of the incident radiation with respect to the scattering plane. The corresponding radar-scattering cross-section can consequently be expressed as

$$\sigma = \frac{4\pi}{k^2}\left(|S_1(\theta)|^2 \sin^2\phi + |S_2(\theta)|^2 \cos^2\phi\right). \qquad (6.3)$$

In Rayleigh scattering, the perpendicular-polarized signal is isotropic in the scattering plane, which is also perpendicular to the induced electric dipole moment. The parallel component, meanwhile, is modified by the cosinusoidal term that appears in the radiation pattern of the ideal dipole antenna. Consequently, S_2 contains a $\cos(\theta)$ dependence that is not in S_1 where θ is the scattering angle.

One way to evaluate the information in the scattering matrix is through graphical representations of the phase function. The phase function is proportional to the terms in parentheses in (6.2) and (6.3). For purposes of plotting, the incident radiation is taken to be unpolarized, the average value is computed, and so the phase function is proportional to $(|S_1(\theta)|^2 + |S_2(\theta)|^2)/2$. In the case of Rayleigh scatter, the phase function is therefore $\cos^2(\theta) + 1$. Phase functions, which are usually normalized to the total scattered power for plotting, are a graphical representation of the relative importance of forward scatter and backscatter. The asymmetry factor, the angle-averaged value of $\cos(\theta)$, is another measure of scattering anisotropy.

6.1.2 Scattering Matrices with Stokes Parameters

$$\begin{bmatrix} I \\ Q \\ U \\ V \end{bmatrix} = \frac{\sigma}{4\pi r^2} \begin{bmatrix} P_{11} & P_{12} & 0 & 0 \\ P_{21} & P_{22} & 0 & 0 \\ 0 & 0 & P_{33} & P_{34} \\ 0 & 0 & P_{43} & P_{44} \end{bmatrix} \begin{bmatrix} I_\circ \\ Q_\circ \\ U_\circ \\ V_\circ \end{bmatrix}$$

Scattering theory is also sometimes cast in terms of the scattering phase matrix, which associates the Stokes parameters of the incident and scattered radiation. In general, the matrix has sixteen independent elements. In the case of scattering from spherical objects, we can note that $P_{22} = P_{11}$, $P_{44} = P_{33}$, $P_{21} = P_{12}$ and $P_{43} = -P_{34}$ so that there are only four independent elements. The term "scattering phase function" refers to $P_{11}(\theta)$ or a normalized version of it. Phase functions are useful for representing Mie scatter, which is described below.

6.2 Mie Scatter

Scattering from dielectric and conducting spheres with sizes comparable to a wavelength is termed Mie, or sometimes Lorenz–Mie scatter in honor of the pioneering work of Ludwig Lorenz and, much later, Gustav Mie. (Peter Debye, a contemporary of Mie, also explored the problem while studying radiation pressure exerted on astrophysical dust.) The theory applies in a broad range of physical contexts, including radar forward and backscatter. It is a building block for other scattering models and can be generalized to cover ellipsoidal targets and targets made of layered material. It is essentially more intricate than Rayleigh scatter. The theory is a vector theory and can accommodate magnetic as well as dielectric materials. A defining characteristic of Mie scatter is a high degree of sensitivity to the scatterer size parameter $x = ka = C_\lambda$, which is sometimes referred to as "morphology-dependent resonance" or MDR.

Presented below is a summary of the theory, culminating in a set of coefficients for computing Mie-scatter cross-sections. While the main ideas behind the theory are presented, many of the mathematical details are omitted in the interest of brevity. More complete treatments of the derivation along with guidance for evaluating the coefficients efficiently can be found in the titles listed under Notes and Further Reading at the end of this chapter. The treatment below is based on that of C. E. Bohren and D. R. Huffman and the references therein.

We seek expressions for the time-harmonic electric and magnetic field that satisfy the homogeneous Helmholtz wave equation everywhere, including in and around the spherical scatterer. The electric and magnetic fields in source-free regions are solenoidal (divergence free) and are related to one another by the curl. It is therefore expedient to introduce the auxiliary vectors

$$\mathbf{M} \equiv \nabla \times (\mathbf{c}\psi) \tag{6.4}$$

$$\mathbf{N} \equiv \frac{1}{k} \nabla \times \mathbf{M}, \tag{6.5}$$

which are also solenoidal and related by a curl. (Here, $\mathbf{c}$, the pilot vector, is either a constant vector or, in the treatment that follows, the radial vector $\mathbf{r} = r\hat{\mathbf{r}}$). It can readily be shown that $\mathbf{M}$ and $\mathbf{N}$ satisfy the homogeneous vector Helmholtz equation if ψ satisfies the homogeneous scalar Helmholtz equation. As the auxiliary variables would seem to meet the requirements for representing the electromagnetic field, solving the vector wave equation has thereby been reduced to the simpler problem of solving the scalar equation. We therefore seek a scalar generating function ψ that solves $\nabla^2 \psi + k^2 \psi = 0$. The solution will also have to satisfy certain boundary conditions at the surface of the spherical scatterer and at infinity, and so it will be expedient to work in spherical coordinates.

Using the method of separation of variables, the scalar field is expressed as the product of three functions $\psi(r, \theta, \phi) = R(r)\Theta(\theta)\Phi(\phi)$. This ansatz is substituted into the Helmholtz equation, and the resulting terms regrouped such that terms involving r, θ and ϕ are isolated. Each of the terms is individually a constant, and so the system constitutes three ordinary differential equations (ODEs) coupled through the separation constants. The ODEs are solved individually for $R(r)$, $\Theta(\theta)$ and $\Phi(\phi)$.

While no one term of the form $R(r)\Theta(\theta)\Phi(\phi)$ will be able to satisfy the boundary conditions in general, a superposition of such terms can. The general solution to the scalar Helmholtz equation is consequently formed from a superposition of basis functions of the form

$$\psi_{emn}(r,\theta,\phi) = \sqrt{\frac{2}{\pi}} \cos(m\phi) P_n^m(\cos\theta) z_n(kr)$$

$$\psi_{omn}(r,\theta,\phi) = \sqrt{\frac{2}{\pi}} \sin(m\phi) P_n^m(\cos\theta) z_n(kr),$$

where n and m are integer indexes (making the terms single-valued) and where e and o refer to solutions with even and odd azimuthal symmetry. The familiar sine and cosine terms are the solutions to the ODE for $\Phi(\phi)$ and are mutually orthogonal. The solutions to the ODE for $\Theta(\theta)$ are the associated Legendre functions of the first kind of degree $n \geq m$ and order $m \geq 0$. The $m = 0$ terms are the Legendre polynomials. The associated Legendre functions are also orthogonal in that

$$\int_{-1}^{1} P_n^m(\beta) P_{n'}^m(\beta) \, d\beta = \frac{2}{2n+1} \frac{(n+m)!}{(n-m)!} \delta_{n,n'},$$

where δ is the Kronecker delta function. Lastly, the functions z_n stand for any of the four spherical Bessel functions j_n, y_n, $h_n^{(1)}$ and $h_n^{(2)}$. These are the solutions to the ODE for $R(r)$. Which functions to use depends on the domain where the solutions are to apply (see below). Recurrence relations for both the associated Legendre functions and the spherical Bessel functions exist that can aid in their eventual manipulation and evaluation.

Any solution to the scalar Helmholtz equation can be expanded in a series of the aforementioned terms, which form a complete basis. The corresponding vector spherical harmonics can be derived from the generating functions using (6.4) and (6.5):

$$\mathbf{M}_{emn} = -\frac{m}{\sin\theta} \sin m\phi P_n^m(\cos\theta) z_n(kr) \hat{\theta} - \cos m\phi \frac{dP_n^m(\cos\theta)}{d\theta} z_n(kr) \hat{\phi}$$

$$\mathbf{M}_{omn} = \frac{m}{\sin\theta} \cos m\phi P_n^m(\cos\theta) z_n(kr) \hat{\theta} - \sin m\phi \frac{dP_n^m(\cos\theta)}{d\theta} z_n(kr) \hat{\phi}$$

$$\mathbf{N}_{emn} = \frac{z_n(kr)}{kr} \cos m\phi \, n(n+1) P_n^m(\cos\theta) \hat{\mathbf{r}} + \cos m\phi \frac{dP_n^m(\cos\theta)}{d\theta} \frac{1}{kr} \frac{d}{dkr}[krz_n(kr)] \hat{\theta}$$
$$- m \sin m\phi \frac{P_n^m(\cos\theta)}{\sin\theta} \frac{1}{kr} \frac{d}{dkr}[krz_n(kr)] \hat{\phi}$$

$$\mathbf{N}_{omn} = \frac{z_n(kr)}{kr} \sin m\phi \, n(n+1) P_n^m(\cos\theta) \hat{\mathbf{r}} + \sin m\phi \frac{dP_n^m(\cos\theta)}{d\theta} \frac{1}{kr} \frac{d}{dkr}[krz_n(kr)] \hat{\theta}$$
$$+ m \cos m\phi \frac{P_n^m(\cos\theta)}{\sin\theta} \frac{1}{kr} \frac{d}{dkr}[krz_n(kr)] \hat{\phi}.$$

154   Scattering

Solutions to the vector Helmholtz equation can similarly be expanded in series of these spherical harmonics. Note that the wavenumber k to be used inside and outside the scatterer is different in accordance with the difference in permittivity (or permeability, in the case of magnetic materials).

We now distinguish between the incident, internal (to the target) and scattered electromagnetic fields in the Mie-scatter problem. In spherical coordinates, the electric field for an incident plane-polarized plane wave can be written (without loss of generality) as

$$\mathbf{E}_i = E_\circ e^{-jkr\cos\theta}\hat{\mathbf{x}}$$
$$\hat{\mathbf{x}} = \sin(\theta)\cos(\phi)\hat{\mathbf{r}} + \cos(\theta)\cos(\phi)\hat{\boldsymbol{\theta}} - \sin(\theta)\hat{\boldsymbol{\phi}},$$

which can be expanded in the infinite series

$$\mathbf{E}_i = \sum_{m=0}^{\infty}\sum_{n=m}^{\infty} A_{emn}\mathbf{N}_{emn} + A_{omn}\mathbf{N}_{omn} + B_{emn}\mathbf{M}_{emn} + B_{omn}\mathbf{M}_{omn}.$$

The values of the undetermined coefficients are found by calculating the inner product of E_i with each of the vector basis functions, e.g.,

$$A_{emn} = \int \mathbf{E}_i \cdot \mathbf{N}_{emn}\, d\Omega \Big/ \int |\mathbf{N}_{emn}|^2\, d\Omega.$$

Due to the orthogonality of the sine and cosine functions, $A_{omn} = B_{emn} = 0$ for all m and n. The only remaining nonzero coefficients, moreover, correspond to $m = 1$, reducing the series to a much simpler form:

$$\mathbf{E}_i = \sum_{n=1}^{\infty} A_{e1n}\mathbf{N}^j_{e1n} + B_{o1n}\mathbf{M}^j_{o1n}.$$

The j superscripts mean that the spherical Bessel function $j_n(kr)$ is being used for z_n here, in view of the fact that the incident field should be finite at the origin of the coordinate system. The coefficients themselves are found by exploiting the recurrence relations for the special functions involved:

$$A_{e1n} = j(-j)^n E_\circ \frac{2n+1}{n(n+1)} \tag{6.6}$$

$$B_{o1n} = (-j)^n E_\circ \frac{2n+1}{n(n+1)}. \tag{6.7}$$

Altogether, and following considerable algebraic manipulation, the incident electric and magnetic field can be written as

$$\mathbf{E}_i = E_\circ \sum_{n=1}^{\infty}(-j)^n \frac{2n+1}{n(n+1)}\left(\mathbf{M}^j_{o1n} + j\mathbf{N}^j_{e1n}\right)$$

$$\mathbf{H}_i = \frac{-k}{\omega\mu}E_\circ \sum_{n=1}^{\infty}(-j)^n \frac{2n+1}{n(n+1)}\left(\mathbf{M}^j_{o1n} - j\mathbf{N}^j_{e1n}\right),$$

where the magnetic field can be found through the application of Faraday's law to the electric field. The complexity of these formulas is a reminder that spherical harmonics is hardly a convenient basis for expanding plane waves!

What remains is to find comparable series expressions for the internal and scattered fields. This requires enforcing boundary conditions at the surface of the target. The relevant boundary conditions are that the tangential components of the electric and magnetic field inside (internal) and outside (incident plus scattered) be equal. It is evident that the only spherical harmonic components involved in the balance will be the $m = 1$ terms. For the scattered fields, the practical choice for $z_n(kr)$ is the spherical Hankel function $h_n^{(2)}(kr)$, an outgoing-wave solution that vanishes at an infinite distance from the origin. The expansions for the scattered electric and magnetic fields can therefore be written as

$$\mathbf{E}_s = E_\circ \sum_{n=1}^{\infty} (-j)^n \frac{2n+1}{n(n+1)} \left(-ja_n \mathbf{N}_{e1n}^h - b_n \mathbf{M}_{o1n}^h \right)$$

$$\mathbf{H}_s = \frac{k}{\omega\mu} E_\circ \sum_{n=1}^{\infty} (-j)^n \frac{2n+1}{n(n+1)} \left(-jb_n \mathbf{N}_{o1n}^h + a_n \mathbf{M}_{e1n}^h \right),$$

where the h superscript denotes the use of the Hankel functions for the radial dependence and where the Mie coefficients a_n and b_n remain to be determined. Similar expansions can be written for the internal fields using additional coefficients, e.g., c_n and d_n. The finiteness of the internal fields at the origin demands the use of the $j_n(kr)$ functions in the corresponding expansions.

The boundary conditions provide sufficient information for determining the Mie coefficients. For each value of n, there are four unknowns (a_n, b_n, c_n and d_n) and four scalar equations enforcing the continuity of the tangential components of the electric and magnetic fields with index n at $r = a$. Solving the system of equations for the coefficients results in

$$a_n = \frac{\mu m^2 j_n(mx)[xj_n(x)]' - \mu_l j_n(x)[mxj_n(mx)]'}{\mu m^2 j_n(mx)[xh_n^{(1)}(x)]' - \mu_l h_n^{(1)}(x)[mxj_n(mx)]'}$$

$$b_n = \frac{\mu_l j_n(mx)[xj_n(x)]' - \mu j_n(x)[mxj_n(mx)]'}{\mu_l j_n(mx)[xh_n^{(1)}(x)]' - \mu h_n^{(1)}(x)[mxj_n(mx)]'},$$

where x is the size parameter $x \equiv ka = C_\lambda$, μ_1 and μ are the permeabilities of the sphere and the surrounding material, respectively, and where m has taken on a new meaning and is the complex index of refraction of the dielectric sphere. (The imaginary part of m is called the index of absorption.)

6.2.1 Scattering Matrix

The preceding equations give sufficient information for evaluating scattering and extinction due to dielectric spheres. The formulas are unwieldy to be sure, but a few simplifications help keep the calculations tractable. It has already been noted that only the $m = 1$ terms in the spherical harmonics play a role in the field expansions. It is therefore expedient to define the angle-dependent functions $\pi_n \equiv P_n^1(\cos(\theta))/\sin(\theta)$ and $\tau_n \equiv dP_n^1(\cos(\theta))/d\theta$ and to reformulate the spherical harmonics accordingly. The shapes of these functions control the shape of the scattered and transmitted radiation patterns. The lobes in the patterns become narrower with increasing order n, and whereas lobes in the forward scatter direction are present for all n, back lobes appear only in every other order. As more orders are necessary to represent accurately the behavior of larger spheres, back lobes will tend to "cancel out" or be suppressed as the scattering parameter x increases.

The asymptotic behavior of the spherical Hankel function in the scattered fields is given by

$$h_n^{(2)}(kr) \approx -\frac{j^n e^{-jkr}}{jkr}, \quad kr \gg n^2.$$

In the radiation zone, we take $kr \gg n_t^2$ where n_t is the largest index used in the truncated approximation of the expansions for the scattered fields found above. Incorporating all of the aforementioned approximations, the scattered radiation field can be written succinctly as

$$\mathbf{E}_s = E_\circ \frac{e^{-jkr}}{jkr} \left(\cos\phi S_2(\theta)\hat{\theta} - \sin\phi S_1(\theta)\hat{\phi} \right),$$

where

$$S_1 = \sum_{n_t} \frac{2n+1}{n(n+1)} (a_n \pi_n + b_n \tau_n)$$

$$S_2 = \sum_{n_t} \frac{2n+1}{n(n+1)} (a_n \tau_n + b_n \pi_n).$$

These are the coefficients that appear in (6.1) (with $S_3 = S_4 = 0$).

A defining characteristic of Mie scatter is the backscatter efficiency ε_s, which was defined already. For Mie scatter, this evaluates as $\varepsilon_s = \sigma/\pi a^2 = 4|S_1(\theta=\pi)/x|^2$. Figure 6.3 is a plot of scattering efficiency versus size parameter for a dielectric sphere with a large index of refraction ($m = 200 + j2$), making it behave like a conductor. At long wavelengths, the scattering efficiency observes the Rayleigh scattering law. The efficiency peaks at $C_\lambda = 1$, however, and oscillates around a value of unity as the wavelength continues to diminish. The oscillations indicate interference between scattered signals from different parts of the target. The behavior in the optical regime holds for spherical targets only.

6.2.2 Scattering, Extinction, and Absorption Efficiency

Other defining characteristics of Mie theory are the scattering, extinction and absorption cross-sections. The theory allows for complex permittivities and therefore for absorption in the material. The extinction cross-section is the total power removed from the incident wave per incident power density, and the scattering cross-section is the total scattered power per incident power density. The difference between these is the absorption

6.2 Mie Scatter   157

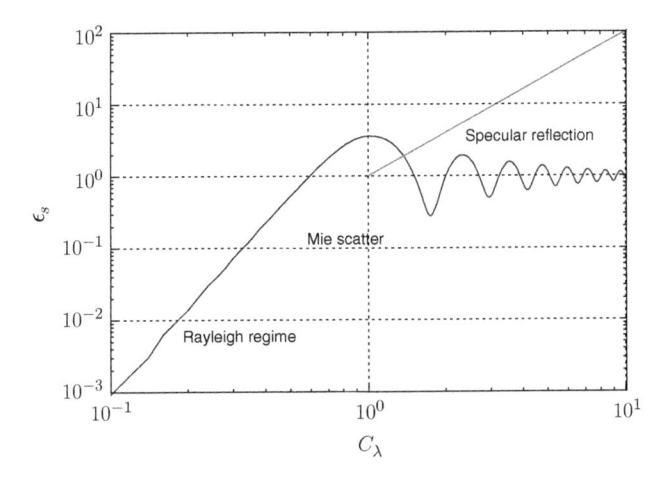

Figure 6.3 Backscatter efficiency ε_s for a spherical conductor versus size parameter $x = C_\lambda$.

cross-section. Dividing these quantities by the physical cross-section of the target gives the associated efficiencies. In terms of the Mie coefficients, the scattering, extinction, and absorption efficiencies are, in that order,

$$Q_s = \frac{2}{x^2} \sum_{n=1}^{\infty} (2n+1)(|a_n|^2 + |b_n|^2)$$

$$Q_t = \frac{2}{x^2} \sum_{n=1}^{\infty} (2n+1)\Re(a_n + b_n)$$

$$Q_a = Q_t - Q_s.$$

Some representative scattering and extinction efficiencies for materials with indices of refraction between 1.3 and 2 are shown in Figure 6.4. The difference between the extinction and scattering efficiencies is due to absorption. Larger absorption is generally accompanied by larger asymmetry. Large oscillations predominate in all of the curves. The period of the oscillations is given by $\delta x = \pi/(n-1)$, where n is the real part of the index of refraction. The large oscillations grow in amplitude with n. They represent interference between the incident and diffracted or scattered radiation. Superimposed on the large oscillations are smaller, secondary oscillations. These have been attributed to interference between waves propagating near the surface of the target and diffracted or scattered radiation.

6.2.3 Mie-Scatter Phase Functions

Finally, Figure 6.5 shows some representative phase functions for dielectric spheres with different complex indices of refraction for two different size parameters $x = C_\lambda$. We consider unpolarized incident radiation and plot the quantity

$$P(\theta) = 2((|S_1(\theta)|/x)^2 + (|S_2(\theta)|/x)^2)/Q_s.$$

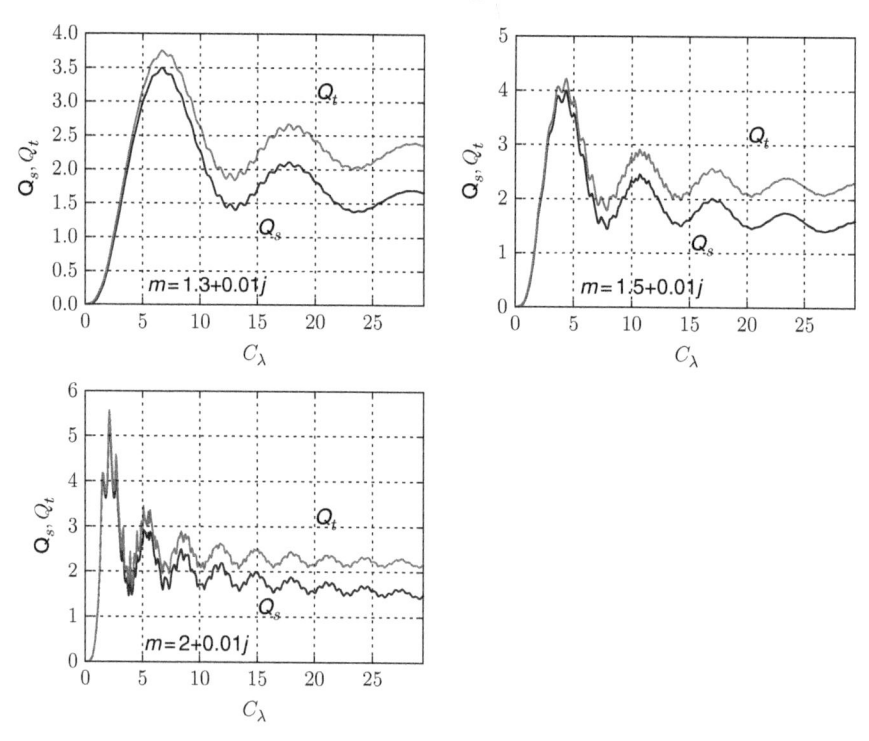

Figure 6.4 Scattering Q_s and extinction Q_t efficiencies for spherical dielectrics with different complex indices of refraction m.

The curves are plotted on a logarithmic scale.

The phase functions resemble antenna radiation patterns with main lobes and sidelobes. The number of lobes in $\theta \in [0, \pi]$ is equal to the size parameter x. The figure illustrates how asymmetry increases with increasing index of absorption. Forward scatter can be much stronger than backscatter in the case of large, lossy targets. This explains why scattered light from smoke in a movie theater is most apparent looking back toward the projector.

6.3 Specular Reflection

The Mie scattering behavior in Figure 6.3 applies only to scatterers that are roughly spherical. Scattering efficiency in the optical regime is a strong function of the shape and orientation of the scatterer. Much more efficient than Mie scatter is backscattering from roughly flat surfaces, which we refer to here as specular reflection or scatter (i.e., as from a mirror). Whereas Rayleigh scatter is related to radiation from elemental dipole antennas, specular reflection is related to emissions from aperture antennas. We can view the flat scatterers as large (compared to a wavelength) apertures illuminated by the incident radiation from the transmitter. The resulting, scattered radiation can be strong and highly directive. Whether strong backscatter is actually received or not

6.3 Specular Reflection 159

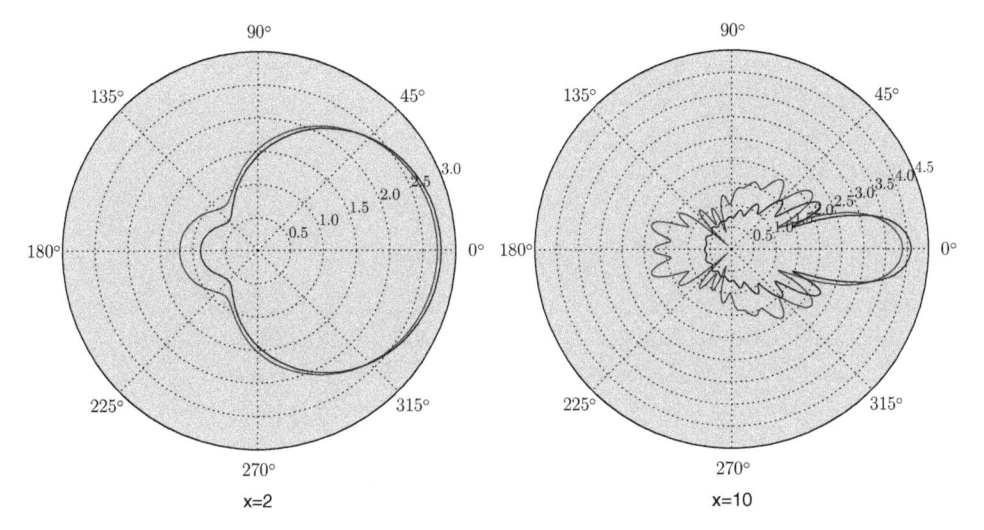

Figure 6.5 Phase functions for spherical dielectrics with size parameters $x = 2$ and 10. The inner curves in each case correspond to $m = 1.5 + j.005$, whereas the outer curves correspond to $m = 1.5 + j0.2$.

depends on whether the aperture is illuminated uniformly and whether the "beams" of the apertures are directed back toward the radar. Both factors are controlled by the scatterer orientation.

Consider backscatter from large, flat disks with $a \gg \lambda$ aligned normally to the radar line of sight. To the extent we can neglect diffraction at the edge of the disks, the total scattering cross-section must be the collecting area or the physical aperture area. The directivity is given by the reciprocity theorem. If we take the effective area also to be equal to the physical area, then the radar cross-section can be estimated to be

$$\sigma = \sigma_t D_s$$
$$= A_{phys} \cdot \frac{4\pi}{\lambda^2} A_{phys}.$$

This implies that the backscatter efficiency is $\varepsilon_s = 4\pi A_{phys}/\lambda^2$ or that $\varepsilon_s = C_\lambda^2$ precisely (with no other constants of proportionality involved). This behavior is also illustrated in Figure 6.3. Targets in the optical regime that are large and optimally oriented can be very efficient scatterers, with radar cross-sections many times larger than their physical areas. Examples are the ionization trails left behind by micro-meteorites, which give rise to strong reflection when the trail is normal to the radar line of sight. If specular targets are misaligned with the radar line of sight, however, little or no backscatter may be received.

As the size of the scatterer approaches the radar wavelength, diffraction around the target periphery must be considered, and the simple relationship between C_λ and ε_s breaks down.

160   *Scattering*

This is the resonance regime, where neither the ideal dipole nor the ideal aperture paradigm applies exactly. Numerical modeling may be required to estimate the radar cross-section in complicated cases. Conductors with sizes approximately equal to half the radar wavelength are particularly strong scatterers. Examples are wire strands and loops deployed in the air (chaff) purposely to confuse radar systems.

6.4 Corner Reflectors

Specular reflection can be intense, but is highly directionally dependent. The directional dependence can, however, be reduced if reflecting surfaces are arranged to meet at right angles. In the plane, a ray that reflects off of two surfaces that meet at a right angle will leave along the same bearing as it entered. The same is true in three dimensions if three reflecting surfaces are brought together at mutual right angles. A ray that returns along the same bearing from which it came guarantees strong backscatter. Figure 6.6 shows two ways that trihedral corner reflectors might be constructed.

The scattering efficiency of a corner reflector scales with target size, the same as an ordinary spectral reflector but with much weaker directionality. That makes corner reflectors bright targets for any radiation that enters through the open quadrant. Multiple corner reflectors can be assembled to expand the directional coverage even further.

Corner reflectors are useful as radar calibration targets. They can be flown in the air to calibrate ground-based radars or placed on the ground to calibrate airborne and space-borne radars. Information about the strength of the backscatter and its location can be useful for calibration. A corner reflector is more directionally dependent than a calibration sphere, but is a brighter target than a comparably sized sphere and will likely be less expensive to construct and deploy.

Corner reflectors can also be used as decoy targets, presenting a scattering cross-section potentially much larger than their physical size. At the same time, the most important

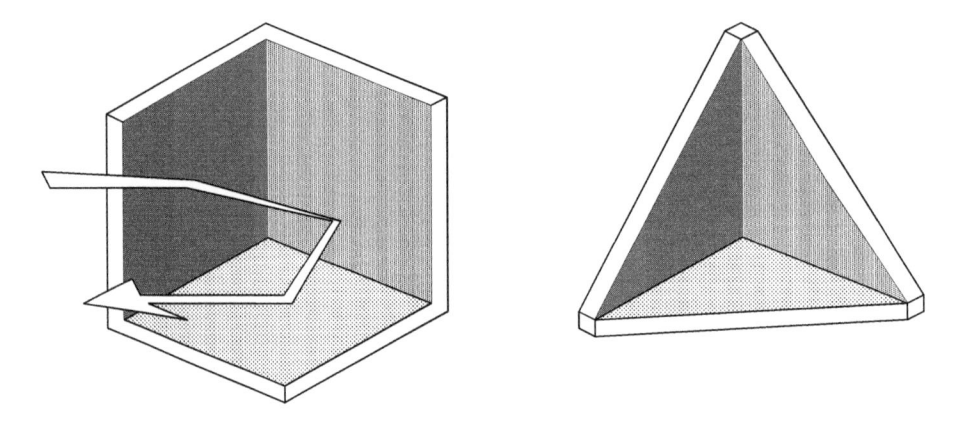

Figure 6.6 Two examples of trihedral corner reflectors.

strategy for building stealthy radar targets is the complete avoidance of flat reflecting surfaces that meet at right angles.

6.5 Thomson Scatter

A simple but illustrative and important problem is Compton scatter or scatter from free electrons. Like dielectric and conducting spheres, free electrons support a current when immersed in electromagnetic radiation, radiate and scatter. By calculating the scattering cross-section, the effective size of an electron can be deduced. By considering scatter from a volume filled with electrons, some additional, general properties of radar backscatter can be found.

Thomson scatter is the low-energy limit of Compton scatter. Whereas Compton scattering may be inelastic in general, Thomson scattering is elastic and leaves the kinetic energy of the electron unchanged. This is the relevant limit for radar.

The Coulomb force causes a free electron illuminated by electromagnetic radiation to accelerate. The acceleration can be found using Newton's kinematic law to be $\mathbf{a} = -(e/m)\,\mathbf{E}_\circ(\mathbf{x}_\circ, t)$, where $E_\circ$ is the electric field of the incident plane wave evaluated at the position $\mathbf{x}_\circ$ of the electron. The electric field produced (scattered) by accelerating electrons (at a vector displacement $\mathbf{r}$) is well known:

$$\mathbf{E}_s(\mathbf{r},t) = -\frac{e}{4\pi\varepsilon_\circ c^2 r^3}\mathbf{r} \times [\mathbf{r} \times \mathbf{a}(t - r/c)]$$

$$= -\frac{e^2}{4\pi\varepsilon_\circ c^2 m}\frac{1}{r}\left\{\begin{array}{c}\cos(\theta) \\ 1\end{array}\right\}\mathbf{E}_\circ(\mathbf{x}_\circ, t - r/c)$$

$$= -\frac{r_e}{r}\left\{\begin{array}{c}\cos(\theta) \\ 1\end{array}\right\}\mathbf{E}_\circ(\mathbf{x}_\circ, t - r/c),$$

where, as in Rayleigh scatter, the $\cos(\theta)$ term applies for parallel propagation but not for perpendicular propagation. Above, we have defined the classical electron radius

$$r_e \equiv \frac{e^2}{4\pi\varepsilon_\circ c^2 m_e}.$$

The radar-scattering cross-section can then be expressed as

$$\sigma = 4\pi r_e^2 \left\{\begin{array}{c}\cos^2(\theta) \\ 1\end{array}\right\}.$$

Note that the numerical value of $4\pi r_e^2$ is 1.0×10^{-28} m^2. The Thomson scattering cross-section is often written as $\sigma = (8\pi/3)r_e^2$, which represents a solid angle-averaged value in the case of unpolarized incident radiation.

6.6 Bragg Scatter

The section following this one describes how to modify the concepts of antenna gain and effective area used in the radar equation to accommodate volume-filling or soft targets. In this section, the concept of Bragg scatter is developed by considering scatter from multiple targets as well. The targets are taken to be free electrons, for the sake of simplicity, although the results are completely general.

Figure 6.7 shows the geometry in question. Let the vector $\mathbf{r}$ represent the relative location of a scatterer within a volume V_s. The vectors R_i and R_s connect the transmitter to the center of the volume to the receiver. The vectors r_i and r_s connect the transmitter and receiver to an individual target at r, i.e., $\mathbf{r}_i = \mathbf{R}_i + \mathbf{r}$, $\mathbf{r}_s = \mathbf{R}_s - \mathbf{r}$. The volume is sufficiently small and distant that the incident wavevector $\mathbf{k}_i$ and the scattered wavevector $\mathbf{k}_s$ are approximately constant throughout the volume. The phase of the wave scattered at $\mathbf{r}$ and received at the receiver will be

$$\phi = \omega t - \mathbf{k}_i \cdot \mathbf{r}_i - \mathbf{k}_s \cdot \mathbf{r}_s$$
$$= \omega t - (\mathbf{k}_i - \mathbf{k}_s) \cdot \mathbf{r} - \mathbf{k}_i \cdot \mathbf{R}_i - \mathbf{k}_s \cdot \mathbf{R}_s,$$

where ω is the frequency of the transmitted wave. The terms involving $\mathbf{R}_i$ and $\mathbf{R}_s$ are set by the experimental geometry, are constant, and can be neglected in the subsequent analysis. With the definition of the scattering wave vector

$$\mathbf{k}_s \equiv \mathbf{k}_i - \mathbf{k}_s,$$

the incident and scattered electric fields are seen to be related by

$$E_s(t) = \frac{r_e}{r_s} \left\{ \begin{array}{c} \cos(\theta) \\ 1 \end{array} \right\} E_\circ e^{j(\omega t - \mathbf{k}_s \cdot r)}.$$

To represent the effects of multiple scatterers in the scattering volume, the result needs to be integrated spatially over the volume

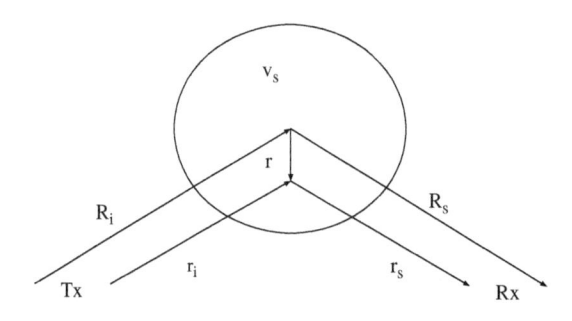

Figure 6.7 Scattering geometry for multiple targets.

6.7 Volume Scatter

$$E_s(t) = r_e \left\{ \begin{array}{c} \cos(\theta) \\ 1 \end{array} \right\} E_\circ e^{j\omega t} \int_{V_s} \frac{1}{r_s} N_e(\mathbf{r},t) e^{-j\mathbf{k}_s \cdot \mathbf{r}} d^3 r$$

$$\sim r_e \left\{ \begin{array}{c} \cos(\theta) \\ 1 \end{array} \right\} E_\circ \frac{e^{j\omega t}}{R_s} N_e(\mathbf{k}_s,t),$$

in which $N_e(\mathbf{r},t)$ is the volume density of scatterers in V_s and $N_e(\mathbf{k}_s,t)$ is the spatial Fourier transform of that quantity, evaluated at $\mathbf{k} = \mathbf{k}_s$. In an electron gas or a plasma, either quantity would be stochastic. Bragg scatter is what is said to occur when the intensity of $\langle |N_e(\mathbf{k}_s,t)|^2 \rangle$ significantly exceeds what would be expected from thermal fluctuations in the electron density.

The result is general and powerful. When backscatter is from a multiplicity of targets, the amplitude of the scattered signal depends on the spatial regularity of the targets. In particular, it depends on structure aligned with the scattering wavevector and with a wavelength given by $2\pi/k_s$. For monostatic scattering where $\mathbf{k}_s = -\mathbf{k}_i$, the scattering wavelength is half the radar wavelength. For bistatic geometries, it is longer. These are the Bragg scattering conditions. Bragg scatter applies not only to regular features in electron gasses but to rough-surface scatter, sea-surface scatter, and clear-air turbulence, to name a few.

6.7 Volume Scatter

The preceding discussion dealt with radar backscatter from a single target. Frequently, multiple targets reside within the *radar-scattering volume*. This is the volume defined in three dimensions by the beam solid angle, the range, and the radar pulse width/ matched filter width (see Figure 6.8). The cross-sectional area of the volume is $r^2 d\Omega$, and the depth of the volume is $c\tau/2$, where τ is the pulse width to which the receiver filter is presumably matched (see next chapter).

The targets in question could be electrons, raindrops, or insects; in the discussion that follows, they are assumed to be identical and randomly distributed throughout the volume. We further assume that the targets are insufficiently dense to cause significant attenuation of the radar signal passing through the volume. This is referred to as the *Born limit*. In this limit, we can define a radar-scattering cross-section per unit volume σ_v as the product of the radar cross-section for each target times the number density of targets per unit volume, $\sigma_v = N\sigma$. A more explicit description of the Born scattering limit is proved at the end of Chapter 11.

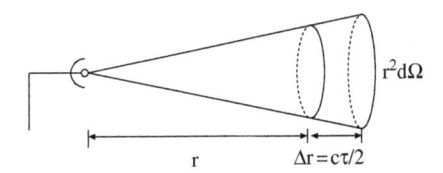

Figure 6.8 Geometry of a radar-scattering volume.

The power density incident on the scattering volume is

$$P_{inc} = P_{tx}\frac{G(\theta,\phi)}{4\pi r^2},$$

where G is the gain and where r, θ, and ϕ are measured from the antenna. The differential power scattered by each sub volume δV within the scattering volume is then given by

$$dP_s = P_{inc}\sigma_v dV.$$

In view of the definition of the radar-scattering cross-section, the corresponding amount of power received from scattering within dV will be

$$dP_{rx} = P_{inc}\sigma_v \frac{A_{eff}(\theta,\phi)}{4\pi r^2}dV.$$

Now, taking $dV = \Delta r r^2 d\Omega = (c\tau/2)r^2\sin\theta d\theta d\phi$ and integrating over the entire scattering volume for this range gate gives

$$\begin{aligned}P_{rx} &= \int P_{tx}\frac{G(\theta,\phi)}{4\pi r^2}\sigma_v\frac{A_{eff}(\theta,\phi)}{4\pi r^2}\frac{c\tau}{2}r^2\sin\theta d\theta d\phi\\ &= \frac{P_{tx}\sigma_v c\tau\lambda^2}{32\pi^2 r^2}\underbrace{\frac{1}{4\pi}\int_0^{2\pi}d\phi\int_0^\pi d\theta\, G^2(\theta,\phi)\sin\theta}_{G_{bs}},\end{aligned}$$

where the reciprocity theorem has been invoked and where the backscatter gain and corresponding backscatter area of the antenna have been defined

$$\begin{aligned}G_{bs} &= \frac{1}{4\pi}\int_0^{2\pi}d\phi\int_0^\pi d\theta\, G^2(\theta,\phi)\sin\theta\\ &\equiv \frac{4\pi}{\lambda^2}A_{bs}.\end{aligned}$$

These are new quantities, not to be confused with the standard antenna gain and effective area. They are constants rather than functions, which quantify signal returns in volume scatter situations. They account for the fact that the antenna gain determines not just the power received from an individual scatterer but also the size of the scattering volume and therefore the number of scatterers involved. It is always true that the backscatter gain is less than the maximum gain and the backscatter area is less than the physical area of the antenna, typically by a factor of about one half.

Altogether, the received power as a fraction of the transmitted power is

$$\frac{P_{rx}}{P_{tx}} = \frac{\sigma_v c\tau\lambda^2}{32\pi^2 r^2}G_{bs}$$

$$= \frac{\sigma_v}{4\pi r^2} \frac{c\tau}{2} A_{bs} \qquad (6.8)$$
$$\propto \frac{A_{bs}}{r^2}.$$

Contrast this with the behavior for single (hard target) backscatter, where the ratio is proportional to $(A_{eff}/r^2)^2$. Since the noise power given by Nyquist's theorem is proportional to the bandwidth, which is the reciprocal of the pulse width, we may further write

$$\frac{S}{N} = \frac{P_{tx}\sigma_v}{2\pi} \left(\frac{\Delta r}{r}\right)^2 \frac{A_{bs}}{K_b T c}$$
$$\propto \left(\frac{\Delta r}{r}\right)^2,$$

where we make use of $\Delta r = c\tau/2$. This expression exemplifies the high cost of fine-rage resolution in volume-scattering situations. In the next two chapters, we will explore how to optimize radar experiments and signal processing in order to deal with weak signals. Pulse compression in particular will be seen to be able to recover one factor of Δr.

6.8 Notes and Further Reading

The fundamental ideas underlying radiation from accelerating charges are explained in the seminal texts by J. D. Jackson and L. D. Landau and E. M. Lifshitz. Lord Rayleigh (John William Strutt) published his theory of scattering of light in a long series of papers that appeared near the end of the nineteenth century. Rayleigh–Mie scattering theory is worked out in detail in the classic electromagnetics text by J. A. Stratton and also in the more contemporary optics texts by H. C. van de Hulst and C. F. Bohren and D. R. Huffman.

The extension of Thomson scatter to scattering from a plasma is described in J. Sheffield's text. A primer for interpreting volume scatter from a plasma quantitatively is presented in the monograph by K. Bowles, et al.

M. Abramowitz and I. A. Stegun. *Handbook of Mathematical Functions with Formulas, Graphs, and Mathematical Tables*. US Government Printing Office, Washington, DC 20402, 1972.

C. F. Bohren and D. R. Huffman. *Absorption and Scattering of Light by Small Particles*. Wiley, New York, 1983.

K. L. Bowles, G. R. Ochs and J. L. Green. On the absolute intensity of incoherent scatter echoes from the ionosphere. *J. of Res. NBS - D. Rad. Prop.*, 66D:395, 1962.

W. H. Bragg and W. L. Bragg. The reflexion of x-rays by crystals. *Proc R. Soc. Lond. A.*, 88(605):428–438, 1913.

H. C. van de Hulst. *Light Scattering by Small Particles*. Wiley, New York, 1957.

J. D. Jackson. *Classical Electrodynamics*, 2nd ed. John Wiley & Sons, New York, 1975.

166   Scattering

L. D. Landau and E. M. Lifshitz. *The Classical Theory of Fields*, 3rd ed. Pergamon Press, New York, 1971.

J. Sheffield. *Plasma Scattering of Electromagnetic Radiation*. New York, 1975.

J. A. Stratton. *Electromagnetic Theory*. McGraw-Hill, New York, 1941.

J. Strutt. On the transmission of light through an atmosphere containing small particles in suspension, and on the origin of the blue of the sky. *Philosophical Magazine*, 47(5):375–394, 1899.

Exercises

6.1 Verify the claim given at the start of section 6.2 that **M** and **N** will satisfy the vector homogeneous Helmholtz equation if ψ satisfies the scalar homogeneous Helmholtz equation.

6.2 A certain 420 MHz radar has a circular aperture antenna with radius 35λ. The radiation pattern is identical to that of a uniformly illuminated circular aperture with a radius of 25λ. The radar is used to observe volume scatter from electrons in the Earth's ionosphere.

   a) Using the uniform aperture approximation, write down expressions for the radiation pattern for this antenna, its maximum directivity, HPBW, and aperture efficiency. Make a plot of the radiation pattern, being careful to demonstrate the HPBW, first sidelobe level, and locations of the first nulls. This part of the problem should be a review for you.

   b) Compute the backscatter gain and backscatter area for this antenna. Compare the former with the maximum gain and the latter with the effective and physical areas.

6.3 There is a considerable amount of space junk or debris in low-Earth orbit. It is made up of everything from spent rocket boosters to flecks of paint. More debris is created all the time as the existing junk undergoes collisions.

   Estimate the smallest scattering cross-section for bits of debris visible to the Arecibo 430 MHz UHF radar at a distance of 300 km. Use the facts that the radar operates at a frequency of 430 MHz and transmits 1 MW of power. The diameter of the Arecibo reflector is precisely 300 m. Take the aperture efficiency to be 70 percent. The radar will be operated in a pulsed mode, and we will see that the pulse width sets the system bandwidth. For our purposes, take the bandwidth to be 100 kHz. The sky noise temperature is about 100 K. For this problem, neglect all other contributions to the system noise temperature. Find the radar-scattering cross-section σ of a target giving rise to a 10 dB signal-to-noise ratio.

   We will see that for targets with a physical cross-section size $\sigma_p \equiv \pi a^2$ and circumference $C \equiv 2\pi a$, a being the physical radius of the target, the ratio $\sigma/\sigma_p \sim (C/\lambda)^4$ for "small" targets that satisfy $C/\lambda \approx < 1$. Does the target you found above satisfy this limit, which is the Rayleigh scattering limit?

7
Signal Processing

What do you do when radar signals arrive? The answer depends on your objective, the sophistication of your equipment, and the era in which you live. Some aspects of signal processing are common to all radar applications. Some are more specialized. New methods of processing radar data and extracting additional information are constantly being developed.

Traditionally, radar signal processing concerned hard target tracking. Is a target in a particular range gate or not? The answer is inherently statistical and carries a probability of correct detection, of missed detection, and of false alarm. Choices are made based on the costs of possible outcomes. Understanding probability and statistics is an important part of the radar art.

More recently, signal processing has expanded to include Doppler processing (related to the speed of a target) and radar imaging. A range of remote sensing problems can now be addressed with different kinds of radars (SAR, GPR, planetary radar, weather radar, etc.). The signal processing can be complicated and demanding in some cases, but statistics remains central to the problem.

What ultimately emerges from a radar receiver is a combination of signal and noise. The noise component is a random process. The signal may be deterministic (as in the case of one or more point targets) or stochastic (as in volume scatter). In either event, the net output voltage is governed by a probability density function controlled by the signal-to-noise ratio (SNR). It is possible to predict the probability density functions for different SNRs. On that basis, one can examine individual voltage samples and make decisions about whether a target is present. The reliability of the decision grows with the number of voltage samples considered. The merits of averaging are discussed below.

7.1 Fundamentals

Most radars are pulsed and can be represented by the block diagram at the top of Figure 7.1. Here, the transmitter and receiver share an antenna through a transmit-receive (T/R) switch. A radar controller generates pulses for the transmitter, controls the T/R switch, controls and drives the receiver, and processes data from a digitizer. The computer also generates some kind of display for real-time data evaluation.

168   Signal Processing

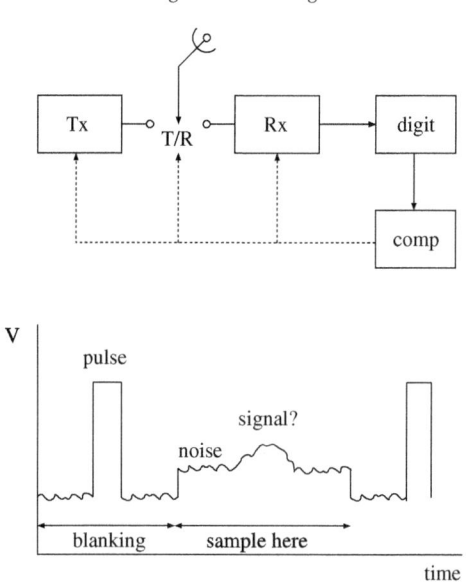

Figure 7.1 Pulsed radar schematic (above) and a-scope output (below).

One of the earliest kinds of radar displays is called an *a-scope* ("a" for amplitude), which is like an oscilloscope except that the vertical deflection represents the modulus of the receiver voltage $|V|$. The horizontal deflection is synchronized to the pulse transmission. A sample a-scope display is depicted at the bottom of Figure 7.1. The transmitted pulse is usually visible even though the receiver is blanked (suppressed) around the time of pulse transmission. Background noise appears in the traces when blanking is discontinued. A signal is indicated by a bump in the trace. Both the oscilloscope phosphor and the eye perform a degree of averaging, making the signal more visible. The time axis represents the target range.

A helpful tool for understanding what the a-scope traces and the underlying voltage samples mean is the *range-time* diagram. The way radar signals vary with time is indicative both of spatial (range) and temporal variations in the backscatter, and a range-time diagram is a guide for sorting out the two. Range-time diagrams are sometimes called Farley diagrams. An example is shown in Figure 7.2.

The ascending diagonal lines represent the path or characteristic of a transmitted pulse of duration τ. The pair indicate the times when the leading and trailing edges of the pulse pass through a given range, or their ranges at a given time. The slope is given by the speed of light, c.

The reflected pulse from a mirror-like target would be bounded by a similar pair of descending lines. Their interpretation is the same, and their slope is $-c$. The width of a reflected, descending pulse is also τ. The receiver filter is matched to the pulse width, and sampling of the filter output takes place at the times shown. Because of the "memory" associated with the matched filter, the filter output at time t includes information received

7.1 Fundamentals

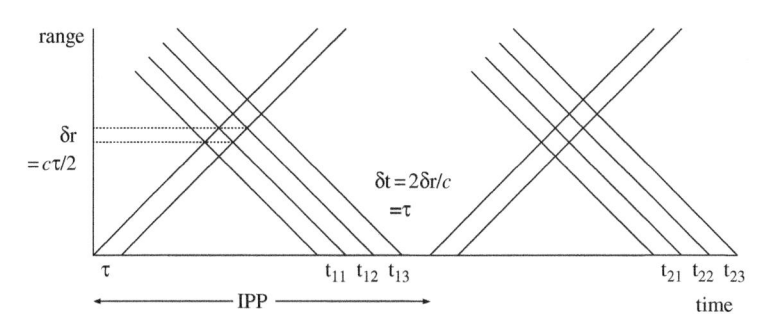

Figure 7.2 Radar range-time diagram.

from earlier times and from a span of altitudes, as indicated by the shaded diamond in Figure 7.2 for sample t_{13}. Samples taken at intervals spaced in time by τ will contain information about ranges spaced by $\delta r = c\tau/2$. The factor of 2 arises from the fact that the pulse must propagate out and back.

After a long time, all the anticipated echoes have been received, and another pulse is transmitted. This is the interpulse period or IPP. Whereas τ may be measured in microseconds, the IPP might be measured in milliseconds. We can often regard short time intervals as designating range and long time intervals as designating time. In fact, time is range and time is time, causing potential ambiguity.

We regard a sample from the ith pulse and jth range gate V_{ij} as having occurred at t_{ij}. Introducing matrix notation, samples acquired over a number of pulses take the form

$$\begin{pmatrix} V_{11} & V_{12} & \cdots & V_{1n} \\ V_{21} & V_{22} & \cdots & V_{2n} \\ \vdots & & & \vdots \\ V_{m1} & V_{m2} & \cdots & V_{mn} \end{pmatrix}.$$

Note that each voltage sample represents scatter from a single range gate with added noise. A row of data arises from a single pulse. A column represents the time history of the backscatter from a particular range gate. Signals from different range gates are normally taken to be uncorrelated, and range gates are therefore analyzed separately. An important point to recognize then is that, whereas data are accumulated by rows, they will be analyzed by columns.

Processing is statistical in nature. There are essentially two ways to process the columns:

1 Incoherent processing. This amounts essentially to averaging power, i.e., $\langle |V|^2 \rangle$.
2 Coherent processing. This makes use of complex voltage samples and generally involves summing them after multiplying by an appropriate phase factor in some cases. Doppler and interferometric information can be extracted in this way. Signal-to-noise ratio improvement is sometimes possible.

Squaring or taking products of voltage samples is sometimes called detection and is a necessary step for computing power. Coherent and incoherent processing really amount to pre-detection and post-detection averaging of one form or another. Practical radar signal processing most often involves coherent processing followed by incoherent processing. An example of incoherent processing is the stereotypical round phosphor radar screen showing target positions versus range and direction. In this case, the electron beam is made to be proportional to the scattered power, and the screen itself performs the averaging, increasing visual detectability somewhat. An example of coherent processing would be to compute the discrete transform (DFT) of a column of samples and then to form an estimate of the Doppler spectrum from it, i.e., $|FT(\omega)|^2$. Each bin represents the power in the corresponding Doppler frequency. An example combining both methods would be to compute Doppler spectra and then average them in time to improve the statistics, i.e., compute $\langle |FT(\omega)|^2 \rangle$.

7.2 Power Estimation and Incoherent Processing

Radar signals are random processes, and the information we extract is statistical in nature (e.g., variances, covariances). We form estimators of sought-after parameters, but these estimators are imperfect and contain errors due to the finite sample sizes involved. The size of the errors relative to the size of the parameters in question decreases with averaging so long as the signal is statistically stationary.

We define P_j as the actual or expected power in range gate j such that $P_j = \langle |V_{ij}|^2 \rangle$. The angle brackets denote an ensemble average, but we can approximate this by a time average if the signal is ergodic. An obvious power estimator is

$$\hat{\mathbf{P}}_j = \frac{1}{k} \sum_{i=1}^{k} |V_{ij}|^2,$$

where k is the number of samples on which the estimate is based. The estimator is unbiased and will converge on the expected power, but there will always be an error due to the finite sample size. What matters is the relative or fractional error. It is important to note that $\hat{\mathbf{P}}_j$ is itself a random variable, whereas P_j is not.

Suppose the estimator is based on a single sample only. The error in that case can be quantified by the estimator variance

$$\sigma_j^2 \equiv \langle (|V_{ij}|^2 - P_j)^2 \rangle.$$

The term σ_j by itself is the standard deviation for a single-sample estimate. Now consider summing multiple samples. If the sample errors are statistically uncorrelated (taken far enough apart in time), the variance of the sum is just the sum of the single-sample variances,

$$\left\langle \left(\sum_{i=1}^{k} (|V_{ij}|^2 - P_j) \right)^2 \right\rangle = k\sigma_j^2.$$

This says that the standard deviation of the sum only grows as $\sqrt{k}$ with increasing samples:

$$\sigma_{sum} = \sqrt{k}\sigma_{1\,sample}.$$

Meanwhile, the mean of the sum is obviously the sum of the means,

$$\left\langle \sum_{i=1}^{k} |V_{ij}|^2 \right\rangle = kP_j.$$

The important measure of statistical error is the standard deviation or RMS error relative to the mean. It is this quantity that determines signal detectability. Compared to the results for a single sample, the result for multiple samples behaves like

$$\left(\frac{\text{RMS error}}{\text{mean}}\right)_{k\,samples} = \frac{\sqrt{k}\sigma_j}{kP_j} = \frac{1}{\sqrt{k}}\left(\frac{\text{RMS error}}{\text{mean}}\right)_{1\,sample}.$$

The quotient term on the right is order unity (single-sample estimators of random processes are very inaccurate). Therefore, the relative error of the power estimator decreases with the *square root* of the number of independent samples. This is a fundamental property of the statistics of random signals. Note that incoherent averaging increases the confidence in our power estimator but does not affect the signal-to-noise ratio, as is sometimes misstated in literature.

Flipping a coin illustrates the principles involved. Consider judging the fairness of a coin by flipping it a predetermined number of times and counting the heads (heads=1, tails=0). For a single toss of a fair coin, the mean outcome is 1/2, the variance is 1/4, and the standard deviation is 1/2. This means the RMS discrepancy between a single outcome (0 or 1) and the expected outcome (1/2) is 1/2. Since the standard deviation is equal to the mean, the relative RMS error for one trial is 1.0.

In 10 flips, the mean and variance both increase by a factor of 10, but the standard deviation increases only by a factor of $\sqrt{10}$, so the relative error decreases by a factor of $\sqrt{10}$. Absolute estimator error grows, but not as fast as the estimator mean, so relative error decreases. After 10 flips, it is likely that the estimator will yield (mean)±(st. dev.) or $10(1/2)\pm\sqrt{10}(1/2)$ or $10(1/2)(1\pm 1/\sqrt{10})$. After 100 flips, the estimator will likely yield $100(1/2)(1\pm 1/\sqrt{100})$ After 10,000 flips, the relative estimator error will fall to 1 percent. It may well take this many flips to determine if a suspect coin is fair. Little could be concluded from a trial of only 10 flips, and nothing could be concluded from a single flip.

7.3 Spectral Estimation and Coherent Processing

The previous treatment made use only of the modulus of the receiver voltage. Early radar equipment sometimes provided only the modulus, and processing was limited to averaging the output power. Modern equipment brought the concept of coherent processing, which involves summing voltages (typically complex or *quadrature* voltages) to improve the signal-to-noise ratio.

The important idea is this: in a pulse-to-pulse radar experiment like the one illustrated in Figure 7.2, the bandwidth of the received signal is determined mainly by the bandwidth of the transmitted waveform and can be quite broad in the case of short or coded pulses. Matched filtering optimizes the overall signal-to-noise ratio of the data matrix. Any one column of the data will contain broadband noise admitted by the matched filter. It will also contain a signal component corresponding to echoes in the given range gate. The spectrum of the echo will typically be much narrower than the spectrum of the noise, which is essentially white. Coherent integration can be viewed as additional matched filtering applied to the data columns, where the filter is matched to the spectrum of the echo in a given range. The objective is additional signal-to-noise ratio improvement. The computation of the Doppler spectrum is an example of coherent integration.

We can decompose the voltages in a column of data into signal (echo) and noise components as

$$V_{ij} = V_{sig,i,j} + V_{noise,i,j}.$$

Let us work with one range gate j at a time and suppress the j subscript in what follows. Consider what happens when we sum voltages over some time interval prior to detection:

$$\left|\sum_{i=1}^{k}(V_{sig,i} + V_{noise,i})\right|^2 = \left|\sum_{i=1}^{k}V_{sig,i} + \sum_{i=1}^{k}V_{noise,i}\right|^2.$$

Assume that V_{sig} varies slowly or not at all in time while the $V_{noise,i}$ are independent random variables. This means that the signal is not fading rapidly and has no significant Doppler shift. The sum over V_{sig} just becomes kV_{sig}, and the expectation of the overall estimator becomes

$$\left\langle \left|kV_{sig,i} + \sum_{i=1}^{k}V_{noise,i}\right|^2 \right\rangle = \left\langle k^2|V_{sig,i}|^2 + kV_{sig,i}\sum_{i=1}^{k}V_{noise,i}^* + c.c. + \left|\sum_{i=1}^{k}V_{noise,i}\right|^2 \right\rangle,$$

where "c.c." means the complex conjugate of the term immediately to the left. The expectation of both cross terms vanish in the event that the noise voltage is a zero-mean random variable. Likewise, all the cross terms involved in the term on the right vanish when the noise voltage samples are statistically uncorrelated. This leaves simply

$$\left\langle \left|\sum_{i=1}^{k}(V_{sig,i} + V_{noise,i})\right|^2 \right\rangle = k^2 \langle |V_{sig,i}|^2 \rangle + k \langle |V_{noise,i}|^2 \rangle$$

$$= k^2 P_{sig} + k P_{noise}.$$

The upshot is that coherent integration has increased the signal-to-noise ratio by a factor of the number of voltage sampled summed, provided that the conditions outlined above (signal constancy) are met over the entire averaging interval. Coherent integration effectively behaves like low-pass filtering, removing broadband noise while retaining narrowband signals.

Coherent integration can also be understood in terms of the random walk principle. Under summation, the signal voltage, which is for all intents and purposes constant, grows linearly with the number of terms being summed. The noise voltage, meanwhile, is stochastic, and grows according to a random walk at the rate of $\sqrt{k}$. Upon detection, we have

$$\left(\frac{P_{sig}}{P_{noise}}\right)_{k\,samples} \sim \left(\frac{kV_{sig}}{\sqrt{k}V_{noise}}\right)^2 \sim k\left(\frac{P_{sig}}{P_{noise}}\right)_{1\,sample}.$$

An important extension of this idea allows coherent addition of signals with Doppler shifts. Such signals can be weighted by terms of the form $e^{-j\omega t}$ to compensate for the Doppler shift prior to summation, where ω is the Doppler frequency. If the Doppler shift is unknown, weighted sums can be computed for all possible Doppler shifts, i.e., $V(\omega_j) = \sum_i V_i e^{-j\omega_j t_i}$. This is of course nothing more or less than a discrete Fourier transform. Discrete Fourier transforms are examples of coherent processing that allows the signal-to-noise ratio in each Doppler bin to be individually optimized. Ordinary (time domain) coherent integration amounts to calculating only the zero-frequency bin of the Fourier transform.

$$\begin{pmatrix} V_{11} & V_{12} & \cdots & V_{1n} \\ V_{21} & V_{22} & \cdots & V_{2n} \\ \vdots & & & \vdots \\ V_{m1} & V_{m2} & \cdots & V_{mn} \end{pmatrix}.$$

Let us return to the data matrix to review the procedure. Data are acquired and filtered (using matched filtering) by rows. Each column is then processed using a discrete Fourier transform (DFT) algorithm. The result is a vector, perhaps of length $k = 2^p$, which represents the signal sorted into discrete frequency bins. Detection leads to a power spectrum for each range gate, with each element representing the power in a given Doppler frequency bin

$$|F(\omega_1)|^2, |F(\omega_2)|^2, \cdots, |F(\omega_k)|^2.$$

Note the following properties of the Doppler spectrum with k bins

- The width of the Doppler bins in frequency, δf, is the reciprocal of the total interval of the time series, $T = k \cdot \text{IPP}$.
- The total width of the spectrum, $\Delta f = k \delta f$, is therefore the reciprocal of the IPP.

For example, if the IPP = 1 ms and $k = 2^p = 128$, $\Delta f = 1000$ Hz and $\delta f \approx 8$ Hz.

Figure 7.3 illustrates how coherent processing optimizes the signal-to-noise ratio in the frequency domain. Here, a signal with a finite (nonzero) Doppler shift is contained entirely within one Doppler bin. The noise, meanwhile, is spread evenly over all k Doppler bins in the spectrum. In the bin with the signal, the signal-to-noise ratio is consequently a factor of k larger than it would be without Doppler sorting.

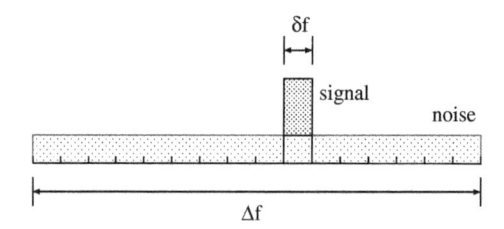

Figure 7.3 Doppler spectrum with signal contained in one Doppler bin.

If we know that the signal is likely to be confined to low-frequency bins in the spectrum, it may be desirable to perform coherent integration in the time domain prior to taking the Fourier transform. This amounts to low-pass filtering and may reduce the signal processing burden downstream. Nowadays, the benefits of this kind of operation, which may degrade the signal quality, may be marginal.

Finally, there is a limit to how large k can be. Because of signal fading, target tumbling, etc., the backscatter spectrum will have finite width. From a SNR improvement point of view, there is no benefit in increasing k beyond the point where the signal occupies a single Doppler bin. (This may be desirable for investigating the shape of the spectrum, but there is no additional SNR improvement.) Following coherent integration and/or spectral processing, additional incoherent integration to improve the signal statistics and detectability is usually desirable. This last operation is limited by time resolution requirements and/or stationarity considerations. The coherent integration time should be limited to intervals that do not permit targets to move between range gates, for example.

7.4 Summary of Findings

The comments above are sufficiently important to warrant repetition. A strategy for designing a radar experiment has emerged.

1. Set the pulse width τ to optimize the tradeoff between range resolution ($c\tau/2$) and system bandwidth ($B \sim \tau^{-1}$) and attendant noise.
2. Set the IPP to fully utilize the transmitter duty cycle. Other issues involving aliasing enter into this decision, however. These issues are discussed below.
3. Set the coherent integration factor k to optimize the tradeoff between spectral resolution δf and SNR improvement.
4. Set the number of incoherent integrations to optimize the tradeoff between time resolution and statistical confidence.

The third factor is perhaps the most subjective and depends on the characteristics of the radar target. There are two way to view the decision:

Time-domain POV Coherent integration works so long as the complex signal voltages (amplitude and phase) remain roughly constant over the averaging interval. Echoes have characteristic correlation times. Beyond this time, coherent integration is not advantageous.

Frequency-domain POV The discrete Fourier transform (DFT) (usually the FFT) increases the SNR in the Doppler bin(s) containing the signal. The effect is greatest when the signal occupies no more than a single Doppler bin. Longer coherent integrations imply narrower bin widths. Improvements in SNR will be seen so long as the bin is as broad as the signal spectral width, which is the reciprocal of its correlation time.

It is important to realize that the spectral width of the target is not related to the bandwidth of the transmitted pulse or received signal. One typically measures and handles very small Doppler widths and Doppler shifts (of the order of 1 Hz) by spectrally analyzing echoes collected from pulses with microsecond pulse widths ($B \sim 1$ MHz), for example in the case of vertical echoes from stratospheric layers. There is no contradiction and no particular technical problem with this.

Following coherent integration, incoherent integration will improve signal detectability by an additional factor of the square root of the number of spectra averaged. One must be careful not to integrate too long, however. The target should not change drastically over the integration time. In particular, it should not migrate between range gates.

With this information in mind, we can begin to explore practical radar configurations and applications. There are essentially two classes of radar: continuous wave (CW) and pulsed. CW radars are relatively simple and inexpensive but provide less information than pulsed radars. A familiar example of a CW radar is the kind used by police to monitor traffic. CW radars typically operate with low peak powers and moderate to high average powers. Pulsed radars require extra hardware not needed in CW radars. They typically operate at high peak power levels, which may necessitate some expensive hardware.

7.5 CW Radars

Schematic diagrams of a CW radar along with two receiver design options are shown in Figure 7.4. The purpose of such a radar is to measure the presence and line-of-sight speed of a target. Because of the *Doppler shift*, the frequency of the received signal f_r differs from the carrier frequency $f_\circ$ according to

$$\begin{aligned} f_r &= f_\circ \frac{1 - v/c}{1 + v/c} \\ &\approx f_\circ \left(1 - \frac{2v}{c}\right), v \ll c \\ &= f_\circ + f_D, \end{aligned}$$

where we take the transmitting and receiving antennas to be collocated. If not, an additional geometric factor must be included. Positive (negative) Doppler shifts f_D imply motion toward (away from) the radar. Target motion transverse to the radar line of sight does not affect the scattered frequency.

The Doppler shift is determined by comparing the frequency of the transmitted and received signals. A common, stable clock (local oscillator) is shared by the transmitter and receiver for this purpose. The middle panel of Figure 7.4 shows an inexpensive design

176   Signal Processing

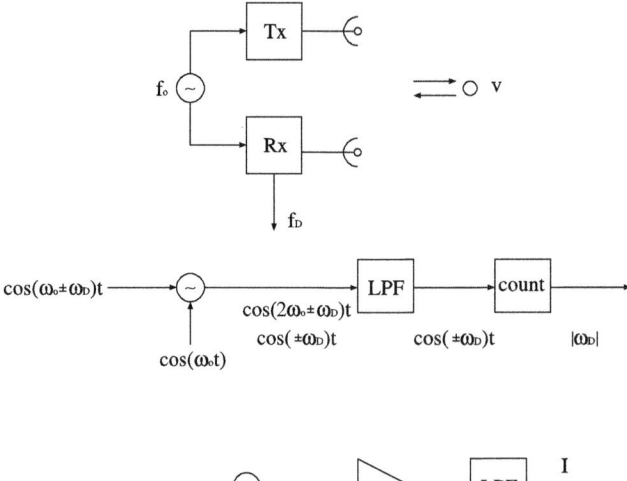

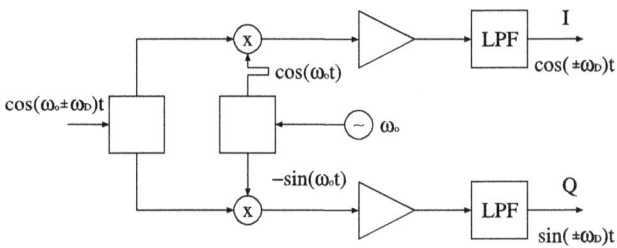

Figure 7.4 Top panel: schematic diagram of a CW radar. Middle panel: simple receiver implementation. Bottom panel: improved receiver implementation.

approach. Here, the output of a single product detector has a frequency equal to the Doppler frequency. The frequency counter gives the magnitude but not the sign of the speed. This is acceptable if the direction of the target is already known, for example in traffic control or baseball. It would not be very useful for tracking, collision avoidance or radar altimetry, where the sign of the Doppler shift is crucial.

A more sophisticated design is shown in the bottom panel of Figure 7.4. In this case, two product detectors are driven with local oscillator signals at the same frequency but 90° out of phase. The so-called "in phase" and "quadrature" output channels together convey not only the Doppler frequency but also the sign of the Doppler shift. These combined outputs can be interpreted as a complex phasor quantity, with I (in-phase component) and Q (quadrature component) representing the real and imaginary parts:

$$\begin{aligned} z(t) &= I + jQ \\ &= x(t) + jy(t) \\ &= C(t)e^{j\phi(t)}, \end{aligned}$$

such that the Doppler shift is the time rate of change of the phase, which can have either sign.

The Doppler spectrum of the radar echoes can be obtained through Fourier analysis of the quadrature signal $z(t)$. The spectrum of a single target with some Doppler shift amid background noise will look something like Figure 7.3, the frequency offset of the peak being the Doppler frequency. If the spectrum were computed from either the I or Q channel alone, the corresponding spectrum would be symmetric, with equal-sized peaks occurring at $\omega = \pm\omega_D$. Symmetric spectra such as this are often an indication that one or the other channel in a quadrature receiver is not functioning properly.

A related issue is the importance of symmetrizing the gains of the I and Q channels. Gain imbalance produces false images in Doppler spectra that can be mistaken for actual targets. Suppose the gains are unbalanced in a particular receiver such that the output from a target with a Doppler shift ω_D is

$$x(t) = \cos\omega_D t$$
$$y(t) = \frac{1}{2}\sin\omega_D t$$
$$z(t) = \frac{3}{4}e^{j\omega_D t} + \frac{1}{4}e^{-j\omega_D t},$$

the third line being obtained from the prior two with the Euler theorem. The net result is like two echoes with Doppler shifts $\omega = \pm\omega_D$ and with an amplitude ratio of 3:1. The first of these is the true echo, while the second is a spurious image. The corresponding power ratio is 9:1. Even though the peak in the spectrum at ω_D will be 9 times more powerful than the image at $-\omega_D$, the latter could cause confusion, depending on the application. As echo power is often displayed in decibels, a factor of nine difference is not as significant as it might at first seem.

7.6 FMCW Radars

CW radars yield information about the presence and motion of targets relative to the radar. They do not give information about the target range, which is what pulsed radars are usually used for. An intermediate class of radar called frequency modulated continuous wave (FMCW) can yield range information through frequency rather than amplitude modulation. While frequency modulated radars can do everything pulsed radars can (given sufficiently sophisticated signal processing algorithms), we focus here on a simple means of extracting range information from an uncomplicated FM waveform using the basic techniques already discussed. The strategy outlined below is often used in radar altimeters on aircraft and spacecraft. General methods for processing FMCW radar signals are outlined later in the text.

The basic principles are illustrated in Figure 7.5. The frequency of the transmitted waveform is ramp-modulated with an amplitude Δf, a period T, and a corresponding ramp frequency $f_m = 1/T$. The received signal is a duplicate of the transmitted one, except that

178   *Signal Processing*

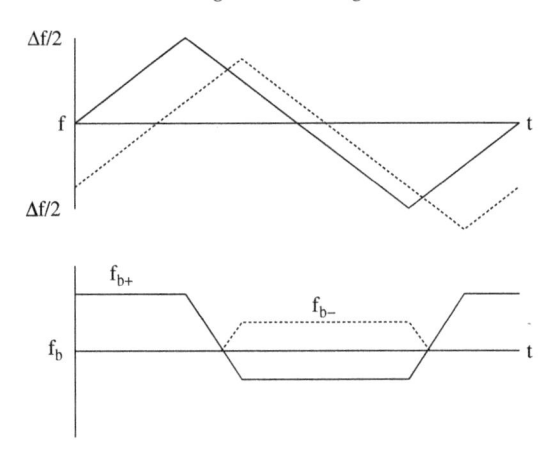

Figure 7.5 FMCW radar. Top panel: transmit (solid) and received (dashed) frequency vs. time. Bottom panel: beat frequency (solid) and magnitude of beat frequency (dashed).

it is delayed in time by $\Delta t = 2R/c$ and also Doppler shifted in frequency. If $\Delta f \ll f_\circ$, then the Doppler shift is just $f_D = -2f_\circ v/c$.

Let the radar receiver be fashioned using the inexpensive implementation from Figure 7.4. The output will be the difference between the transmitted and received frequency or the beat frequency

$$f_b = |f_{tx} - f_{rx}|,$$

or rather the modulus of the difference. We define f_{b+} and f_{b-} as the beat frequency during the upswing and downswing of the transmitted ramp and ignore intervals when the beat frequency is in transition. If there is only one target, not much interference, and the range and Doppler shift of the target are not too great, it will be possible to measure curves like the bottom panel of Figure 7.4 and to determine f_{b+} and f_{b-} reliably and accurately.

Knowing the two beat frequencies gives enough information to determine the target range and Doppler shift. The received frequency is equal to the transmit frequency at some time in the past, shifted by the Doppler frequency:

$$f_{rx}(t) = f_{tx}(t - \Delta t) + f_d$$
$$\Delta t = 2R/c,$$

where Δt is the time required for the signal to go out and back to the target. Since the frequency ramp is linear, we can write

$$f_{tx}(t - \Delta t) = f_{tx}(t) - \Delta t f'(t)$$
$$f'(t) = \pm \Delta f/(T/2) = \pm 2\Delta f f_m.$$

The beat frequency is then

$$f_b = |f_{tx}(t) - f_{rx}(t)|$$
$$= |\pm 2\Delta f f_m \Delta t - f_D|,$$

giving

$$f_{b+} = 2\Delta f f_m \Delta t - f_D$$
$$f_{b-} = 2\Delta f f_m \Delta t + f_D.$$

We have two equations and two unknowns, $\Delta t = 2R/c$ and f_D, for which we can solve. In practice, successful determination of range and Doppler shift is only possible for single, strong targets and requires a good initial guess. Over-the-horizon (OTH) radars use frequency modulated waveforms to observe multiple targets over enormous volumes in marginal SNR conditions, but only with the adoption of sophisticated and computationally expensive signal-processing algorithms. Approaches to processing radar signals using arbitrary modulation schemes will be discussed in Chapter 8.

7.7 Pulsed Radars

Simultaneous determination of the range and Doppler shift of multiple targets is usually accomplished with pulsed radars incorporating amplitude and perhaps frequency and/or phase modulation. Such radars must generally operate at very high peak power levels to maintain sufficient average power. This and the necessity of transmit/receive (T/R) switches and a complex radar controller make pulsed radars relatively costly compared to CW radars. Figure 7.6 shows a block diagram of a typical pulsed radar.

Radar pulses are most often emitted periodically in time at intervals of the interpulse period (IPP). This arrangement simplifies hardware and software design as well as data processing and is, in some sense, optimal. Ideas pertaining to the Nyquist sampling theorem can be applied directly to the radar signals that result. However, periodic sampling leads to certain ambiguities in signal interpretation. Much of the rest of the chapter addresses these ambiguities, which are called *range aliasing* and *frequency aliasing*. Aperiodic sampling can be used to combat aliasing, but this is beyond the scope of the text.

7.7.1 Range Aliasing

Figure 7.7 shows an A-scope trace for a hypothetical radar experiment. An echo with a range delay (measured from the prior pulse) Δt is evident in the trace. What is the target's range? The answer cannot be uniquely determined, since the echo could correspond to any of the pulses in the periodic train. In fact,

$$R \stackrel{?}{=} \frac{c\Delta t}{2} \stackrel{?}{=} \frac{c(\Delta t + T)}{2} = \frac{c(\Delta t + nT)}{2},$$

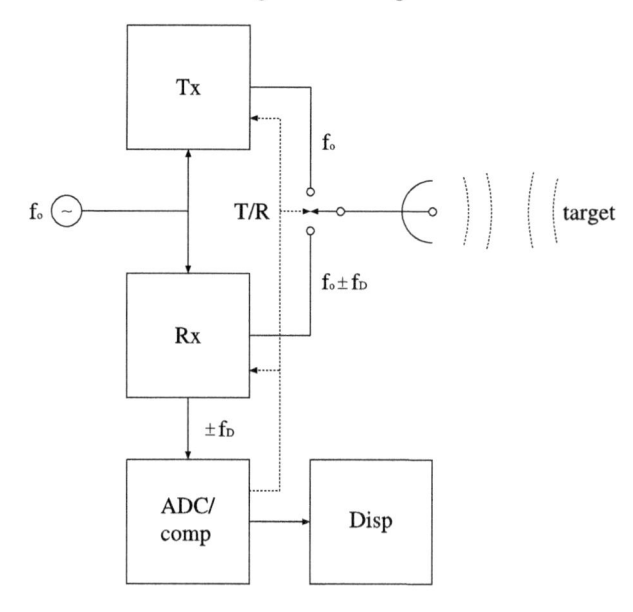

Figure 7.6 Schematic diagram of a pulsed radar.

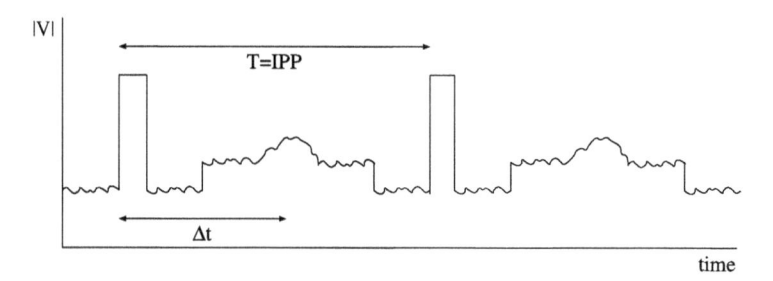

Figure 7.7 A-scope output for a pulsed radar experiment.

where n is an integer. The ambiguity implied is serious and can lead to misleading results. There is a infamous example of lunar echoes being detected by a Ballistic Missile Early Warning System (BMEWS), causing an alert. The echoes aliased to ranges consistent with potentially threatening airborne targets (range aliasing always leads to an underestimate of range). Only later was the true nature of the echoes discovered and the problem mitigated. Another infamous story has the radio altimeter on a Russian Venera spacecraft misjudging the altitude of mountain peaks on Venus because of range aliasing. The altimeter had been based on a design for a jet aircraft and used an IPP too short for planetary radar applications.

A vivid example of range aliasing is depicted in Figure 7.8, which shows echoes from a geosynchronous satellite measured with the 50 MHz Jicamarca Radio Observatory, looking vertically. That the satellite drifted into and out of the radar beam is an indication that

7.7 Pulsed Radars

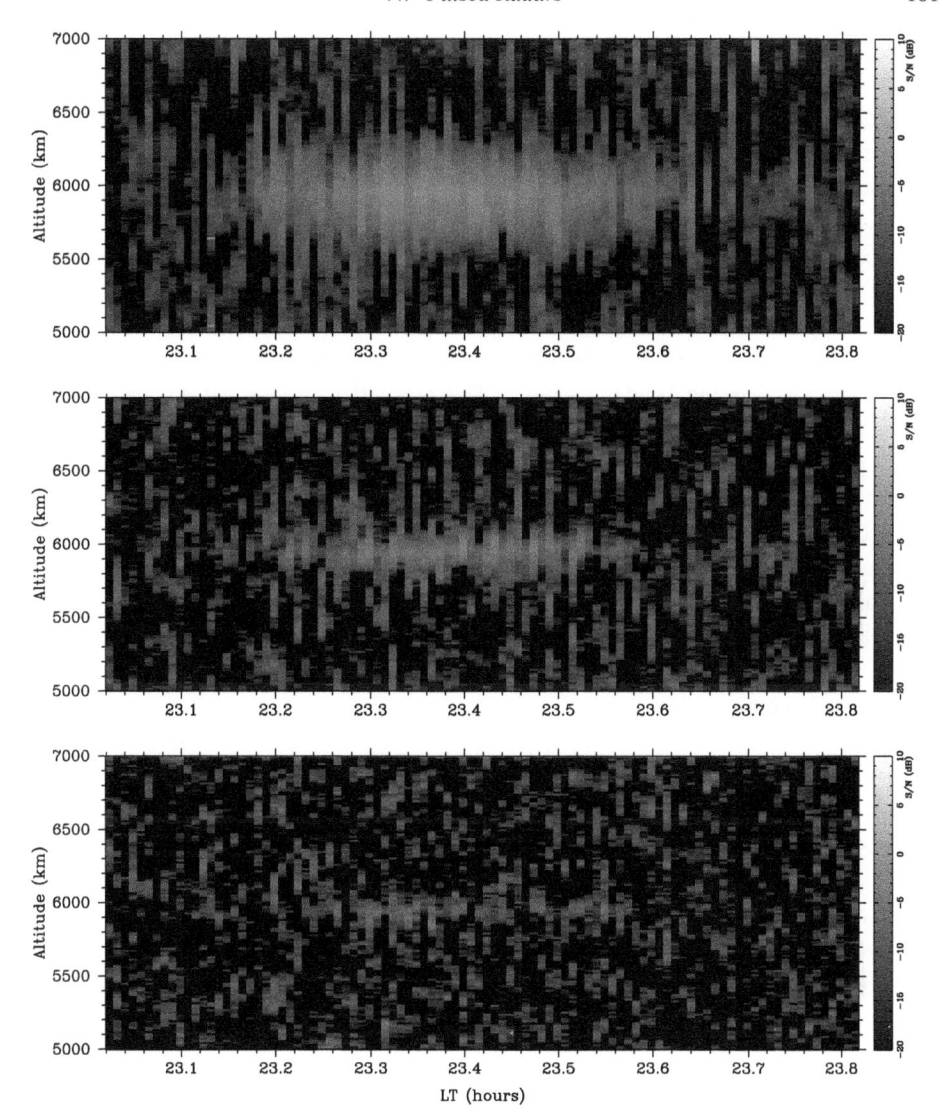

Figure 7.8 Backscatter from a geosynchronous satellite observed at the Jicamarca Radio Observatory near Lima, Peru. The upper, center and bottom panels show echoes received using 4 ms, 2 ms and 1 ms pulses, respectively. Grayscales indicate the signal-to-noise ratio on a dB scale. The interpulse period was 100 ms. (The data were acquired with the help of Jicamarca's director Dr. Marco Milla.)

it was in geosynchronous and not geostationary orbit. The interpulse period was 100 ms, implying range aliasing for targets more distant than 15,000 km. In fact, the target was at a range of nearly 36,000 km = 2 × 15,000 km + 6000 km. (The precise altitude for geosynchronous orbit is 35,786 km, indicating that the target was in a slightly higher "graveyard" orbit.)

The satellite was observed using pulse lengths of 4 ms, 2 ms and 1 ms, and the three panels show the effects of different pulse lengths on detectability and range resolution. The latter issue will be discussed in subsequent chapters in the context of radar experiment optimization. Insofar as calculating the expected signal-to-noise ratio, we can do this using the radar equation for hard targets using information about the radar.

The Jicamarca radar uses a square phased-array antenna with dimensions 300×300 m and an aperture efficiency of 0.63. The wavelength is 6 m, and the transmitted power for the measurements shown was 2 MW. The noise temperature at the time in question was about 8000 K. For the scattering cross-section we estimate $\sigma \approx 10$ m^2 and presume that the scattering efficiency was about unity. Given these numbers, the signal-to-noise ratio should have been about 3 in the 4-ms pulse mode. This estimate is in reasonable agreement with the measurements.

Range aliasing can be avoided by choosing an interpulse period sufficiently long that all the echoes are recovered for each pulse before the next one is transmitted. This may be impractical, however, for reasons discussed later in the chapter. A way of detecting range aliasing is to vary the IPP slightly. If the apparent range to the target changes, the echoes are aliased. The idea is illustrated in Figure 7.9. In this case, when the IPP was changed from T_1 to T_2, the apparent range to the target changed from Δt_1 to Δt_2. On the basis of the first pulse scheme, the true delay must be $\Delta t_1 + nT_1$. On the basis of the second, it must be $\Delta t_2 + mT_2$. Equating gives

$$\Delta t_1 + nT_1 = \Delta t_2 + mT_2.$$

We have one equation and two unknowns, n and m; since these are integers, one or more discrete solutions can be found. In the case of the example in Figure 7.9, the echoes arrive from the pulse sent just before the most recent one, and $m = n = 1$ is a solution. In the event that multiple plausible solutions exist, a third IPP might be tried to resolve the ambiguity.

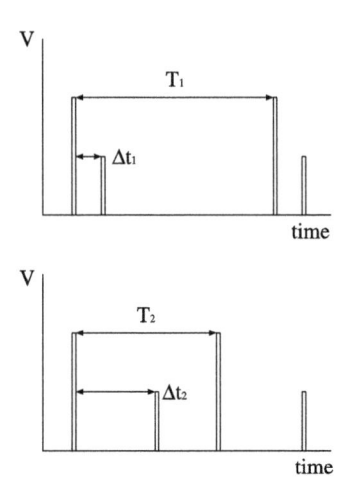

Figure 7.9 Effect of changing the IPP on range aliasing.

An application in which range aliasing is certain to be an issue is planetary radar. Consider as an example using the Arecibo radar to probe the surface of Venus. The round-trip travel time for radio waves reflected from Venus at its closest approach is about 5 minutes. An IPP of 5 minutes is necessary to completely avoid range aliasing, but this is too long for a practical experiment. A more efficient use of the radar would be to transmit pulses at 40 ms intervals (with a 25 Hz PRF). Since the radius of Venus R_v is about 6000 km, $2R_v/c \sim 40$ ms, meaning that all of the echoes from a given pulse will be received before echoes from the next pulse in the train begin to appear. Absolute range information is lost, but relative range information (relative to the subradar point, for example) is recovered unambiguously. The radar can transmit a continuous pulse sequence for 5 min. and then switch to receive mode for the next 5 min. This illustrates that the condition for avoiding range aliasing is therefore actually

$$T \geq \frac{2L}{c},$$

where L is the span of ranges from which echoes could conceivably come.

In fact, radar studies of Venus at Arecibo have been conducted with an IPP of about 33 ms for reasons that will be explained below. Since echoes from the periphery of the planet are much weaker than echoes from the subradar point, information from the periphery is effectively lost in such experiments. Range ambiguity is not a problem, however.

7.7.2 Frequency Aliasing

Frequency aliasing in the radar context is no different from frequency aliasing in the usual sampling problem and is governed by the Nyquist sampling theorem. It is important to keep in mind, however, that the radar data have been decimated; since we are only considering the samples corresponding to a given range gate, the effective sampling rate is the pulse repetition frequency (PRF) which is much less than the matched filter bandwidth. The noise, which has a bandwidth B determined by the matched filter, will always be undersampled and therefore aliased. Whether the signal is aliased depends on the fading or correlation time of the radar target.

Figure 7.10 illustrates the sampling process in the time (upper row) and frequency (lower two rows) domains. In the time domain, sampling the signal is equivalent to multiplying it by a train of delta functions separated in time by the interpulse period T. In the frequency domain, that pulse train is another series of delta functions, spaced in frequency by $\Delta f = 1/T$. Multiplication in the time domain transforms to convolution in the frequency domain. Consequently, the spectrum of the sampled time series is a sum of spectra identical to the original, only replicated periodically at frequency intervals Δf. The spectrum that emerges from discrete Fourier analysis of the data is the portion of the spectrum enclosed in the frequency window $\pm \Delta f/2$. This is equivalent to the original spectrum so long as there were no spectral components in the original at frequencies outside the window. Any spectral features originally outside the window

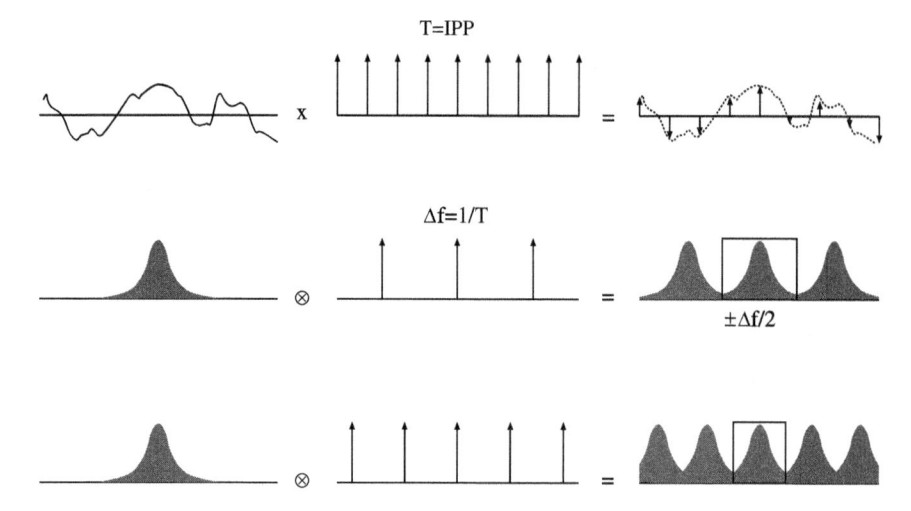

Figure 7.10 Sampling and frequency aliasing. The bottom row shows the effect of increasing the IPP, which narrows the width of the spectral window $\pm\Delta f/2$ and leads to frequency aliasing.

will be mis-assigned or aliased in frequency when folded back into the window by convolution. This represents distortion, which can lead to misinterpretation of the radar data.

For example, suppose a target with a Doppler frequency of 120 Hz is observed with an interpulse period of 5 ms. The spectrum resulting from discrete Fourier analysis will encompass frequencies falling between $\pm$ 100 Hz. Due to aliasing, the echo will appear in the -80 Hz frequency bin. Thus, a target with with a positive Doppler shift signifying approach will appear to have a negative Doppler shift, signifying retreat. Imagine the consequence for a collision avoidance radar!

Frequency aliasing can be prevented by choosing an interpulse period small enough to satisfy the Nyquist sampling theorem for the anticipated targets. As with range aliasing, however, this constraint may be impractical. If frequency aliasing is suspected, it can be resolved by changing the IPP and observing the change in the apparent Doppler shift. The procedure is similar to that described in the previous section.

It should be obvious that it may be impossible to simultaneously avoid both range and frequency aliasing, since the former (latter) requires a long (short) IPP. The conditions for avoiding both can be expressed as

$$2|f_D|_{max} \leq \text{PFR} \leq \frac{c}{2L},$$

where the pulse repetition frequency (PRF) is the reciprocal of the interpulse period (IPP) and where $|f_D|_{max}$ is the modulus of the largest anticipated Doppler shift. Recall that $f_D = 2v/\lambda$, where v is the approach speed of the target. If a range of existing PRFs satisfy both conditions, the target is said to be *underspread*. Otherwise, it is *overspread*. The outcome

7.8 Example – Planetary Radar

depends on the correlation time of the target and also on radar frequency, the condition being easier to satisfy at lower frequency. There are means of dealing with overspread targets. These methods will be discussed in Chapter 10.

7.8 Example – Planetary Radar

We revisit the planetary radar example, which is illustrative of both underspread and overspread targets. Backscatter from the rough surfaces of planets can be used to construct detailed images, even from Earth. It is important to realize that the images do not come from beamforming; the entire planet is well within the main beam of even the largest-aperture radar. Instead, detailed images arise from our ability to make extremely accurate measurements of time. From time, we derive information about both range and Doppler shifts.

The geometry of a planetary radar experiment is shown in Figure 7.11. Echoes arrive only from the rough surface of the planet. Surfaces of constant range are planes of constant x. The intersection of these planes with the planet's surface are circles, the largest having a radius equal to that of the planet. The surface echoes will have Doppler shifts arising from the planet's rotation. Surfaces of constant Doppler shift are planes of constant y. To see this, note that the Doppler shift depends only on the line-of-sight projection of motion, $v_x = -v \sin\theta$, where $v = r\dot\theta = r\Omega$, r being the distance from a point on the surface to the axis of rotation,

$$f_D = \frac{2v_x}{c}f_\circ \;=\; -\frac{2v\sin\theta}{c}f_\circ \;=\; -\frac{2r\sin\theta\,\Omega}{c}f_\circ \;=\; -\frac{2y\Omega}{c}f_\circ,$$

where $f_\circ$ is the radar frequency and Ω is the planetary rotation rate in radians/s. The planet's surface, a plane of constant range, and a plane of constant Doppler frequency intercept at two points in the northern and southern hemispheres, respectively. Presuming that the hemispheric ambiguity can somehow be resolved, each point in a range-Doppler spectrogram represents the backscatter power from a point on the planet's surface. (The ambiguity

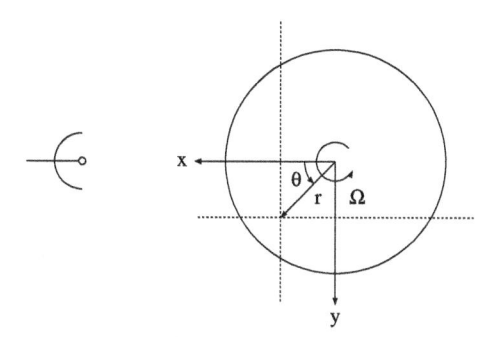

Figure 7.11 Schematic diagram of a planetary radar experiment. The intersection of the dashed lines is a point on the surface with cylindrical coordinates (r, θ). The radius of the planet is R.

can most easily be resolved by comparing results from experiments in different seasons or using interferometry to suppress one of the regions.)

Since y is limited by the planet's radius, the range of Doppler shifts that will be measured is given by

$$-\frac{2R\Omega f_\circ}{c} \leq f_d \leq \frac{2R\Omega f_\circ}{c},$$

with the extreme limits being reached only when $x = 0$, i.e., only for echoes received from the extreme edges of the planet. To avoid frequency aliasing, we require the Nyquist frequency (the largest frequency in the spectral window) to exceed this figure, or

$$\frac{1}{2T} \geq \frac{2R\Omega f_\circ}{c}$$

$$T \leq \frac{c}{4R\Omega f_\circ}.$$

Meanwhile, range ambiguity is avoided by making $T > 2L/c = 2R/c$. Combining these conditions gives

$$\frac{2R}{c} \overset{?}{\leq} T \overset{?}{\leq} \frac{c}{4R\Omega f_\circ},$$

which can only be satisfied if

$$f_\circ \leq \frac{c^2}{8R^2\Omega}.$$

Whether a planet is overspread therefore depends on its size, rotation rate and the radar frequency. In the case of Venus, $R_v \sim 6000$ km, the rotation period is about 250 days, and the cutoff frequency is therefore $f_\circ \sim 1$ GHz. In fact, the apparent rotation period of Venus as viewed from the Earth is somewhat greater than 250 days, and the planet is nearly underspread for observations with Arecibo's 2.38 GHz S-band radar. Frequency aliasing can be avoided by using an IPP of 33 ms, which permits the mapping of all but the most distant parts of the planet's surface. On the other hand, with its ~24.6 hour day, Mars is strongly overspread at S-band, and only a narrow strip of the planet near its equator can be mapped from the Earth.

7.9 Range and Frequency Resolution

We now know enough to design a competent radar experiment. Four parameters related to four different experimental timescales must be set in a particular order. Table 7.1 shows how each parameter is set to strike a balance between competing demands. Satisfying all the constraints can sometimes be difficult, and compromise is often necessary. The questions arise: How well can one do? What ultimately limits the range and frequency resolution of an experiment?

For example, consider transmitting a series of $\tau = 1\mu s$ pulses. The corresponding range resolution is $\Delta r = c\tau/2 = 150$ m, and the corresponding bandwidth is $B = 1/\tau = 1$ MHz.

7.9 Range and Frequency Resolution

Table 7.1 *Tradeoffs in Radar Experiment Design*

Parameter	Balance
pulse length	range resolution vs. bandwidth
IPP	range aliasing vs. frequency aliasing
coherent processing	signal-to-noise ratio vs. correlation time
incoherent processing	statistical confidence vs. time resolution

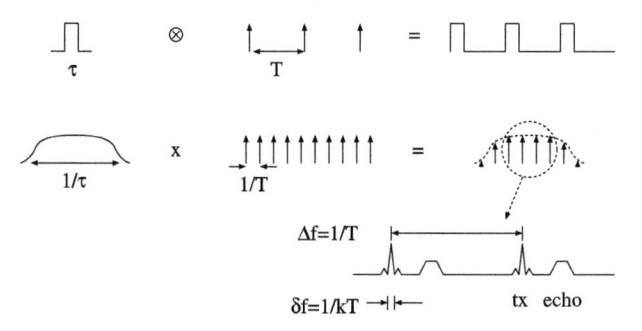

Figure 7.12 Representation of a radar pulse train in the time domain (top row) and frequency domain (middle row). An exploded view, showing both the transmitted and received spectra, is shown in the bottom row.

Several factors affecting signal detectability are thus determined. Now suppose that the echoes from the target have a Doppler width and Doppler shift of just a few Hz. Is this detectable, in view of the broad bandwidth of the transmitted signal? In fact it is, perhaps surprisingly, and the problem is not even technically challenging. The spectrum of the transmitted signal and the final echo Doppler spectrum are essentially independent. This is a consequence of coherent pulse-to-pulse detection, made possible by the accuracy of the clocks involved in the measurement timing, and untrue of remote sensing with lasers, for example.

The key idea is that a train of pulses rather than a single pulse is being transmitted, and a train of echoes is received and coherently processed. This is represented schematically in Figure 7.12. The width of each pulse sets the system bandwidth, but the length of the train determines the spectral resolution. We can think of the transmit pulse train as the convolution of a single pulse of width τ with a train of delta functions spaced by the IPP T. In the frequency domain, the pulse has a spectrum of width $B = 1/\tau$ and the pulse train becomes an infinite train of delta functions spaced in frequency by $\Delta f = 1/T$. Convolution transforms to multiplication. Consequently, the spectrum of the transmitted pulse train is a series of very narrow, periodically spaced spectral peaks contained in an envelope of bandwidth B.

If the original pulse train had an infinite extent in time, the spectral peaks just mentioned would be infinitely narrow Dirac delta functions. In fact, data from a finite pulse train of

k pulses are analyzed during coherent processing. The finite duration of the pulse train in the time domain translates to a finite spectral width for the peaks in the transmitted spectrum. That width is the reciprocal of the total transmission time, or $\delta f = 1/kT$. Given an interpulse period measured in tens of ms and given pulse train lengths numbering in the hundreds, it is not difficult to design experiments with $\delta f \leq 1$ Hz.

The received echo spectrum will be a Doppler-shifted and Doppler-broadened copy of the transmitted spectrum, as shown in the last row of Figure 7.12. When we decimate the radar samples and examine one range gate at a time, we undersample the received spectrum, which still has a bandwidth B. In so doing, all the spectral components of the echoes will alias to so as to coalesce into a single copy of the spectrum, in what is called *baseband*. That spectrum will be a faithful representation of the target's Doppler spectrum (with added white noise) so long as the target spectrum is wider than δf, which is to say that its correlation time is shorter than kT. Our ability to measure precisely the width of the Doppler spectrum, differentiate between the spectra of two different echoes and measure a Doppler shift are all limited by $\delta f = 1/kT$. Nowhere in this analysis does the pulse length enter.

Finally, the Doppler spectrum can be incoherently averaged to improve the experimental statistics. This can continue, so long as the target remains in the same range gate and its spectrum doesn't change. Target stationarity and the need for good time resolution ultimately limit incoherent integration time.

The topic can also be appreciated in the time domain. Consider transmitting a series of pulses with an interpulse period T and receiving the corresponding series of echoes from a hard target. For a particular range gate, this produces voltage samples of the form

$$V_i = V_{i-1} e^{j\omega_D T},$$

where ω_D is the angular Doppler frequency. This is the rate of change of the phase of an echo from a target in motion, i.e.,

$$\Delta \phi = \frac{2\pi}{\lambda} 2(\Delta x = vT)$$
$$= \omega_D T.$$

To measure the Doppler shift, we require that kT be long enough such that $\omega_D T$ is measurable (while T is small enough to prevent frequency aliasing). The ability to measure ω_D accurately depends on the length of the time series and the accuracy of the clock. The transmitter pulse width is essentially irrelevant.

7.10 MTI Radars

One final application of coherent signal processing is found in so-called Moving Target Indicator (MTI) radars. The purpose of such systems is to remove radar clutter from stationary targets, leaving behind only radar signals from the moving targets of interest.

7.10 MTI Radars

The targets in question could be aircraft or other kinds of vehicles. The stationary background clutter, meanwhile, could arise from the land or sea. In many instances, the clutter could be much stronger than the signal, and removing the former is paramount to detecting the latter.

One way to distinguish the clutter from the signal is coherent processing and spectral analysis using DFTs followed by incoherent integration as described above. While this approach is perfectly satisfactory, it could represent overkill if the objective is simply to identify the presence or absence of a moving target amid background clutter. A small variation on the basic coherent integration technique can simplify the problem and reduce the amount of hardware and signal processing necessary to implement it.

Even a "stationary target" gives rise to backscatter with a small range of Doppler shifts. Doppler spreading can arise from minuscule target motion (e.g., sea surface waves, swaying trees), from the motion of the radar antenna, from jitter in the radar frequency and from index of refraction variations along the propagation path. Each of these contributors can be described by a spectral variance. Since the various contributors are statistically uncorrelated, these variances add. The MTI notch filter bandwidth must be tuned to the overall clutter bandwidth to be effective.

In the simplest implementation of coherent processing, sample voltages from a given range gate were added in phase so that slowly varying signal components added constructively. In the simplest version of an MTI radar, adjacent sample voltages are added 180 degrees out of phase, and so slowly varying signal components add destructively and vanish. Conceptually, the single-stage MTI filter implies the use of feedback. Mathematically, the filter acts like

$$y_1(t) = x(t) - x(t-T)$$
$$Y_1(\omega) = X(\omega)[1 - \exp(-j\omega T)],$$

where T is the IPP. With some reorganization, we can express the filter frequency response function $H(\omega) = Y(\omega)/X(\omega)$ as

$$|H_1(\omega)| = 2|\sin(\omega T/2)|.$$

This function, which is shown in Figure 7.13, has a null at zero frequency and maxima at $\omega/2\pi = f = \pm 1/2T$, i.e., at the extreme boundaries of the spectral window across which it is periodic. It therefore behaves like a high-pass or notch filter, albeit a rudimentary one, with little attenuation away from zero frequency.

By itself, a single-stage MTI filter may have too narrow a notch for many applications and would tend to pass much of the clutter along with the signal. The obvious generalization is the multi-stage MTI filter, which incorporates another stage of feedback,

$$y_2(t) = y_1(t) - y_1(t-T)$$
$$= x(t) - 2x(t-T) + x(t-2T)$$

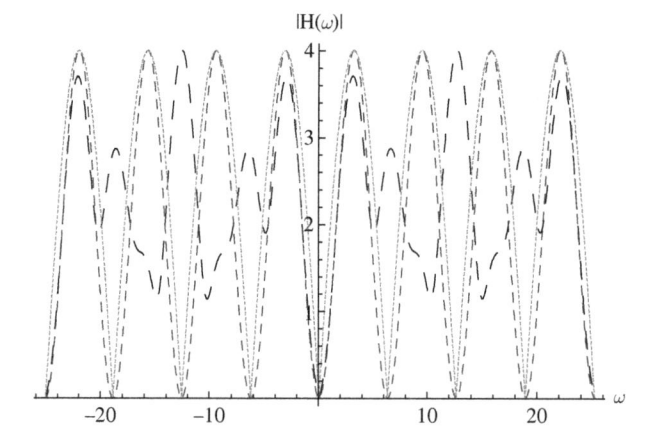

Figure 7.13 Frequency response function $|H(\omega)|$ for three MTI radar configurations calculated with $T = 1$; (solid) $2|H_1(\omega)|$, (dash) $|H_2(\omega)|$, (long dash) $|H(\omega)|$ from (7.1) with $a = 3/4$.

$$Y_2(\omega) = X(\omega)\left[1 - 2\exp(-j\omega T) + \exp(-j2\omega T)\right]$$
$$= X(\omega)\left[1 - \exp(-j\omega T)\right]^2,$$

so that the frequency response function becomes

$$|H_2(\omega)| = 4\left|\sin(\omega T/2)\right|^2,$$

which represents a broader notch filter than the single-stage version (see again Figure 7.13). Still higher-order feedback produces wider notch filters with commensurate orders. If the calculation seems familiar, it is essentially the same one used to calculate the radiation pattern of a binomial antenna array.

By using additional stages and incorporating recursion and gain, it is possible to design feedback networks with filter stop-bands that precisely match the clutter spectrum. The feedback weights can be tuned "on the fly" for optimal performance as the signal and clutter characteristics change. This could be done adaptively, using a generalization of the adaptive array formalism developed back in Chapter 3. Fixed and adaptive filter design is an enormous field of study and is beyond the scope of the present discussion. Beyond some level of complexity, however, the benefits of MTI become dubious, and full spectral analysis becomes attractive.

The MTI filters described above suffer from a pathology due to frequency aliasing. Along with the zero-frequency components, components with frequencies near integer multiples of T^{-1} are also attenuated. This renders moving targets with speeds near certain "blind speeds" invisible to the radar. The same problem exists for ordinary spectral analysis, where targets with Doppler shifts that alias to low apparent frequencies could easily be mistaken for ground clutter.

The techniques mentioned earlier for frequency aliasing mitigation also apply to the MTI problem. One popular strategy involves using slightly staggered IPPs to break the symmetry that underlies frequency aliasing. If the IPPs are transmitted with a small jitter, signals with broad ranges of nonzero Doppler shifts will be able to pass through the MTI filter with little attenuation. This can be seen by reformulating the multi-stage MTI filter response function in a more general way:

$$|H(\omega)| = \left| \sum_{n=0}^{N} C_n \exp(-j\omega t_n) \right|.$$

For the two-stage filter problem considered above, we had $N = 2$, $C_n = \{1, -2, 1\}$ and $t_n = \{0, T, 2T\}$. For illustrative purposes, let us modify the pulsing scheme and associated filter by making $t_1 = aT$. The new filter response function then becomes

$$|H(\omega)|^2 = \left[(1 - 2\cos a\omega T + \cos 2\omega T)^2 + (-2\sin a\omega T + \sin 2\omega T)^2 \right]. \quad (7.1)$$

The filter still has a null at zero frequency but does not in general have nulls (blind frequencies) at integer multiples of $1/T$, depending on the value of a. With some experimentation, it can be shown that, for values of a of the form $a = (m-1)/m$, where m is an integer not equal to 3, blind frequencies appear at integer multiples of $f_{\text{blind}} = m/T$. (If $m = 3$, $f_{\text{blind}} = 3/2T$.) In the example shown in Figure 7.13, we have set $a = 3/4$ giving $f_{\text{blind}} = 4/T$.

IPP staggering produces a filter shape with regular maxima and minima along with some degradation in the vicinity of the main nulls. This can be corrected to some degree by adjusting the weights C_n above. Even so, the filter shape may prove problematic when it comes to interpreting and analyzing the final Doppler spectrum, which will tend to exhibit spurious ripples. This is the main reason that staggered pulses are infrequently used in general-purpose radar applications. The situation is more tolerable in simple MTI systems where the purpose may be merely to detect the presence or absence of a moving target. Literature discussing tradeoffs in this area is extensive, but advances in signal processing and radar hardware may have rendered the technique largely obsolete.

7.11 Notes and Further Reading

The first appearance of range-time-intensity diagrams may have been in D. T. Farley's 1969 article on radar correlation-function measurement, and many of the issues and techniques introduced in this chapter were clearly described in Farley's article on signal processing for MST (mesosphere-stratosphere-troposphere) radars. A comprehensive text on radar signal processing with an emphasis on stochastic processes can be found in the text by V. Tazlukov. Additional historical and contemporary references on signal-processing fundamentals appear in the reference list below.

An approachable primer to planetary radar was published in the early days by I. I. Shapiro, and a review of the current capabilities of the planetary radar system at Arecibo with an up-to-date reference list was presented recently by D. B. Campbell et al. MTI radars are discussed in most textbooks on radars including N. L. Levanon's.

B. D. O. Anderson and J. B. Moore. *Optimal Filtering*. Springer-Verlag, New York, 1979.

R. B. Blackman and J. W. Tukey. *The Measurement of Power Spectra*. Dover Press, New York, 1958.

J. W. Cooley and J. W. Tukey. An algorithm for the machine computation of complex Fourier series. *Math. Comput.*, 19:297–301, 1965.

D. B. Campbell, J. K. Harmon and M. C. Nolan. Solar system studies with the Arecibo planetary radar system. General Assembly and Scientific Symposium, 2011 XXXth URSI, August 13–20, 2011. doi: 10.1109/URSIGASS.2011.6051217.

D. T. Farley. Incoherent scatter correlation function measurements. *Radio Sci.*, 4:935–953, 1969.

On-line data processing for MST radars. *Radio Sci.*, 20:1177–1184, 1985.

F. J. Harris. On the use of windows for harmonic analysis with the discrete Fourier transform. *Proc. IEEE*, 66:51, 1978.

L. H. Koopmans. *The Spectral Analysis of Time Series*. Academic Press, New York, 1974.

N. L. Levanon. *Radar Principles*. Wiley, New York, 1988.

A. Papoulis. *Probability, Random Variables, and Stochastic Processes*. McGraw-Hill, New York, 1984.

Jr. P. Z. Peebles, Jr. *Radar Principles*. Wiley, New York, 1998.

C. M. Rader. An improved algorithm for high-speed autocorrelation with application to spectral estimation. *IEEE T. Acoust. Speech*, AU-18(4):439–441, 1970.

M. A. Richards. *Fundamentals of Radar Signal Processing*. McGraw-Hill, New York, 2005.

I. I. Shapiro. Planetary radar astronomy. *IEEE Spectrum*, March:70–79, 1966.

V. Tazlukov. *Signal Processing in Radar Systems*. CRC Press, New York, 2013.

P. M. Woodward. *Probability and Information Theory, with Applications to Radar*, 2nd ed. Pergamon Press, Oxford, 1964.

Exercises

7.1 The text alluded to (in the context of target detection, false alarms, and missed detections) the fact that, if the probability density functions (PDFs) for raw receiver voltages $V = x + jy$ are Gaussian, the PDF for the magnitude of those voltages $|V|$ is something different.

Consider the output of a quadrature receiver receiving noise only. Let the PDFs of the in-phase and quadrature components of the output voltages have Gaussian distributions so that

Exercises

$$P(x)dx = \frac{1}{\sqrt{2\pi}\sigma} e^{-\frac{x^2}{2\sigma^2}} dx$$

$$P(y)dy = \frac{1}{\sqrt{2\pi}\sigma} e^{-\frac{y^2}{2\sigma^2}} dy.$$

Since the two components are statistically independent, $P(x,y) = P(x)P(y)$, and so the probability of producing a voltage in a given dx, dy interval is $P(x,y)dxdy = P(x)P(y)dxdy$.

Now, we can change variables from x and y to $|V|$ and ϕ where $|V|^2 = x^2 + y^2$ and $\phi = \text{atan}(y/x)$. Converting everything to polar coordinates and recalling that $dxdy = |V|d|V|d\phi$, we can rewrite the PDF for the receiver output as

$$P(|V|,\phi)d|V|d\phi = \frac{1}{2\pi\sigma^2} e^{-\frac{|V|^2}{2\sigma^2}} |V|d|V|d\phi.$$

Integrating over all phase angles ϕ then, we can write the PDF for V as

$$P(|V|) = \frac{|V|}{\sigma^2} e^{-\frac{|V|^2}{2\sigma^2}},$$

which is a Rayleigh distribution.

Now the question: suppose that, in addition to noise, a deterministic signal is also present that makes a constant added contribution to the output voltage of the form $V_\circ = a + ib$, with $V_\circ^2 = a^2 + b^2$ and $\phi_\circ = \text{atan}(b/a)$. Find the resulting PDF $P(|V|)$. Sketch it for the cases where the deterministic signal is much weaker than and much stronger than the noise.

Hint: $\int_0^{2\pi} \exp(c\cos(\phi - \phi_\circ)) = 2\pi I_\circ(c)$ where $I_\circ$ is the modified Bessel function of the first kind of order zero, $I_\circ(0) = 1$, and $I_\circ(x) \sim \exp(x)/\sqrt{2\pi x}$ for large x. The result you find is called a Rician distribution.

7.2 A 300 MHz radar transmits 20 μs pulses, giving a range resolution of 3 km, toward a target 75 km away. The IPP is 2 ms, and we assume that range aliasing is not occurring. The radar cross-section of the target is such that the SNR for each individual pulse is -20 dB. The target is approaching the radar at 40 m/s, and the Doppler spread caused by the target tumbling is about 2 Hz. Design a data processing scheme, consisting of coherent FFT processing (discrete Fourier transform of 2^n samples, giving 2^n frequency bins) followed by incoherent averaging of the resulting power spectra, that will give the best output SNR. In other words, what is the largest useful value of n, and how many spectra can you average incoherently? What is the resulting output SNR, roughly? The idea is to make the FFT such that all the signal is more or less in one frequency bin, giving the best SNR in that bin, and then to incoherently average as many of these spectra as possible. How many can you average, given that you want the target to remain in one range gate during the averaging?

7.3 Range aliasing. A radar searching for aircraft transmits a long series of pulses with an IPP of 500 μs. An echo is observed at delay of 300 μs. The radar operator suspects that the target may actually be range aliased, however, and checks this by changing the IPP to 400 μs. The echo delay changes to 200 μs. What is the smallest possible target range?

7.4 Frequency aliasing. Suppose that, in the above example, the observed Doppler shift of the target also changed from +400 Hz for the original IPP to +900 Hz for the second IPP. What is the smallest possible (in absolute value) true Doppler shift consistent with these observations?

8
Pulse Compression

In this chapter, we discuss strategies for modulating radar pulses to strike an optimal balance between range resolution and sensitivity while fully utilizing a transmitter's duty cycle capabilities. Most high-power pulsed transmitters are peak-power rather than average-power limited and can generate longer pulses than might be desired, in view of the fact that range resolution degrades with pulse length. (Recall that the width of a range bin is $c\tau/2$.) As things stand, this can lead to very inefficient radar modes. For example, MW-class vacuum tube transmitters typically have maximum duty cycles of about 5 percent. Suppose you wanted to use such a transmitter to observe a target with 150 m range resolution with an IPP of 10 ms. (We suppose that this IPP avoids range and frequency aliasing.) The required pulse length, $1\mu s$, is only 0.01 percent of the IPP, meaning that the average power of the transmitter is drastically underutilized, and sensitivity is compromised.

The key idea is that the fine range resolution of the short pulse stems from its broad bandwidth and the fact that its matched filter output (a simple triangle) is narrow. It's the width of a pulse's matched filter output, or its autocorrelation function, that determines range resolution. However, a pulse need not be short for its autocorrelation function to be narrow. A suitably modulated long pulse can have a broad bandwidth and a narrow autocorrelation function while still utilizing a transmitter's full duty cycle. The modulation is called pulse compression since it makes a long pulse behave like a shorter pulse with higher peak power. In effect, modulated pulses contain more information than unmodulated ones, and it is from this information that finer range resolution can be obtained. Pulse compression is related to spread spectrum communications and to the pulse coding schemes used in GPS.

There are two main classes of pulse compression: phase coding and frequency chirping. The former (latter) is most often implemented digitally (with analog components). The two are equivalent, and implementation is a matter of expediency.

8.1 Phase Coding

Phase coding is usually implemented with binary phases, where the phase is either 0° or 180°, although quadriphase codes (0°, 90°, 180°, 270°) and other polyphase codes are sometimes used. (There are small advantages to the latter in practice, although the execution is more complicated.) The code sequences can be random or preset. An important

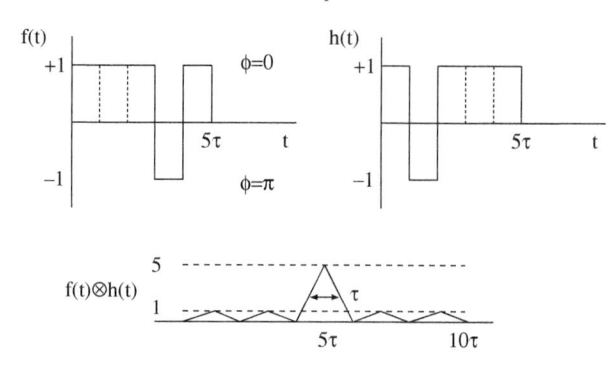

Figure 8.1 Example of a 5-bit binary phase code $f(t)$. This particular example belongs to the class of Barker codes. Also shown is the matched filter impulse response function, $h(t)$ and the convolution of the two, $f(t) \otimes h(t)$.

class of codes are Barker codes, of which an example is shown in Figure 8.1. The bit or "chip" width in this example is τ, and the pulse width is 5τ or 5 bits. Because of the phase modulation, the bandwidth of the coded pulse is approximately τ^{-1} instead of $(5\tau)^{-1}$. The duty cycle is proportional to 5τ nonetheless.

Also shown in Figure 8.1 is the impulse response function $h(t)$ of a filter matched to $f(t)$. Such a filter could be implemented with analog circuitry using delay lines or, more directly, with a digital filter. Finally, Figure 8.1 shows the convolution of $f(t)$ and $h(t)$, which is the matched filter output. This is equivalent to the autocorrelation function (ACF) of $f(t)$. Notice that the main lobe of the ACF has a width $\sim \tau$ rather than 5τ. In this sense, the coded pulse performs like an uncoded pulse of length τ while more fully utilizing the transmitter duty cycle. The main lobe of the ACF is associated with ideal pulse compression. In this case, compression is not ideal, and the ACF has four sidelobes. These are undesirable and represent an artificial response (range sidelobes) to a point target.

What distinguishes Barker codes like this one is that the sidelobe level is at most unity everywhere. The main lobe, meanwhile, has a voltage amplitude equal to the number of bits in the code (n). The bandwidth of the coded pulse is again approximately τ^{-1} rather than $(n\tau)^{-1}$. Barker codes can be found for code sequences with $n = 2, 3, 4, 5, 7, 11$ and 13 only. The Barker codes are given in Table 8.1. Many useful codes with more bits can be found, but none have sidelobes limited to unity.

An expedient method for calculating the matched filter response to a binary coded pulse with a range-time diagram is illustrated by Figure 8.2. The figure graphically represents filtered data associated with a single range of interest, signified here by the broken horizontal line. The $+/-$ signs along the bottom of the figure represent the transmitted pulse code (lower left) and the time-reversed impulse response function (lower right). The two sequences are of course identical. The intersections of the upgoing and downgoing characteristics are populated with the products of these signs. Summing across rows yields the impulse response, shown here with a main lobe and some sidelobes. We can interpret the result in two ways, which are really equivalent. It shows at once (1) the response to echoes

8.1 Phase Coding

Table 8.1 *The Barker codes.*

Length	Code
2	++, +−
3	++−
4	+++−, ++−+
5	+++−+
7	+++−−+−
11	+++−−−+−−+−
13	+++++−−++−+−+

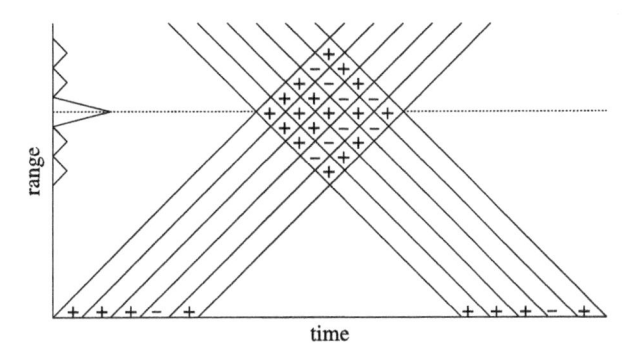

Figure 8.2 Range-time diagram calculation of matched filter response to a coded pulse.

from the intended range as well as from others nearby and (2) the response nearby range gates will have to echoes from this range. The sidelobes are thus an avenue of power leakage between adjacent range gates. We call this leakage *radar clutter*. In this case, since the sidelobes are reduced by a factor of 5 from the main lobe, the clutter power will be reduced by a factor of 25, or about 14 dB.

Pulse compression increases the signal-to-noise ratio of the echoes without degrading range resolution. We say that an n-bit code of length $n\tau$ has a compression ratio of n compared to an uncoded pulse of length τ. Assuming that the echoes remain phase coherent throughout the duration of the pulse, pulse compression increases the signal voltage by a factor of n and the signal power by a factor of n^2. Noise voltage meanwhile grows by a factor of $\sqrt{n}$ (random walk), and noise power by a factor of n. The process is like coherent integration, only the summing is taking place along the rows rather than the columns of the data matrix this time. Ultimately, the signal-to-noise ratio grows by a factor of n at the expense of added clutter. The longest Barker code has a compression ratio of 13 and a sidelobe level of $(1/13)^2$ or -22.3 dB. This may or may not be acceptable, depending on the application.

Besides Barker codes, a number of other binary phase codes with low sidelobe levels are sometimes used. Long codes can be constructed by nesting the Barker codes, for example,

Table 8.2 *Maximum-length codes for sidelobe levels between 1–5.*

Bits	SLL	Code (hex)
13	1	1F35
28	2	DA44478
51	3	71C077376ADB4
69	4	1D9024F657C5EE71EA
88	5	9076589AF5702502CE2CE2

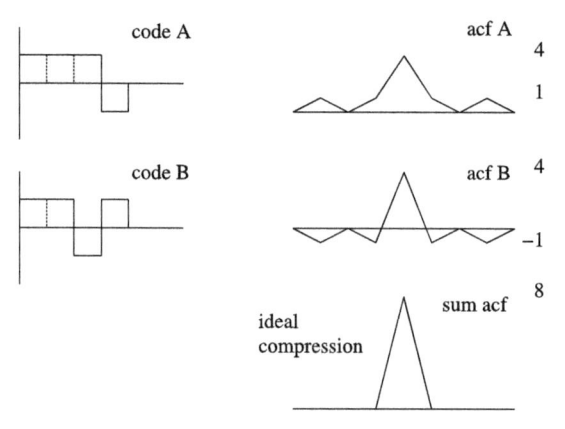

Figure 8.3 Complementary code pair leading to ideal compression after summation. A key requirement here is that the correlation time of the scatterer be longer than the interpulse period.

and others have been found through numerical searches or by exploiting number theory (see Table 8.2). There is a 28-bit code with a sidelobe level of 2 ($(2/28)^2 = -22.9$ dB), a 51-bit code with a sidelobe level of 3 (-24.6 dB) and a 69-bit code with a sidelobe level of 4 (-24.7 dB). An 88-bit code used at Arecibo for ionospheric research has a sidelobe level of 5 (-24.9 dB). Even lower sidelobe levels are possible when signals from different pulses are combined coherently. Some of the possibilities are described below.

8.1.1 Complementary (Golay) Codes

There are code sequences that have no sidelobes at all, which are referred to as "perfect codes." The most common examples of these are complementary codes. Complementary codes are code pairs used to modulate successive pulses. The resulting echoes are matched-filtered and summed across the pair via coherent integration prior to detection. The codes are complementary in that the sidelobes can be made to cancel identically. An example of a 4-bit pair is shown in Figure 8.3. After summing the matched-filter outputs of the two sequences, the main lobes add, but the sidelobes cancel.

8.1 Phase Coding

Whereas the codes discussed earlier could only be used when the correlation time of the echoes was greater than the pulse length, complementary codes can only be used when the correlation time is greater than the interpulse period. In applications where this highly restrictive constraint can be met, complementary coding approaches ideal pulse compression. Complementary codes are often used in radar studies of the lower and middle atmosphere, where the correlation times can be quite long compared to a typical interpulse period. They are impractical in applications involving rapidly moving or fading targets like vehicles or highly turbulent media.

Complementary codes come in lengths of 2^n and 10×2^n bits. A simple generating formula for long codes exists. If the sequences A, B make up a complementary code pair, so do the sequences $AB, A\overline{B}$. The shortest complementary code pair is 1, 1. Complementary codes can be arbitrarily long and are ultimately limited only by the maximum pulse length allowed by the transmitter.

8.1.2 Cyclic Codes

Sometimes called *linear recursive sequences, linear feedback shift register codes*, or pseudo-random noise (PRN) codes, these are pseudo-random sequences generated using shift registers with exclusive-or feedback. Such a device is shown in Figure 8.4. Upon each clock tick, the states stored in the registers shift one bit to the right, and the leftmost bit is set according to an exclusive-or operation on the previous state of the 6th and 7th registers. This represents feedback. The rightmost bit is the output. While a device with p registers has 2^p possible states, the state with all zeros is inaccessible since the output would be all zeros thereafter. Therefore, a p-stage device produces a non-repeating sequence up to $n = 2^p - 1$ in length. If the sequence produced is actually this long, the sequence is called a *maximal length* or m-sequence. Even longer non-repeating sequences are possible if the feedback configuration is changed from time to time. The sequence produced by the 7-stage device in Figure 8.4 is an m-sequence and repeats every 127 transitions.

Long repeating sequences of cyclic codes are processed by correlating them with a single, non-repeating sequence. (Formally, this is an example of unmatched filtering, since the impulse response function is matched to only a fraction of the emitted waveform.) A main lobe with a voltage amplitude n will occur in the correlation every n tics. In this way, the sequence repetition time serves as an effective interpulse period. Between the main lobes, sidelobes will appear with a uniform value of -1. The overall relative sidelobe power level is therefore $1/n^2$, as it is for a Barker code. That the sidelobes are uniform will make them less visible in radar imaging applications, for example.

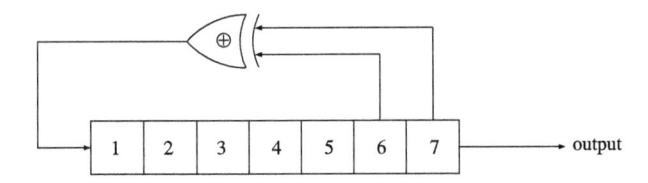

Figure 8.4 Seven-stage shift register configuration for cyclic code generation.

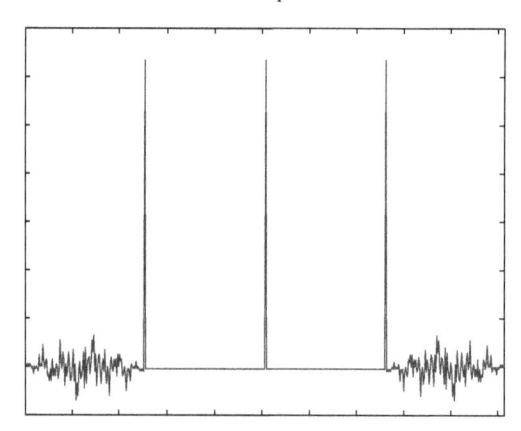

Figure 8.5 Correlation of the 127-bit cyclic code produced by the device in Figure 8.4 with a sequence made by repeating that same code three times. The main lobes have amplitudes of 127 and are separated by 127 state transitions. The sidelobes between them have a uniform amplitude of -1 except at the edges of the sequence.

What makes cyclic codes particularly attractive is the ease of generation of very long sequences using minimal extra hardware. The planetary radar studies at Arecibo discussed earlier utilized an $n = 8191$ length sequence. The compression ratio of 8191 made the Arecibo 1 MW peak-power transmitter perform as if it were an 8 GW transmitter! A 4 μs bit width afforded a range resolution of 600 m. The cyclic code was transmitted continuously in a 5 min. on / 5 min. off cycle. Since it repeats every $8191 \times 4\mu s =\sim 33$ ms, the transmitted waveform had an effective 33 ms interpulse period. Echoes were received in the intervening 5 min. intervals.

8.1.3 Gold Codes

Maximal-length or m-sequences codes were introduced in the discussion about shift register codes above. These sequences are also the building blocks for another set of codes called Gold codes (named after their inventor). Gold codes or sequences have bounded small mutual cross-correlations. They are useful in pulse-compression scenarios where multiple transmitters are operating simultaneously in the same frequency band. Notable examples include GPS and CMDA-based cellular communications. Gold codes could also be used gainfully in multistatic radar networks.

Gold sequences can be generated from pairs of m-sequence codes from different shift-register configurations with the same size p. For certain pairs, the codes derived from them have cross-correlations that take on one of three values: -1, $-t$, and $t - 2$, where

$$t = \begin{cases} 2^{(p+1)/2} + 1, & p \text{ odd} \\ 2^{(p+2)/2} + 1, & p \text{ even}. \end{cases}$$

Such pairs are called "preferred pairs."

From two distinct preferred m-sequence codes of length $n = 2^p - 1$ can be generated n new Gold sequences. Call the preferred m-sequence codes in the pair c_1 and c_2. Then the corresponding gold sequences are

$$c_1, c_2, c_1 \oplus c_2, c_1 \oplus Tc_2, c_1 \oplus T^2c_2, \cdots, c_1 + T^{n-1}c_2,$$

where $\oplus$ is the exclusive-or operator and where T^k is the cyclic bit-shift operator applied k times. The maximum correlations are smaller than those for two different ordinary m-sequences, and the improvement increases rapidly with increasing p.

8.1.4 Polyphase Codes

Phase codes do not have to be binary but can involve discrete phase jumps that are any integer factor of 360°. Quaternary phase codes (with 90° jumps) are one example. The sequence $(1, j, -1, -j)$ is analogous to a 4-bit Barker code in that the modulus of its sidelobes are limited to values of unity. Other advantageous sequences are known although the space of polyphase codes is not nearly so well explored as the space of binary phase codes.

Polyphase codes have some small advantages over binary phase codes. One is that they are less prone to a phenomenon known as straddle loss. When bit transitions in the received signal occur in between digital samples, matched-filtering of the individual bits causes the signal to be attenuated. Polyphase codes exhibit less straddle loss than binary phase codes. They also invite smaller losses in the event that the filter is not perfectly matched to the code.

In practice, the difficulty of generating accurate polyphase codes generally outweighs the benefits outlined above. Few operational systems employ polyphase codes in the manner described here, although the widespread availability of digital waveform synthesis may be altering the situation. Polyphase codes are used widely in digital implementations of frequency chirping applications as described below, however.

8.2 Frequency Chirping

Ultimately, it is the spectral broadening of the coded pulse that permits better range resolution than is possible with an uncoded pulse of the same duration. Another approach to broadening the bandwidth is to introduce a frequency variation or chirp using either analog or digital circuitry. Chirped waveforms are widely used in radar tracking applications.

The simplest form of frequency chirping is linear frequency modulation, or LFM, wherein the frequency varies linearly from f_1 to f_2 over the time interval $[t_1, t_2]$ of pulse transmission. We can express the transmitted waveform in this time interval as

$$s(t) = \cos[(\omega_\circ + \mu t)t]$$

$$\mu = \frac{d\omega}{dt} = 2\pi \frac{f_1 - f_2}{t_1 - t_2} = 2\pi \frac{\Delta f}{T}.$$

We will see later in this chapter that the matched filtered output power of such a pulse is given approximately by

$$P(t) \approx \left| \frac{\sin(\pi \Delta f t)}{\pi \Delta f t} \right|^2,$$

reminiscent of the radiation pattern of a rectangular aperture. Like the binary phase-coded pulses, the chirped pulse has an output with a main lobe and several range sidelobes (the largest down just 13.5 dB). The width of the main lobe in this case is approximately Δf^{-1}. We can therefore define a compression ratio as the ratio of the pulse width to the main lobe width:

$$c = \frac{T}{\Delta f^{-1}}$$
$$= (t_2 - t_1)(f_2 - f_1).$$

The compression ratio, which has the same meaning here as in the context of phase-coded pulses, is therefore just the time-bandwidth product of the chirped pulse. The longer the pulse and the greater the frequency swing, the larger the compression. Compression ratios of the order of 10,000 are possible using relatively inexpensive (albeit highly specialized) surface acoustic wave (SAW) devices.

For reasons we will soon discover, it turns out that the LFM chirp has a severe pathology that makes it unsuitable for applications where both the range and Doppler shift of the target are unknown. The problem arises from the regular rate of chirp. At any instant in time, the frequency of the received radar signal depends on the range delay of the target and also its Doppler shift in a manner that cannot be unambiguously distinguished. Displacing the target slightly in range or Doppler frequency has nearly the same effect on the received signal.

8.2.1 Frank Codes

Before discussing how to mitigate the pathologies associated with the LFM chirp, it should be noted that frequency modulation is often implemented with digital hardware using discrete phase and frequency jumps rather than continuous frequency chirps. Such codes can serve as a discrete approximation of the continuous LFM chirp or can be applied more generally to the pulse compression problem. An example of the former case are Frank codes. These are polyphase codes with quadratic phase shifts that specify the desired frequencies required to approximate LFM chirps.

Frank codes have lengths $n = l \times l$ that are perfect squares. Consider first the $l \times l$ Frank polyphase matrix F below:

$$F = \begin{pmatrix} 0 & 0 & 0 & \cdots & 0 \\ 1 & 2 & 3 & \cdots & l-1 \\ 2 & 4 & 6 & \cdots & 2(l-1) \\ 3 & 6 & 9 & \cdots & 3(l-1) \\ \vdots & \vdots & \vdots & \ddots & \vdots \\ (l-1) & 2(l-1) & 3(l-1) & \cdots & (l-1)^2 \end{pmatrix}. \tag{8.1}$$

Each row of the matrix represents phases corresponding to certain fixed frequencies. The frequencies of the rows are integer multiples of the fundamental frequency in increasing order. The phases to be transmitted are then found by reading out the n elements of the matrix from left to right, top to bottom, multiplying each by $2\pi/l$. The results is not so much frequency chirping as it is frequency hopping, as illustrated in the left panel in Figure 8.6. When l is sufficiently large and the code is sufficiently granular, its performance approaches that of LFM.

8.2.2 Costas Codes

We return now to the issue of ambiguity in LFM chirped pulses. Analogous ambiguity is present in the digital approximation of LFM implemented with Frank codes. Similar ambiguity arose in FMCW radar experiments but was resolved through the combination of upward and downward frequency swings. Another way to resolve the ambiguity involves changing the radar frequency in an irregular way rather than progressively. This is most readily implemented in the digital domain, as represented in Figure 8.6. By randomizing the frequency steps, the symmetry that lead to the ambiguity in the progressive case can be broken. An optimal frequency stepping code, called a Costas code, is one in which no time-frequency displacement appears more than once in a figure like Figure 8.6. A more comprehensive explanation of Costas codes will be given later in the chapter, after the mathematical machinery necessary to analyze them has been introduced.

Throughout most of this chapter, it was assumed that the Doppler shift of the target was sufficiently small that the phase of the received signal was constant during decoding. If not, pulse compression fails. This is easily seen in the case of binary phase codes. If the phase of the received signal changes by an appreciable fraction of π over the length of the pulse T, the coherent addition taking place in the main lobe of the filter response will become

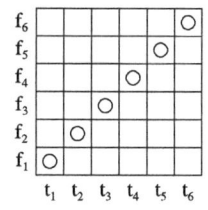

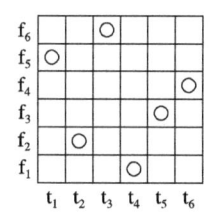

Figure 8.6 Illustration of frequency hopping. (left) Frank code with $n = 6$. (right) Costas code with $n = 6$.

coherent subtraction, and the response will be partially or almost totally destroyed. Sidelobe levels will also rise. Similar effects occur for chirped pulses. In order to understand the phenomena more fully and to appreciate the merits and pitfalls of phase codes and chirped pulses, we must study radar *ambiguity functions*.

8.3 Range-Doppler Ambiguity Functions

Ambiguity functions are used to isolate and analyze the performance of radar pulse pattern/filter pairings. They are analogous to impulse response functions or Green's functions in that they give the response of a radar system to a single mirror reflector. They provide the mathematical basis for describing phenomena heretofore discussed only qualitatively.

The definition of the range-Doppler ambiguity function springs directly from the fundamentals of radar signal matched filtering. Suppose that the transmitted waveform is

$$u(t)e^{-j\omega_o t},$$

where ω_o is the carrier frequency and $u(t)$ is the modulation envelope. The signal received by a mirror reflector will be, after removal of the carrier frequency, a time-delayed, Doppler-shifted copy,

$$r(t,\nu) = u(t-t_o)e^{j2\pi\nu t} + n(t),$$

where ν is the Doppler frequency, $t_o = 2R/c$ is the time delay, and $n(t)$ is the system noise. Take the filter to be matched to $u(t)$. Then the output of that filter is

$$m(\tau - t_o, \nu) = \int_{-\infty}^{\infty} r(t,\nu) h(\tau - t_o - t)\,dt$$
$$= \int_{-\infty}^{\infty} u(t-t_o) u^*(t-\tau) e^{j2\pi\nu t}\,dt + \text{noise term}.$$

Without loss of generality, we can take $t_o = 0$ and consider a target at zero range. Then,

$$m(\tau,\nu) = \chi(\tau,\nu) + \text{noise term} \tag{8.2}$$
$$\chi(\tau,\nu) \equiv \int_{-\infty}^{\infty} u(t) u^*(t-\tau) e^{j2\pi\nu t}\,dt, \tag{8.3}$$

which is a function of the Doppler shift ν and the delay time τ relative to the optimal detection time, which we have taken to be $t_o = 0$. The function $|\chi(\tau,\nu)|^2$ is called the radar ambiguity function, which can be used to analyze the performance of a pulse envelope/matched filter pair. By convention, the modulation envelope is usually normalized to have unity power, i.e., $\int_{-\infty}^{\infty} u(t) u^*(t)\,dt = 1$ so that $|\chi(0,0)|^2 = 1$.

The radar ambiguity function is governed by the following four theorems:

a) $|\chi(\tau,\nu)|^2 \leq 1$.
b) $\int_{-\infty}^{\infty} d\tau \int_{-\infty}^{\infty} d\nu\, |\chi(\tau,\nu)|^2 = 1$.

c) $|\chi(\tau,\nu)| = |\chi(-\tau,-\nu)|$.
d) If $u(t) \to \chi(\tau,\nu)$, then $u(t)e^{j\pi k t^2} \to \chi(\tau,\nu+k\tau)$.

Theorem a) is a consequence of the normalization condition and can be proved with the help of the Schwartz inequality. Define $f = u(t)\exp(j2\nu t)$ and $g = u^*(t-\tau)$. Then, according to (5.10), we have $|\chi(\tau,\nu)|^2 \leq (\int_{-\infty}^{\infty}|u(t)|^2 dt)^2$. The right side of the inequality is unity, in accordance with the normalization convention.

Theorem b) shows that the volume under the 2-D surface is invariant and only the shape of the ambiguity function can be changed. The proof is somewhat involved. Call the integral in question I. Then

$$I = \int_{-\infty}^{\infty} d\tau \int_{-\infty}^{\infty} d\nu \left| \int_{-\infty}^{\infty} \underbrace{u(t))u^*(t-\tau)}_{P(t,\tau)} e^{j2\pi\nu t} dt \right|^2$$

$$= \int_{-\infty}^{\infty} d\tau \int_{-\infty}^{\infty} d\nu \, |\tilde{P}(\nu,\tau)|^2,$$

where we have defined the function $P(t,\tau)$ and recognized its Fourier transform. According to Parseval's theorem, the integral of $\tilde{P}^2(\nu)$ over all frequencies is equal to the integral of $P^2(t)$ over all times, which can be viewed as an expression of energy conservation. Consequently,

$$I = \int_{-\infty}^{\infty} d\tau \int_{-\infty}^{\infty} dt \, |P(t,\tau)|^2$$

$$= \int_{-\infty}^{\infty} dt \int_{-\infty}^{\infty} d\tau \, |u(t)|^2 \, |u(t-\tau)|^2.$$

Finally, defining $t' = \tau - t$ and carrying out the appropriate change of variables proves the theorem

$$I = \int_{-\infty}^{\infty} dt \, |u(t)|^2 \int_{-\infty}^{\infty} dt' \, |u(t')|^2$$

$$= 1.$$

Theorem c) is a statement of odd symmetry. It can be proven by replacing ν and τ with their negatives in (8.3). Defining a new variable $t' = t+\tau$ and substituting then yields

$$\chi(-\tau,-\nu) = e^{-j2\pi\nu\tau} \int_{-\infty}^{\infty} u^*(t'-\tau) e^{-j2\pi\nu t'} u(t') dt'.$$

We can take the complex conjugate of χ without affecting its modulus, which is also unaffected by the phase term to the left of the integral. The variable t' is a dummy variable, and the theorem is thus proven.

Finally, theorem d) is a stretching property that will be useful for understanding linear FM chirps. Its proof is obvious, since the new quadratic phase term associated with the chirp can be incorporated into the existing phase term in (8.3) by redefining $\nu \to \nu + k\tau$.

The ambiguity function spells out the response of the radar to echoes with different Doppler frequencies at different times surrounding the nominal response time. Ideally, we desire a "thumbtack" response, highly peaked near the origin, small at small frequencies and delays, and zero further out. Peaks well away from the origin represent artifacts that could lead to misinterpretation. Various pathologies can be detected using ambiguity functions, which are therefore an important design tool.

As an example, consider a (properly normalized) lone rectangular pulse of width t_p and the associated ambiguity function. To start, define a rectangular function $R(t)$ with unit width

$$R(t) = \begin{cases} 1, & |t| \leq \frac{1}{2} \\ 0, & |t| > \frac{1}{2} \end{cases}.$$

Using this notation, the pulse envelope becomes

$$u(t) = \frac{1}{\sqrt{t_p}} R\left(\frac{t}{t_p}\right)$$

$$\chi(\tau, \nu) = \frac{1}{t_p} \int_{-t_p/2+\tau}^{t_p/2} e^{j2\pi\nu t} dt, \tau \geq 0$$

$$= \frac{1}{t_p} \int_{-t_p/2}^{t_p/2+\tau} e^{j2\pi\nu t} dt, \tau < 0.$$

The integrals are elementary and lead to the result

$$|\chi(\tau, \nu)| = \begin{cases} \left| \left(1 - \frac{|\tau|}{t_p}\right) \frac{\sin\left[\pi t_p \left(1 - \frac{|\tau|}{t_p}\right)\nu\right]}{\pi t_p \left(1 - \frac{|\tau|}{t_p}\right)\nu} \right|, & |\tau| \leq t_p \\ 0, & |\tau| > t_p \end{cases}.$$

Plots of the ambiguity function along with illustrative cuts through it are shown in the first column of Figure 8.7. The $\nu = 0$ and $\tau = 0$ cuts are the familiar functions

$$\chi(\tau, 0) = 1 - \frac{|\tau|}{t_p}, |\tau| \leq t_p$$

$$\chi(0, \nu) = \left| \frac{\sin \pi t_p \nu}{\pi t_p \nu} \right|.$$

The first is just the matched-filter response to a square pulse of duration t_p, which we calculated earlier assuming no Doppler shift. The ambiguity function reveals the time interval over which the echo will be spread. The second function, a sinc function, gives the frequency response of the matched filter. The bandwidth of the filter is given approximately by t_p^{-1}. Echoes with Doppler shifts within the bandwidth will be detected, along with noise.

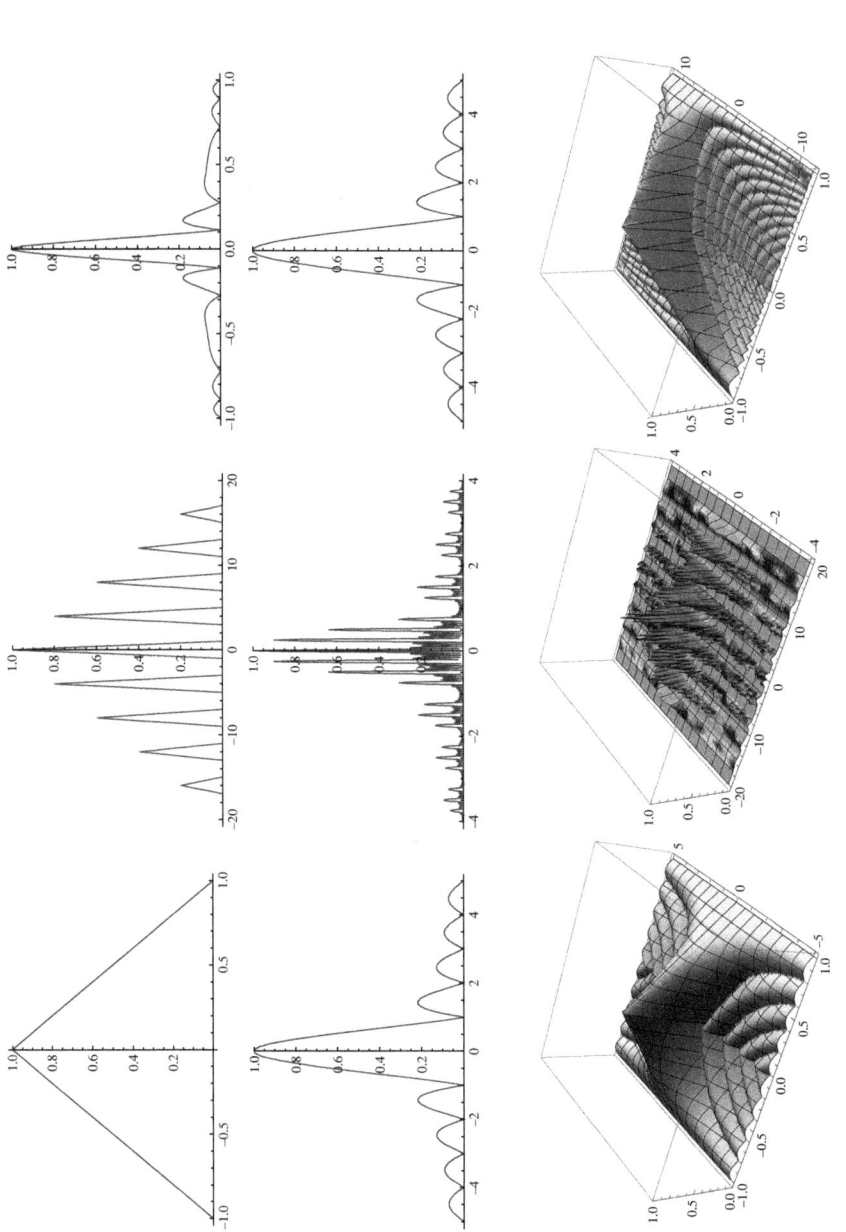

Figure 8.7 Radar ambiguity functions. The first, second and third columns apply to a square pulse, a train of square pulses and a linear FM chirp pulse, respectively. The top, middle and bottom rows represent the time axis cut, frequency axis cut and full 2-D ambiguity functions, respectively.

The full 2-D ambiguity function extends these concepts, showing the temporal spread of echoes with different Doppler shifts or, equivalently, the frequency response of the filter at different times.

It should be reiterated that the frequency response discussed above has little to do with the Doppler spectrum recovered from pulse-to-pulse data processing, for reasons discussed at the end of the previous chapter. To analyze pulse-to-pulse processing, one needs to specify a train of pulses with $u(t)$ rather than a single pulse. This problem, too, has been investigated in the second column of Figure 8.7. Implementing a pulse train turns the ambiguity function into a "bed of nails," as shown. The various peaks are all quite narrow, depending on the length of the pulse train, implying that both the range and Doppler shift of the echo can be known rather precisely. The multiplicity of peaks corresponds to range and frequency aliasing. There is therefore still ambiguity in range and frequency, but it is quantized. Using an irregular pulse train would destroy many of the peaks and reduce the ambiguity, but this topic is beyond the scope of the current discussion. Note that the ambiguity function shown here does not take into account data processing with a DFT and merely reflects the results for the zero Doppler bin.

A final illustrative example is the linear FM chirp waveform. Here, we take the envelope to be

$$u(t) = \frac{1}{\sqrt{t_p}} R\left(\frac{t}{t_p}\right) e^{j\pi k t^2},$$

so that the instantaneous frequency is kt Hz, which is bounded by $[\pm(1/2)kt_p]$. Applying the last rule of ambiguity functions shows that

$$|\chi(\tau, \nu)| = \begin{cases} \left|\left(1 - \frac{|\tau|}{t_p}\right) \frac{\sin\left[\pi t_p \left(1 - \frac{|\tau|}{t_p}\right)(\nu + k\tau)\right]}{\pi t_p \left(1 - \frac{|\tau|}{t_p}\right)(\nu + k\tau)}\right|, & |\tau| \leq t_p. \\ 0, & |\tau| > t_p \end{cases}$$

The third column of Figure 8.7 shows this function along with the $\nu = 0$ and $\tau = 0$ cuts. The latter of these is unchanged from the rectangular pulse example, which is to say that the matched filter frequency response through the main vertical of the ambiguity function is unchanged. The former now looks like

$$\chi(\tau, 0) = \left|\left(1 - \frac{|\tau|}{t_p}\right) \frac{\sin\left[\pi \Delta f \tau \left(1 - \frac{|\tau|}{t_p}\right)\right]}{\pi \Delta f \tau \left(1 - \frac{|\tau|}{t_p}\right)}\right|, \quad |\tau| \leq t_p,$$

where the frequency chirp $\Delta f = kt_p$ has been defined. We find that the triangular impulse response function has been modified by the multiplication of a sinc function with an argument that first increases and then decreases with lag τ. The first null of the sinc function occurs where $\tau \approx \Delta f^{-1} = t_p/c$, where c is the compression ratio. This is now the effective width of the impulse response, which was formerly just t_p. As with phase coding, range sidelobes (at the ~ -14 dB level and below) accompany frequency chirping.

8.3 Range-Doppler Ambiguity Functions

Table 8.3 *Difference table for digital FM code sequence shown on the right side of Figure 8.6.*

a_j	5	2	6	1	3	4
$i=1$	−3	4	−5	2	1	
2	1	−1	−3	3		
3	−4	1	−2			
4	−2	2				
5	−1					

The full 2-D ambiguity function shows that the narrowing of the impulse response has come about as a result of the shearing of the overall function. While this is fine for the zero Doppler frequency case, it introduces a bias in cases with finite Doppler shifts. Note that the function maximum falls on the line $\tau_{max} = -v/k$. Given finite Doppler shifts, the maximum occurs here rather than at $\tau = 0$. The associated bias would lead to misleading results if it were not somehow corrected. The effect can be compensated for once the Doppler shift of the echo has been ascertained. Even better, it can be avoided with the use of Costas codes, for which the ambiguity function is more like the optimal thumbtack.

8.3.1 Costas Codes Revisited

Having studied the ambiguity function for the LFM chirp pulse, we are in a better position to understand its pitfalls and also the benefits of frequency hopping and digital FM code sequences like the one shown on the right side of Figure 8.6. Some digital FM codes, like some binary phase codes, are better than others. What criteria should be applied in the search for an optimal sequence?

In FM radar, a Doppler-shifted echo from some range will always masquerade as an unshifted echo from another range. The range error is proportional to the offset in time that shifts the transmit frequency just so as to account for the Doppler shift. With a linear FM chirp, that offset in time is always the same throughout the entire transmission interval. The range error is therefore reinforced. The reinforcement is evident in the left side of Figure 8.6, where the same time-frequency jumps are repeated again and again throughout the transmitted waveform. The LFM chirp is, in fact, the worst digital FM sequence possible in this family!

A digital FM code sequence will only generate an ambiguity function with a "thumbtack" response if there are no repeated time-frequency jumps. There's a simple, graphical way to check whether this is true. Consider the example shown in the right side of Figure 8.6. First, we can represent the entire sequence as $\{a_j\} = 5,2,6,1,3,4$. Next, we can construct a so-called "difference table" from this vector (see table above). Each number in the table is $d_{i,j} = a_{i+j} - a_j$. The first "−3" in the first row says that there is a pair of code elements spaced by −3 frequency steps and one time step. The "1" in the next row says that there is a pair of elements spaced by 1 frequency step and two time steps. The code is an optimal (Costas) code if no number occurs twice on the same row of the difference table. This example meets the requirement.

Upon reception, signals that differ in frequency by more than the reciprocal of the length of the signal are statistically uncorrelated. This condition generally applies to the "bits" in a digital FM code sequence. The only way signals emitted at different frequencies will give rise to correlated echoes is if the Doppler shift brings them into congruence. Roughly speaking, the ambiguity function of a digital FM sequence can therefore be estimated by calculating the 2-D autocorrelation function of the sequence represented graphically in the manner of Figure 8.6. The tilted ridge in the FM chirp ambiguity function in the right column of Figure 8.7 is a direct consequence of the line of dots in the left panel of Figure 8.6. An ambiguity function with a thumbtack response follows, meanwhile, from the autocorrelation of the Costas code on the right panel of Figure 8.6.

8.3.2 Volume Scatter Revisited

The last section in Chapter six concerned the volume scattering cross-section σ_v and a calculation of the power received from a scattering volume full of soft targets. The scattering volume was shown to be defined by the transmitting antenna gain, which controls the angular size of the illuminated region, and the antenna effective area, which determines the sensitivity of the radar to scatter from different directions. The two functions were combined into a single parameter, the backscatter gain/area, which encompasses both effects and specifies the solid angle of the scattering volume $\Omega_{bs} = 4\pi/G_{bs} = \lambda^2/A_{bs}$.

In the aforementioned calculation, the depth of the scattering volume illuminated by a rectangular pulse was given in terms of the radar pulse width $\Delta r = ct_p/2$. This is a crude approximation, which can be improved upon by incorporating the ambiguity function. The ambiguity function specifies the radar response to targets at different ranges and so defines the actual depth of the scattering volume. That depth depends on both the transmitted pulse width and the receiving impulse response function in a manner analogous to the way the backscatter gain depends on both the antenna gain and antenna effective area. There is no new mathematics to be performed here, since all the pieces of the calculation have already been worked out.

Equation (6.8) predicted the received power for soft-target (volume-filling) backscatter as fraction of the transmitted power:

$$\frac{P_{rx}}{P_{tx}} = \frac{\sigma_v}{4\pi r^2} \frac{ct_p}{2} A_{bs}. \tag{8.4}$$

Neither the range nor the frequency dependence of the volume-scattering cross-section σ_v was considered. Generalizing this result, $ct_p/2$ should be replaced by an integral involving the square of the ambiguity function, which gives the output power arising from a target at a given range and Doppler frequency:

$$\frac{P_{rx}(t)}{P_{tx}} = \frac{A_{bs}}{4\pi r^2} \int dr \int d\nu\, \sigma_\nu(r,\nu) |\chi(t-2r/c,\nu)|^2, \tag{8.5}$$

where $t - 2r/c$ is the time delay relative to the anticipated arrival time for echoes from range r and where the contributions of echoes from all possible Doppler frequencies are considered. In writing (8.5), it is assumed that the ambiguity function is sufficiently narrow in the delay dimension for the r^2 term in the denominator to be treated as a constant and left outside the integral. The volume-scattering cross-section σ_ν written here has dimensions per unit length per unit frequency.

For example, consider a volume-filling target that is homogeneous in range and has a narrow spectrum such that $\sigma_\nu(r,\nu)$ can be approximated by $\sigma_\nu \delta(\nu)$ in (8.5). Using a rectangular pulse and ideal matched filtering, (8.5) becomes in this case

$$\frac{P_{rx}}{P_{tx}} = \frac{\sigma_\nu}{4\pi r^2} \frac{ct_p}{3} A_{bs}. \tag{8.6}$$

The result resembles (8.4), only multiplied by a factor of $L_r = 2/3$ (or reduced by about 1.76 dB). This can be regarded as a loss factor arising from the incomplete use of echoes from different ranges in the scattering volume. In the event that a filter with a much broader bandwidth than the matched filter were used, the factor would approach unity at the expense of increased noise and degraded SNR. The fact that the backscatter area is less than the antenna effective area can be thought of as an analogous loss factor having to do with the incomplete use of echoes coming from across the radar beam.

8.4 Passive Radar

Passive radar refers to utilizing existing radiation sources (sources of opportunity) rather than operating dedicated radar transmitters. The sources in question are typically broadcast radio and television stations, although they could equally well include cellular towers and existing radar installations. Passive radar is sometimes referred to as *stealth radar* because it can be conducted unobtrusively. This terminology also is used in conjunction with spread spectrum radar, which is a related but distinct topic.

In passive radar, we rely on the characteristics of the source of opportunity to provide pulse compression. Some sources are more appropriate than others. In order to understand what constitutes an appropriate radiation source, we consider a simple configuration and an intuitive signal processing strategy. The same strategy could be applied to signals collected by traditional radar installations. The analysis that follows can therefore be applied to almost any kind or radar experiment or mode.

A passive radar configuration is illustrated in Figure 8.8. A target is equidistant from a ratio transmitter and a radar receiver, which are otherwise isolated by some form of obstruction (a mountain). This prevents the receiver from being overwhelmed by signals coming directly from the transmitter. Interpreting the scattered data requires complete knowledge of the transmitted waveform, which therefore must be sampled and forwarded

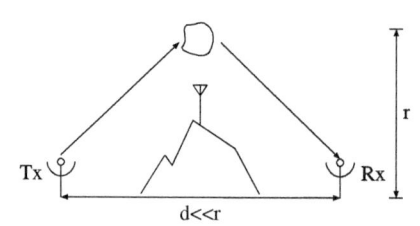

Figure 8.8 Geometry for a particular kind of passive radar experiment. The distance between the transmitter and receiver is assumed here to be small compared to the range to the target.

by an intermediate receiver, shown here at the top of the mountain. The signal received by the receiver ($y(t)$) is related to the transmitted waveform ($x(t)$) by

$$y(t) = \int_0^\infty x(t - 2r/c)\sigma(r, t - r/c) dr, \qquad (8.7)$$

where $\sigma(r,t)$ is the scattering amplitude of the target, including the appropriate distance scaling. This function is a random variable, and meaningful information about it is contained in its autocorrelation function,

$$R_\sigma(r, \tau) \equiv \langle \sigma(r,t)\sigma^*(r, t - \tau) \rangle.$$

It is this autocorrelation function, or its Fourier transform (the Doppler spectrum), that we seek to extract from the data. Consider the estimator

$$\hat{Q}(r, \tau) \equiv \frac{1}{T} \int_T y(t) x^*(t - 2r/c) y^*(t - \tau) x(t - 2r/c - \tau) dt,$$

where T is the incoherent integration interval. We can evaluate the meaning of the estimator by substituting $y(t)$ from above. In so doing, we replace the dummy variable r with p and q in the two instances where $y(t)$ appears:

$$\hat{Q}(r,\tau) = \frac{1}{T} \int_T dp\, dq\, dt\, x_1(t - 2p/c) \sigma_1(p, t - p/c)$$
$$x_2^*(t - 2r/c) x_3^*(t - 2q/c - \tau) \sigma_2^*(q, t - q/c - \tau) x_4(t - 2r/c - \tau),$$

where subscripts have been added to help keep track of the various terms. This is obviously a very complicated expression. As in the case of pulse coding, however, we rely on the properties of the transmitted signal to simplify things, permitting the estimator to revert to something like the desired autocorrelation function. Symbolically, we have

$$\hat{Q} = \int dp\, dq\, \langle \sigma_1 \sigma_2^* \rangle \langle x_1 x_2^* x_3^* x_4 \rangle,$$

where we regard the time average as a proxy for the expectation and note that the scattering amplitude and transmitted signal are statistically uncorrelated. Concentrate first on the term involving the scattering amplitudes, which is itself nearly the desired ACF:

$$\langle \sigma_1 \sigma_2^* \rangle = R_\sigma(p,\tau)\delta(p-q).$$

The delta function arises from the assumption that the scatterers in different volumes (ranges) are statistically uncorrelated. This is a common assumption for volume-filling targets, but should probably be examined in the case of hard targets.

The remaining term can be handled with the fourth moment theorem for Gaussian random variables, which is how we regard the transmitted waveform (see Chapter 1):

$$\langle x_1 x_2^* x_3^* x_4 \rangle = \langle x_1 x_2^* \rangle \langle x_3^* x_4 \rangle + \langle x_1 x_3^* \rangle \langle x_2^* x_4 \rangle + \langle x_1 x_4 \rangle \langle x_3^* x_2^* \rangle$$
$$= R_x(2(r-p)/c)R_x^*(2(r-q)/c) + R_x(2(q-p)/c+\tau)R_x^*(\tau).$$

The last of the three terms on the first line is obviously zero, since the random phase terms involved do not cancel. The remaining two can be expressed in terms of the autocorrelation function of the transmitted waveform. Exactly how R_x evaluates will be different for different kinds of broadcast services. Some are better suited for passive radar than others. To see this, we substitute back into the equation for $\hat{Q}$ and integrate, making use of the Dirac delta:

$$\hat{Q}(r,\tau) = \int dp R_\sigma(p,\tau) \left(|R_x(2(r-p)/c)|^2 + |R_x(\tau)|^2 \right)$$
$$= \int dp R_\sigma(p,\tau) \underbrace{|R_x(2(r-p)/c)|^2}_{\sim \delta(r-p)} + \int dp R_\sigma(p,\tau) \underbrace{|R_x(\tau)|^2}_{\sim \delta(\tau)}.$$

A suitable broadcast signal is one with a singly, sharply peaked autocorrelation function, much as was the case for radar pulse codes. An example of a particularly bad signal is standard broadcast (NTSC) television, which has regular sync pulses that lead to a multiple-peaked autocorrelation function. (FM radio and HDTV are much better.) To the extent R_x can be approximated by a Dirac delta function, the estimator becomes

$$\hat{Q}(r,\tau) \approx R_\sigma(r,\tau) + \text{const}\,\delta(\tau),$$

where the term with the constant is an artifact that contaminates the estimator in the zero lag only. This really represents a kind of transmitter-induced clutter that has to be dealt with, in some way. The zero lag of the ACF translates to the noise floor of of the spectrum in the frequency domain. The artifact may be easy to distinguish, estimate and remove, leaving behind a faithful representation of the backscatter spectrum/ACF.

Passive radar analysts must still contend with range and Doppler aliasing. Although there is no interpulse period as such, the operator imposes an effective IPP with the choice of the length of the lags of the ACF that are actually computed. Since this can be decided after the data are collected, the analyst has considerable flexibility and should be able to strike an optimal balance.

The main difficulty posed by passive radar is neither clutter nor aliasing, but rather, practical implementation. Problems involved with data transport and processing can be daunting. Furthermore, the correlations involved in the estimate can be computationally expensive. Fast Fourier transforms can be incorporated to reduce the computational burden, as can be significant coherent integration. Coarse quantization and specialized hardware can be used to improve the performance of practical systems further.

8.5 Notes and Further Reading

Most all of the introductory radar texts cited in this book discuss range-Doppler ambiguity functions and pulse compression. A few notable references are given below. Original source material from R. H. Barker, M. J. E. Golay, J. P. Costas, R. L. Frank, and R. Gold is also cited below. The maximum-length codes (MLCs) discussed in this chapter were worked out by A. M. Kerdock et al. A recommended general text on coding theory is the one by R. E. Blahut.

The treatment of passive radar in this chapter followed that of J. D. Sahr. and F. D. Lind. Additional information about passive bistatic radars can be found in the book chapter by P. E. Howland et al.

R. H. Barker. Group synchronization of binary digital systems. In W. Jackson, editor, *Communication Theory*. Academic, New York, 1953.

R. E. Blahut. *Algebraic Codes for Data Transmission*, 2nd ed. Cambridge University Press, Cambridge, UK, 2003.

J. P. Costas. A study of a class of detection waveforms having nearly ideal range-Doppler ambiguity properties. *Proc. IEEE*, 72(8):996–1009, 1984.

R. L. Frank. Polyphase codes with good nonperiodic correlation properties. *IEEE Trans. Inform. Theory*, IT-9:43–45, 1963.

M. J. E. Golay. Notes on digital coding. *Proc. IRE*, 37:657, 1949.

Complementary series. *IRE Trans. Inform. Theory*, 7(2):82–87, 1961.

P. E. Howland, H. D. Griffiths and C. H. Baker. Passive bistatic radar systems. In M. Cherniakov, editor, *Bistatic Radar: Emerging Technology*. John Wiley & Sons, Chichester, UK, 2008.

A. M. Kerdock, R. Mayer, and D. Bass. Longest binary pulse compression codes with given peak sidelobe levels. *Proc. IEEE*, 74(2):366, 1986.

N. L. Levanon. *Radar Principles*. Wiley, New York, 1988.

F. E. Nathanson. *Radar Design Principles*. McGraw-Hill, New York, 1969.

J. D. Sahr and F. D. Lind. The Manashtash Ridge radar: a passive bistatic radar for upper atmospheric radio science. *Radio Sci.*, 32(6):2345–2358, 1997.

Exercises

8.1 Costas codes can be generated in a couple of ways. For example, consider the two prime numbers $n = 2$ and $m = 5$. Generate $m - 1$ numbers by evaluating $2^1 \% 5 = 2$,

$2^2\%5 = 4$, $2^3\%5 = 3$, and $2^4\%5 = 1$ (here, % represents the modulo operator, the remainder you get after dividing the first argument by the second). Together, they form the sequence (2,4,3,1), which can be represented in tabular form as shown below. This is a Costas code because its 2-D autocorrelation function, and consequently the range-time ambiguity function for a FM-coded waveform, has the form of a thumbtack.

$$
\begin{array}{c|cccc}
f_4 & & & & \circ \\
f_3 & & & \circ & \\
f_2 & \circ & & & \\
f_1 & & & & \circ \\
\hline
& t_1 & t_2 & t_3 & t_4
\end{array}
$$

For this problem, calculate a Costas code for the case $n = 3$ and $m = 7$. Plot the 2-D autocorrelation function for the code is such a way that it clearly has the form of a thumbtack.

8.2 In addition to binary phase codes, one can also transmit quadriphase codes that have much the same effect. Show that the code with the complex phase envelope $(1, j, 1, -j)$ will give a Barker code-like output from a matched filter receiver. Note that for complex signals, the impulse response function of a matched filter is the complex conjugate of the time reverse of the anticipated signal. Sketch the output of the matched filter for an echo from a target at some range R. Take the baud length of the code to be *tau* and assume it was transmitted at a time $t = 0$.

8.3 A receiver is matched to a *single* 5-baud binary Barker code $(+ + + - +)$.

   a) What will be the output of the filter if the input consists of contiguous ("touching") Barker codes, each of length 5, assuming no Doppler shift?

   b) Repeat a) for the case when the two codes are separated by exactly one baud length and the second is inverted in sign, i.e., the sequence is $(+ + + - + 0 - - - + -)$. The receiver decoder (filter) remains the same as for a). You should notice that some of the range sidelobes vanish for the second case. This sort of code provides two samples of echos from a given target separated by a very short time interval. This kind of information could be useful in probing overspread targets.

8.4 Complementary codes are pairs of binary phase codes that work together to produce no range sidelobes. The idea is to transmit the first pulse in the pair, wait for an IPP, and then transmit the second pulse. Matched filter decoding is performed separately for the two pulses, and then the results are added together prior to detection (squaring of the voltages). So long as the echoes are slowly fading and remain phase coherent for an entire IPP, the range sidelobes of the two pulses in the pair cancel identically.

   a) The shortest complementary code pair is $(1,1)$ followed by $(1,-1)$. Verify that this is indeed a complementary phase code pair.

Longer complementary codes can be assembled from the basic code in part (a). To form the first code in the pair, concatenate together both of the previous codes (i.e., get $(1,1,1,-1)$). To form the second code in the pair, take the first one you just calculated and reverse the signs of the bauds in the second half of the code (i.e., get $(1,1,-1,1)$). Repeat as necessary.

b) Find the complementary code with 8-baud pulses and verify that it produces no range sidelobes.

8.5 Write a program to implement the shift register code generator shown in Figure 8.4. Verify that it produces a code sequence 127 bits long before repeating. Construct a cyclic code from three such sequences concatenated end-to-end. Correlate this code with a single sequence and reproduce the result shown in Figure 8.5. Verify that the main peaks have amplitudes 127 times that of the sidelobes.

8.6 Barker codes can be combined to produce much longer codes. The idea is to modulate the sub-pulses of a Barker coded pulse with another Barker code. For example, if $B_4 = (++-+)$ and $B_5 = (+++-+)$, then $B_{54} = (+++-+,+++-+,---+-,+++-+)$, where the commas aren't breaks but are just intended to help illustrate what's going on. Note that the 20-bit code that results is not itself a Barker code.

a) Write down the sequence for B_{43}.
b) Calculate the matched filter output for this code.
c) What are the sidelobe level and compression ratio?

8.7 Due to its slow rotation rate, Venus is an underspread target (or nearly so) at the Arecibo S-band radar frequency of 2.38 GHz. This is not true for Mars, which rotates with a period of 1.03 Earth days. If we ignore echoes from portions of Mars sufficiently distant from the subradar point, however, we can make the planet effectively underspread. This means that only a narrow strip of the surface near the equator can be mapped, however.

a) Verify that Mars is overspread at 2.38 GHz. The radius of Mars is about 3400 km. Is Mars also overspread for the 430 MHz radar at Arecibo?
b) Find the size of the region of Mars that can be mapped without frequency aliasing.
c) Suppose that we use a long cyclic code with a repetition period (effective IPP) of $n = 2^p - 1$ bauds, where the baud length is 4 μs. What period (what integer value of p) should you choose to give the appropriate IPP? Ignore the rotation of Earth for this problem, even though it is not really negligible.

8.8 Consider the ambiguity function for the 5-bit Barker coded pulse $(+++-+)$.

a) With the help of a computer, generate representative plots of the ambiguity function $|\chi(\tau,v)|$. Plot the $\tau = 0$ and $v = 0$ cuts and also the entire ambiguity function (as a contour plot or in 3-D) over a representative range of delays and frequencies.

b) By examining the plots, show that the range resolution of the pulse is related to the baud length rather than the pulse length of the coded pulse. Show that the same is true of the bandwidth.
c) It should be clear from the contour or 3-D plot that the pulse code fails to give anything like ideal pulse compression for targets with substantial Doppler shifts. Indicate this breakdown on your plot.

9

Propagation

The discussion so far has neglected mention of how radar signals propagate to and from the target. In a homogeneous medium, radio wave propagation is by direct line of sight. The Earth's atmosphere and ionosphere constitute an inhomogeneous medium, however, and spatial variations in the index of refraction (as well as ground effects) must be considered. The curvature of the Earth limits the distance over which line-of-sight communication occurs, but this can be extended by diffraction, refraction, reflection and scattering in the atmosphere and ionosphere and by ground effects. The extension may be advantageous, as in the case of distant-object tracking, or may simply present additional range clutter, as in the case of aircraft or weather system tracking.

In the lower atmosphere, variations in the index of refraction n at commonly used radar frequencies arise from variations in the neutral temperature, pressure, water vapor content and liquid water content according to the formula

$$N \equiv (n-1) \times 10^6 \quad = \quad 77.6 \frac{P}{T} + 3.73 \times 10^5 \frac{P_w}{T^2} + 1.4W, \tag{9.1}$$

where N is called the refractivity, P is atmospheric pressure in mbar, T is the Kelvin temperature, P_w is the partial pressure of water vapor and W is liquid water content in grams per cubic meter. The first term on the right side of (9.1) is associated with the polarizability of air molecules, the second with permanent dipole moments of water vapor molecules, and the last with scattering from water droplets of some representative size. Except during extreme meteorological conditions, deviations in the index of refraction are generally limited to a few hundred parts per million. Nevertheless, it is possible for waves in the VHF and UHF bands to undergo sufficient refraction in the lower atmosphere and to propagate over paths significantly longer than line-of-sight propagation would allow. From a radar perspective, this can permit the detection of targets at much greater ranges than might otherwise be possible. This can lead to confusion, particularly if the echoes are range aliased.

In contrast, variations in the index of refraction that occur in the upper atmosphere and ionosphere can be substantial (up to 100 percent) at HF frequencies. This is what permits shortwave radio signals to propagate around the Earth. So-called "over the horizon" (OTH) radars operate at HF frequencies and utilize long propagation paths to track aircraft

Propagation 219

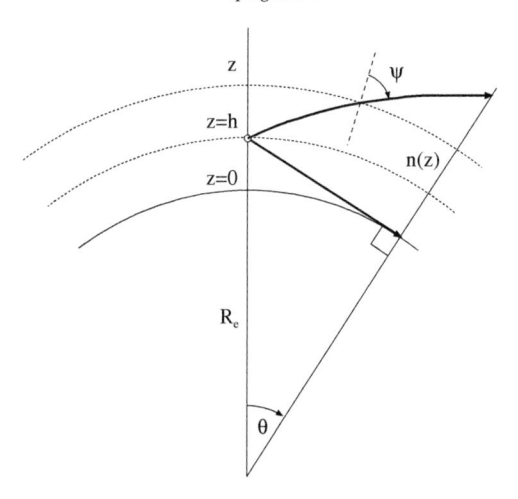

Figure 9.1 Radio wave propagation geometry. Two ray paths are shown: one along a straight path terminating at the horizon, and the other along a refracted path. The latter must have a turning point ($\psi = \pi/2$) if the ray is to return to the ground past the transmitter's horizon.

and ships at distances of thousands of kilometers. The propagation paths taken by OTH radar signals are complicated and highly variable. Predicting them requires large numerical models driven by routine and extensive ionospheric measurements from the ground and from space. The problem of inverting OTH radar signals is an active area of contemporary research.

We can calculate the greatest distance a radio wave can travel along a straight line before encountering the horizon with the aid of Figure 9.1. Here, R_e is the Earth's radius, and $h \ll R_e$ is the transmitter height above ground. Solving for the angle θ gives

$$\cos\theta = \frac{R_e}{R_e + h} = \frac{1}{1 + h/R_e}$$

$$1 - \frac{\theta^2}{2} \approx 1 - \frac{h}{R_e},$$

or $\theta \approx \sqrt{2h/R_e}$. The corresponding arc length on the ground is $R_e\theta = \sqrt{2R_e h}$. For example, radio signals launched from a $h = 100$ m tower will encounter the horizon at a distance of about 35 km. This is about the limit for FM radio and television signals, which propagate mainly by line of sight. The range can be extended by elevating the radio source further, as in the case of airborne radar (AWACS), which can be about 100 times higher and have ten times the range of ground-based radar installations. Space-borne radar can cover still greater range. The range of communications links can also be extended by incorporating high-altitude scatterers. Turbulent scatter in the troposphere and stratosphere and meteor scatter in the mesosphere and lower thermosphere have been utilized in this way. Communications satellites have rendered these techniques largely obsolete, however.

9.1 Tropospheric Ducting

The index of refraction in the Earth's troposphere varies with altitude, mainly because of varying water vapor concentration, and radio waves propagate along curved lines as a result. This can greatly increase the distance that radio signals travel, particularly over water when meteorological conditions are favorable. Radio and television signals sometimes propagate across the Great Lakes and, rarely, across even larger bodies of water in so-called tropospheric ducts, which function like weakly guiding dielectric waveguides.

In a spherically stratified medium, Snell's law applied along a radio ray path can be shown to take the form (Bouger's law):

$$(R_e + z)n(z)\sin\psi(z) = \text{const}, \tag{9.2}$$

where $n(z)$ is the index of refraction, z is the height above ground, ψ is the zenith angle shown in Figure 9.1 and the radius of curvature of the dashed line is $R_e + z$. After differentiation with respect to z and noting that $dz \approx \cot\psi (R_e + z)d\theta$, we have

$$\frac{1}{(R_e+z)}\frac{d\psi}{d\theta} = -\left(\frac{1}{n}\frac{dn}{dz} + \frac{1}{R_e+z}\right), \tag{9.3}$$

where the term on the left is the rate of change of zenith angle with lateral displacement and the trailing term in the parentheses on the right is a purely geometric effect associated with propagation over a spherical body. In order for the ray to travel horizontally, exactly following the curvature of the Earth, we require $d\psi/d\theta = 0$ and therefore

$$\frac{dn}{dz} = -\frac{n}{R_e + z} \approx -\frac{1}{R_e}, \tag{9.4}$$

where we note again that the index of refraction is typically nearly unity throughout the troposphere, differing by no more than a few hundred ppm. In fact, under standard atmospheric conditions, $dn/dz \approx -1/4R_e$. This means that the gradient in the index of refraction is usually insufficiently strong to maintain horizontal radio ray paths, that $d\psi/d\theta < 0$, and that the radio waves will exhibit some curvature but ultimately escape into space.

Examining (9.3), we see that the rate of change of the wave zenith angle under standard atmospheric conditions is identical to what it would be if there were no refraction and if the radius of the Earth were $a_e \approx (4/3)R_e$. Consequently, the height of the ray above ground would be the same in either case. A common practice for incorporating the effects of atmospheric refraction in propagation calculations is to assume straight-line propagation, only replacing R_e with a_e. This has the minor effect of increasing the distance estimates from the previous section by a factor of $\sqrt{4/3}$.

When $|n'(z)|$ is greater than (less than) about $1/4R_e$, we have superstandard (substandard) refraction. If it is greater than $1/R_e$, ducting may occur, and radio waves can be refracted back to Earth. This can happen at ground level (a surface duct) or well above the surface (an elevated duct) during unusual meteorological conditions, particularly over water. It can also occur where surface heating or other meteorological conditions create a temperature inversion with associated effects on the refractivity profile. Surface ducts tend to be lossy because they incorporate scattering off the ground, which sends radiation off in all directions. Elevated ducts act like dielectric waveguides created by perturbed refractivity layers at altitude. They can convey radiation efficiently over long distances, but coupling radiation into and out of them is difficult. How well confined a radio wave is to a duct and over what distance it can propagate depends on the duct thickness and steepness (the refractivity gradient).

We can model an elevated duct as a thin layer where the index of refraction has a parabolic shape:

$$n(z) = n_\circ + n_1 z(h-z)(4/h^2), \, 0 \leq z \leq h,$$

where $n_\circ$ and n_1 are the indices of refraction at the edges and the center of the duct, respectively. Let the radius of curvature of the guided wave be sufficiently small when it is in the duct that we can neglect the Earth's gradual curvature by comparison. Suppose a wave is launched into the midpoint of the duct at $z = h/2$ and at an angle ψ_i with respect to the vertical. Snell's law for a vertically stratified medium is just $n \sin \psi = \text{const}$, or

$$n(z) \sin \psi = n(h/2) \sin \psi_i.$$

At the turning points, the wave propagates horizontally, and $n(z) = n(h/2) \sin \psi_i$. Solving from the model, the heights of the turning points must be

$$z = z_t = \frac{h}{2} \left[1 \pm \left(1 + \frac{n_\circ}{n_1} \right)^{1/2} (1 - \sin \psi)^{1/2} \right].$$

The question then is whether or not the turning points are inside the layer. If they are, the wave will remain within the waveguide. Otherwise, it will penetrate the upper boundary and escape. This happens if the initial zenith angle is too steep. The critical angle is the initial zenith angle giving a turning point precisely at $z = h$, viz,

$$\sin \psi_c = \frac{1}{1 + (n_1/n_\circ)}.$$

For example, taking $n_\circ = 1$ and $n_1 = 1 \times 10^{-4}$ (corresponding to a 100 ppm maximum perturbation in the index of refraction across the duct) gives the critical angle $\psi_c = 89.2°$, showing that waves must be launched nearly horizontally to be ducted. This means that it is easier to couple radio waves into surface ducts than elevated ducts. Of those waves with shallow enough zenith angles to be confined, only wave modes with $\lambda < h$ will actually be confined to and guided by the waveguide, and only those with $\lambda \ll h$ are likely to be remain confined over long distances. Perturbations in the index of refraction caused by water vapor are most significant for waves in the microwave band, but these also tend to suffer the most absorption. In practice, waves in the VHF and UHF bands are the most prone to ducting.

9.2 Ionospheric Reflection and Refraction

Radio waves fall into one of three categories: ground waves, space waves or sky waves. Ground waves are guided by the dielectric waveguide formed at the air/ground interface. This is a lossy waveguide, and only low-frequency waves can propagate long distances without severe attenuation. Low-frequency signals, including AM radio signals, can propagate several tens of kilometers this way, permitting reception even after the transmitting tower falls below the horizon.

VHF and higher frequency waves (along with HF waves launched at high elevation angles) propagate along almost straight lines into space and are known as space waves. HF and lower frequency waves meanwhile are refracted as they pass through the Earth's ionosphere, a layer of the upper atmosphere that is partially ionized and that represents a population of free electrons. Whereas the index of refraction is only perturbed slightly in atmospheric ducts, it can vary drastically in the ionosphere. If both the radio wave frequency and elevation angle are low enough, signals can return along wide arcs to ground level at great distances (over the horizon) from their origin. These are sky waves. The gap where neither ground waves nor sky waves are detected is called the *skip distance*. At sufficiently low frequencies, ionospheric refraction becomes reflection, the wide arcs become sharp angles, and there is no skip distance.

Roughly put, waves between 10 and 40 MHz can propagate long distances via ionospheric refraction, whereas waves between 2 and 10 MHz can do so via ionospheric reflection. Below about 2 MHz, strong absorption in the lowest layer of the ionosphere (the D region) prohibits long-distance propagation during the day, but reflection still occurs at night. This phenomenology can be discerned simply by listening to AM radio stations at different times of day.

Figure 9.2 shows representative daytime and nighttime electron density profiles for the Earth's ionosphere. Actual profiles vary with location, local time, season, and solar cycle. There are three main ionospheric layers: the D, E and F layers. These layers are generated by photoionization from solar radiation and shaped by chemical and dynamical (winds, electric fields) processes. Above the F layer, the neutral atmosphere is too tenuous for significant photoionization to occur. Below the D layer, meanwhile, most of the ionizing radiation from the sun has been absorbed. The E layer ("E" stands for electric) was discovered first because of the strong electric currents that flow in it. The F layer is the densest and is associated with long-distance radio wave propagation. During the day, it splits subtly into two layers (although the F_1 layer is really more of a ledge). The D layer exists only during the day. Collisions with the dense neutral atmosphere cause this layer to be very lossy, and radio signals that interact with it are strongly attenuated. At night, chemical recombination removes much of the lower ionosphere, but the upper ionosphere remains largely intact.

9.2 Ionospheric Reflection and Refraction

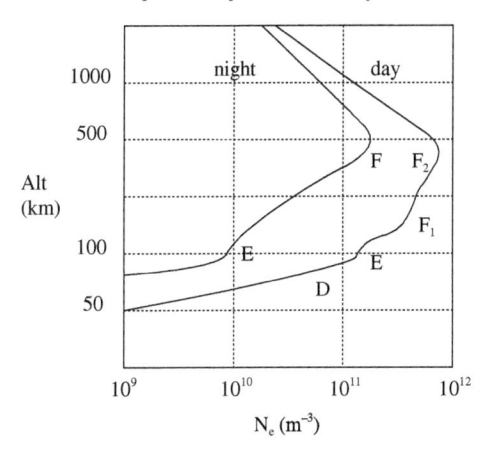

Figure 9.2 Electron number density in the Earth's ionosphere.

The ionosphere is everywhere very nearly charge neutral, with electrons and ions occurring in equal numbers. Electrons and ions respond differently to electric fields, however, and so conduction current can flow. Conduction current combines with displacement current and modifies the index of refraction of the medium, causing radio waves to reflect, refract and exhibit other interesting and important phenomena.

Some of these phenomena can be understood my modeling the ionospheric plasma as a uniform background of ions which are massive and immobile with an equal number of mobile electrons intermixed. We will neglect the Earth's magnetic field **B** for the moment, although this will turn out to be important, as we will see in Section 9.4, The electrons obey Newton's second law:

$$m_e \frac{d\mathbf{v}}{dt} = -e\mathbf{E} - m_e \nu \mathbf{v}.$$

Reading from left to right, the terms in this equation for the electron momentum represent inertia, the Coulomb force and collisional drag, with ν representing the collision rate between electrons and neutral species. Applying phasor notation and solving for the electron velocity yields

$$\mathbf{v} = \frac{-e\mathbf{E}}{(j\omega + \nu)m_e},$$

where ω is the frequency of the electric field applying force to the electrons. Notice that the electron drift is generally out of phase with the applied Coulomb force because of the effect of inertia. Associated with this electron drift is a current density $\mathbf{J} = -N_e e \mathbf{v}$, where N_e is the number density of electrons. This current is also out of phase with the electric field. The current can be incorporated into Ampère's law:

$$\nabla \times \mathbf{H} = \mathbf{J} + i\omega\varepsilon_\circ \mathbf{E}$$
$$= i\omega\varepsilon \mathbf{E}$$
$$\varepsilon(\omega) \equiv \varepsilon_\circ \left(1 - \frac{e^2 N_e/\varepsilon_\circ m_e}{\omega(\omega - j\nu)}\right).$$

Notice how the conduction current has been absorbed here into an effective, frequency-dependent dielectric constant $\varepsilon(\omega)$. This redefinition of the dielectric constant affords a simple means of assessing the effect of conduction currents in a plasma on radio wave propagation, since the index of refraction in a dielectric is defined by $n^2 \equiv \varepsilon/\varepsilon_\circ$. A plasma is not a dielectric *per se*, since it has mobile charge carriers, but the analogy is a powerful and useful one.

Plasmas exhibit a characteristic resonance frequency called the *plasma frequency*

$$\omega_p = 2\pi f_p = \sqrt{\frac{e^2 N_e}{\varepsilon_\circ m_e}}$$
$$f_p \approx 9\sqrt{N_e} \quad \text{(MKS units)},$$

which corresponds to the natural oscillation frequency of the electrons with finite inertia and a Coulomb restoring force. The index of refraction for electromagnetic waves in an unmagnetized plasma is then given by

$$n^2 = 1 - \frac{\omega_p^2}{\omega(\omega - j\nu)} \tag{9.5}$$
$$\approx 1 - \frac{\omega_p^2}{\omega^2}, \quad \nu \ll \omega.$$

When the wave frequency is comparable to both the plasma frequency and the collision frequency, the index of refraction has a significant imaginary component, and attenuation occurs. This happens to MF (medium frequency) waves during the daytime. In the collisionless limit, when the wave frequency is less than the plasma frequency, the index of refraction becomes imaginary and wave propagation ceases. Radio waves reflect at altitudes where this occurs. Since the peak plasma number density is seldom much greater than 1×10^{12} m^{-3}, reflection is usually limited to frequencies below about 10 MHz. (The plasma frequency associated with the plasma density at the F layer peak is called the F-layer *critical frequency* f_c, which is the highest reflection frequency.) Finally, the index of refraction can depart significantly from unity even when the wave frequency is above the plasma frequency. This is the regime of ionospheric refraction. Note that, for $\nu \ll \omega$, ω_p plays the same role as the cutoff frequency for waves propagating in a waveguide. In that way, the plasma can act like a high-pass filter.

9.2.1 Vertical-Incidence Radio Sounding

The earliest radar experiments involved sending pulses upward into the ionosphere and estimating the shape of the electron density profile on the basis of the time it took to

9.2 Ionospheric Reflection and Refraction

receive echoes at different radar frequencies. A special kind of radar used to to probe the ionosphere by transmitting HF signals over a broad range of frequencies and measuring the return times of the echoes is called an *ionosonde*. The record of propagation times versus frequency it produces is called an *ionogram*. In the event the electron density is vertically stratified (a good assumption most of the time), radio waves launched vertically from the ground will travel vertically upward, reflect at the critical height (where the radar frequency matches the local plasma frequency), and travel vertically downward, back to the radar. The delay time depends on the group speed of the waves along the propagation path and so is an integrated quantity.

A dispersion relation for electromagnetic waves in a collisionless plasma without a background magnetic field can be derived from (9.5) by noting that $n^2 \equiv k^2 c^2/\omega^2$:

$$\omega^2 = \omega_p^2 + k^2 c^2.$$

The group speed can be found from the dispersion relation by evaluating $v_g = \partial \omega/\partial k$ implicitly:

$$v_g = \frac{c^2}{\omega/k} = \frac{c^2}{v_p},$$

where $v_p \equiv c/n$ is the phase speed of the waves. This implies the remarkable conclusion that $v_p v_g = c^2$, that the product of the phase speed and the group speed is c^2. Since the phase speed is always greater than c, the group speed is always less than c, as we should expect.

The time it takes for a wave to travel upward to the reflection height h and down again is

$$T(f) = 2 \int_0^h \frac{dz}{v_g(z)} > \frac{2h}{c}, \tag{9.6}$$

where the group speed is a function of the electron density and therefore of altitude z and where the frequency dependence f enters through the reflection height h, the altitude where the wave frequency matches the plasma frequency. Since $v_g < c$, $T > 2h/c$. For frequencies greater than the critical frequency, $f > f_c$, no reflection occurs, and the reflection time is infinite.

It is customary to define the virtual height as $h' = cT(f_c)/2$. This is the height of a hypothetical reflector that would give an echo at the time $T(f_c)$ in vacuum. By definition, $h' > h$. The concept of virtual height can be generalized with the definition of critical heights for the E, F_1, and F_2 layers. The table below gives representative virtual heights for different ionospheric layers: Virtual heights are easily measured and are recorded with ionosondes and tabulated at sites around the globe. As shown below, the virtual-height parameter turns out to be useful for radio propagation calculations. Measuring electron density profiles is more difficult and requires the inversion of (9.6). Note that since signals are never received from altitudes above the F-layer peak, only the bottom side of electron density profiles can be measured with ionosondes.

226   Propagation

Table 9.1 *Representative ionospheric virtual heights.*

layer	h′
E	110 km
F_1	200–250 km
F_2	250–400 km
F (night)	300 km

The propagation of radio waves through the ionosphere at off-vertical incidence angles is more complicated and usually requires ray-tracing calculations to interpret completely. The ray paths followed by the signals are not straight lines in general but curved arcs, and the phase and group speeds depart significantly from the speed of light, particularly near turning points. A few simple ideas can give some insight into the oblique sounding problem, however. These ideas date back to the earliest days of radio science.

9.2.2 Oblique Sounding and the Theorem of Breit and Tuve

This surprising and useful theorem says that the time it takes for a wave to travel along an actual, curved ray path with apex height h in a lossless, isotropic ionosphere is equal to the time it takes for a wave to travel along two straight-line ray paths with the same initial zenith angle and joining at the virtual height h' in vacuum. The theorem makes it possible to analyze ionospheric propagation using simple geometric constructions along with information about virtual heights.

The time it takes for a wave in vacuum to travel along straight lines from the ground to the apex height h' and back to the ground is

$$\begin{aligned} t &= \frac{2}{c}(d/2)\csc\psi_i \\ &= (d/c)\csc\psi_i, \end{aligned}$$

where d is the horizontal distance traveled and ψ_i is the zenith angle, which remains constant throughout. The time it takes for an ionospheric wave to propagate a distance ds along its ray path while simultaneously propagating a horizontal distance dx is

$$\begin{aligned} dt &= \frac{ds}{v_g} \\ &= \frac{dx}{v_g \sin\psi}, \end{aligned}$$

where v_g is the groups speed, ψ is the zenith angle and both quantities vary along the ray. In an isotropic plasma, $v_g = cn$. According to Snell's law, $n\sin\psi = \sin\psi_i$ where ψ_i is the initial zenith angle at ground level where the index of refraction n is unity. Consequently,

$$dt = \frac{dx}{c \sin \psi_i}$$

$$t = \frac{1}{c} \csc \psi_i \int_0^d dx$$

$$= (d/c) \csc \psi_i,$$

which proves the theorem.

9.2.3 Maximum Usable Frequency (MUF)

Consider next the condition for which a radio wave launched from the ground obliquely into the ionosphere has a turning point and returns to the ground. While the Earth's curvature cannot be neglected completely in the analysis of what happens, it is generally much more gradual than the curvature of a sky wave propagating in the ionosphere. We can therefore assume straight-line propagation below the ionosphere and make use of Snell's law for a vertically stratified medium within,

$$n_i \sin \psi_i = n \sin \psi, \tag{9.7}$$

where $n_i = 1$ and $\psi = \psi_i$ at the base of the ionosphere (see Figure 9.3). At a turning point, $\psi = \pi/2$ and we have the condition $\sin \psi_i = n = \sqrt{1 - \omega_p^2/\omega^2}$. The condition depends on the zenith angle of entry, the wave frequency, and the plasma frequency. Of great interest is the highest frequency for which a turning point exists, called the maximum usable frequency, f_{max}. The turning point in question will occur at the electron density peak (the critical layer) where the plasma frequency is highest ($f_p = f_c$). This gives

$$1 - \frac{f_c^2}{f_{max}^2} = 1 - \cos^2 \psi_i$$

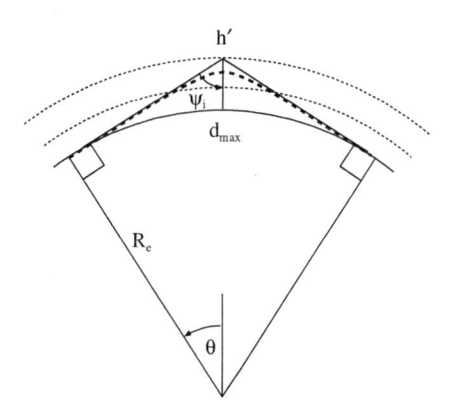

Figure 9.3 Radio wave propagation at the MUF.

$$f_{max} = f_c \sec \psi_i. \tag{9.8}$$

This is the highest frequency for which refracted waves will return to the Earth. Typical daytime values are 25–40 MHz, depending on the solar cycle, season and location. Predictions of the MUF are tabulated and published regularly. Signals propagating at the MUF have the highest arcs and longest propagation paths of all the sky waves.

It might appear as if the MUF could be arbitrarily large if ψ_i is sufficiently close to $\pi/2$. What limits ψ_i is the curvature of the Earth, which must now be taken into account. The MUF and the largest ψ_i possible will clearly occur for radiation launched horizontally at the ground. Figure 9.3 shows a radio wave emitted horizontally, refracting at the critical layer, and intercepting the ground horizontally a distance $d_{max} = 2\theta R_e$ away. Although the ray path is curved, we can define a virtual reflection h' as the height of the interception of the two tangent lines shown using the theorem of Breit and Tuve.

From the discussion at the beginning of the chapter, we have $d_{max} = 2\sqrt{2R_e h'}$ and $\theta = \sqrt{2h'/R_e}$. To complete the calculation of the MUF, we make use of the following geometric equation,

$$\left(1 + \frac{h'}{R_e} - \cos\theta\right) \csc\theta = \cot\psi_i, \tag{9.9}$$

which is true for any initial and final elevation angle (takeoff angle) and not just for radio waves launched and received horizontally. For example, taking $h' = 400$ km gives $d_{max} = 4500$ km (2500 mi). Assuming $N_e = 10^{12} m^{-3}$, we have $f_c = 9$ MHz. Applying (9.9) gives $\psi_i = 74°$ and $f_{max} = 32$ MHz, which is a reasonable estimate for the MUF during the daytime at solar maximum.

In practice, radio waves are not launched quite horizontally, but (9.9) can still be used to calculate ψ_i for a finite takeoff angle (angle above the horizon). This is equivalent to calculating ψ_i and the associated MUF for path lengths shorter than d_{max}. The procedure is as follows: begin by specifying a path length $d = 2\theta R_e$ less than d_{max}. Associated with this foreshortened path length is an angle θ and also a takeoff angle $\pi/2 - \theta - \psi_i$ which will be greater than zero. Both ψ_i and then the takeoff angle can be calculated using (9.9) for a specified virtual height. From this ψ_i, an f_{max} for the calculated takeoff angle and path length and the given virtual height can be found. Note that the normal definition of the MUF in the absence of any qualifiers refers to zero takeoff angle and the maximum path length, however.

One sometimes finds the MUF calculated using Bouger's law for spherically stratified media (9.2) to trace the ray all the way from the ground directly to the ionospheric turning point,

$$R_e \sin\psi_{i0} = (R_e + h)n,$$

where the initial height of the ray above ground is zero, ψ_{i0} refers to the zenith angle at ground level and h is the true height of the turning point. Solving for the MUF as before gives

$$f_{\max} = \frac{f_c}{\sqrt{1 - \frac{R_e^2}{(R_e+h)^2}\sin^2\psi_{i0}}}, \quad (9.10)$$

which should be equivalent to what we found using (9.8) and (9.9) except formulated in terms of the true height of the critical layer instead of the virtual height. While this formulation is somewhat easier to use, true height measurements are not as prevalent as virtual height measurements in the ionosphere. Comparing results obtained from the two approaches is a means of estimating one from the other, however.

Finally, keep in mind that propagation paths longer than d_{max} can be achieved through multiple hops. Round-the-world propagation is not difficult to achieve in this way when conditions are favorable.

9.3 Generalized Wave Propagation in Inhomogeneous Media

Here, we develop the basic mathematics underlying atmospheric and ionospheric refraction in more detail. The discussion is limited to linear, isotropic inhomogeneities with spatial scales large compared to a wavelength. In terms of radio wave propagation in the ionosphere, we are still neglecting the effects of the Earth's magnetic field, which can be important (see Section 9.4). The analysis presented here goes by several names: geometric optics, eikonal theory, the method of characteristics, ray tracing and even classical mechanics. Wave propagation in inhomogeneous media is a deep and compelling topic with numerous applications, including radar and sonar. Interested readers should have no trouble finding material for further study.

Combining Faraday's and Ampère's laws in phasor notation gives the following wave equation

$$\nabla \times \nabla \times \mathbf{E} - \omega^2 \mu \varepsilon(\mathbf{x})\mathbf{E} = 0,$$

where the spatial dependence of the permittivity is the new, crucial feature. Let us rewrite $\nabla \times \nabla \times \mathbf{E} = \nabla(\nabla \cdot \mathbf{E}) - \nabla^2 \mathbf{E}$ and also $\mu\varepsilon \equiv n^2/c^2$, where n is the index of refraction. Also, since $\nabla \cdot (\varepsilon \mathbf{E}) = 0$ in a charge-free region by Gauss' law, we can write $\nabla \cdot \mathbf{E} = -(1/\varepsilon)\nabla\varepsilon \cdot \mathbf{E} = -\nabla \ln \varepsilon \cdot \mathbf{E}$. The wave equation therefore becomes

$$\nabla^2 \mathbf{E} + \nabla\left[\nabla\left(\ln n^2\right) \cdot \mathbf{E}\right] + k_\circ^2 n^2 \mathbf{E} = 0,$$

where $k_\circ = \omega/c$ as usual. To solve for the fields in terms of a known, spatially varying index of refraction, we assume the form of solution

$$\mathbf{E}(\mathbf{x}) = \mathbf{E}_\circ(\mathbf{x})e^{-jk_\circ \phi(\mathbf{x})}, \quad (9.11)$$

where $\mathbf{E}_\circ$ is a slowly varying amplitude and ϕ is a rapidly varying phase term or *eikonal*. The eikonal replaces a term like $\exp(jn\mathbf{k}_\circ \cdot \mathbf{x})$ that describes plane waves in homogeneous media. Surfaces of constant phase are surfaces of constant ϕ now rather than planes, and

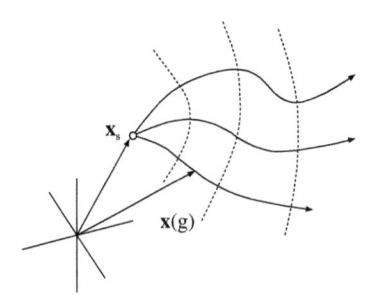

Figure 9.4 Ray tracing geometry. The paths of phase propagation (solid curves) are normal to surfaces of constant phase (dashed curves). Waves are launched at $\mathbf{x}_s$ and follow parametrized trajectories $\mathbf{x}(g)$.

the paths along which waves propagate, which are normal to the constant phase surfaces, are no longer straight lines in general.

Substituting (9.11) into the wave equation leads to a complicated expression involving the unknowns $\mathbf{E}(\mathbf{x})$, $\phi(\mathbf{x})$ and their spatial derivatives. Multiplying the various terms are different powers of $k_\circ$. In the small wavelength limit of interest, $k_\circ$ is a large term compared to the ∇ operator, for example. We treat $k_\circ$ as an ordering parameter and equate separately terms that are the same order in $k_\circ$. The first and most important of the resulting equations, called the eikonal equation, comes from equating terms of order $k_\circ^2$:

$$\nabla\phi \cdot \nabla\phi = n^2. \tag{9.12}$$

Given $n(\mathbf{x})$, we can solve for the phase term everywhere with this equation. From this, we can trace the propagation paths of the waves. Having solved for $\phi(\mathbf{x})$, we can then turn to the equation formed from terms of order $k_\circ^1$,

$$-\nabla^2\phi \mathbf{E}_\circ - 2\nabla\phi \cdot \nabla\mathbf{E}_\circ - \nabla\ln n^2 \cdot \mathbf{E}_\circ \nabla\phi = 0,$$

which gives sufficient information for determining the amplitude of the waves $\mathbf{E}_\circ$. Note that this last step need not be performed if one is only interested in ray tracing, which is the limit of our interest here.

We can solve the partial differential (9.12) using the method of characteristics. We seek curves $\mathbf{x}(g)$ parametrized by the scalar g and representing propagation paths from a starting point $\mathbf{x}_s$. The paths are everywhere normal to the surfaces of constant phase:

$$\frac{d\mathbf{x}}{dg} = \nabla\phi.$$

According to (9.12), we then have

$$\frac{d\mathbf{x}}{dg} \cdot \frac{d\mathbf{x}}{dg} = n^2. \tag{9.13}$$

Differentiating with respect to g yields

$$\frac{d^2\mathbf{x}}{dg^2} = n\nabla n. \qquad (9.14)$$

What was once a PDE has become three uncoupled, 2nd order, scalar ODEs, one for each vector component of $\mathbf{x}(g)$. Note also the form of the equations. With the appropriate scaling, (9.14) has the form of Newton's second law, with the $n\nabla n$ term playing the role of a generalized force. Ray paths evidently follow ballistic trajectories, and solutions to (9.14) can be highly intuitive as a result. Of course, the path depends on the initial conditions as well as on n, just as the initial position and velocity of a ballistic object, together with the applied force, determine its trajectory.

What is g? It is related to path length, but what is the relationship? Note that, by definition, one may always write

$$\frac{d\mathbf{x}}{ds} \cdot \frac{d\mathbf{x}}{ds} = 1,$$

where $ds^2 = dx^2 + dy^2 + dz^2$ and ds is a physical path length element. Comparison with (9.13) shows that $ds/dg = n$, or that dg is linearly related to ds by the index of refraction. We call g the optical path length, which is proportional to the path length measured in wavelengths rather than fixed units.

9.3.1 Stratified Media

Consider what happens to (9.14) in vertically stratified media. In the horizontal (x) direction, we have

$$\frac{d^2x}{dg^2} = \frac{d}{dg}\left(\frac{dx}{dg}\right) = n\frac{d}{ds}\left(n\frac{dx}{ds}\right) = 0.$$

Integrating yields

$$n(z)\frac{dx}{ds} = c,$$

where c is a constant of integration. If the zenith angle of the ray at any instant is θ, then dx/ds is just $\sin\theta$, meaning

$$n(z)\sin\theta = c = \sin\theta_\circ, \qquad (9.15)$$

where we take the index of refraction to be unity at the origin of the ray path, where $\theta = \theta_\circ$. Equation (9.15) is of course just Snell's law in a stratified medium. While this is sufficient information to trace the ray, we can develop an explicit equation for the ray trajectory by manipulating the vertical (z) component of (9.14):

$$n\frac{d}{ds}\left(n\frac{dz}{ds}\right) = n\frac{dn}{dz} = \frac{1}{2}\frac{d(n^2)}{dz}.$$

To this, we apply Snell's law in the form $ds = (n/c)dx$ to find

$$\frac{d^2z}{dx^2} = \frac{1}{2c^2}\frac{d}{dz}(n^2).$$

Note further that, along a line segment, $d/dx = (dz/dx)d/dz$. Therefore, the left side of the preceding equation is $(dz/dx)d/dz(dz/dx)$, or

$$\frac{1}{2}\frac{d}{dz}\left(\frac{dz}{dx}\right)^2 = \frac{1}{2c^2}\frac{d}{dz}(n^2).$$

Integrating once in z gives

$$\left(c\frac{dz}{dx}\right)^2 = n^2(z) + B,$$

where B is another constant of integration. Recalling that we have taken $n(0) = 1$ and that $c = \sin\theta_\circ$ as a consequence allows the assignment $B = -\sin^2\theta_\circ$ (since dz/dx at the origin is $\cot\theta_\circ$). Taking the reciprocal of the square root of the resulting equation and integrating then gives the desired equation for the ray trajectory:

$$x(z) = \int_0^z \sqrt{\frac{\sin^2\theta_\circ}{n^2(z) - \sin^2\theta_\circ}}\,dz.$$

In a homogeneous medium with $n = 1$, this obviously gives the equation for a straight line, $x = z\tan\theta_\circ$. In inhomogeneous media, it specifies a ray that may or may not return to the $z = 0$ interface, depending on the steepness of the index of refraction gradient and the initial zenith angle $\theta_\circ$.

9.3.2 Radius of Curvature

It is often useful to formulate the ray tracing problem in curvilinear coordinates, as in the case of a ray bent into a nearly circular arc. Care must be taken here since the coordinate axes themselves vary with position. Returning to the ballistic equation, we have

$$\frac{d^2\mathbf{x}}{dg^2} = n\nabla n \quad \text{with} \quad \frac{ds}{dg} = n$$

$$n\frac{d}{ds}\left(n\frac{d\mathbf{x}}{ds}\right) = n\nabla n.$$

In polar coordinates, $\mathbf{x} = r\mathbf{e}_r$ and $ds = rd\theta$ on a circular arc. Furthermore, $d\mathbf{e}_r/d\theta = \mathbf{e}_\theta$ and $d\mathbf{e}_\theta/d\theta = -\mathbf{e}_r$. Altogether,

$$\frac{d}{ds}(n\mathbf{e}_\theta) = \nabla n$$

$$\frac{dn}{ds}\mathbf{e}_\theta - \frac{n}{r}\mathbf{e}_r = \nabla n,$$

where r has been held constant throughout. Taking the dot product with $\mathbf{e}_r$ then gives

$$\frac{1}{r} = -\nabla \ln n \cdot \mathbf{e}_r,$$

which relates the radial gradient in the index of diffraction to the radius of curvature of a refracted ray. For example, we can determine how the index of refraction must vary in order for the ray to be bent into a circular arc of radius R_e:

$$-\frac{dr}{R_e} = d \ln n$$

$$-\frac{r}{R_e} + C = \ln n$$

$$n = n_\circ e^{-r/R_e}.$$

This shows that the radius of curvature of a circular ray is the scale height of the index of refraction in a radially stratified medium, in agreement with statements made earlier. Note that the calculations become more cumbersome if deviations from a circular path are allowed.

9.4 Birefringence (Double Refraction)

Reflection and refraction are hallmarks of radio wave propagation in inhomogeneous media. Birefringence, meanwhile, is an additional effect characteristic of waves propagating in anisotropic media. An anisotropic medium is one that responds differently depending on the direction of forcing. Many crystals have this property because of asymmetries in their lattice structure. In the ionosphere, anisotropy arises from the Earth's magnetic field and the effect of the $\mathbf{J} \times \mathbf{B}$ component of the Lorentz force on the currents driven by electromagnetic waves. This effect has been neglected so far but is quite important for understanding ionospheric radio wave propagation.

Mathematically, the dielectric constant in an anisotropic medium is generally described by a matrix rather than a single constant. The consequence of this is that the equation governing the index of refraction is quadratic and has two viable solutions or modes. Two distinct kinds of waves may therefore propagate in any given direction in the medium. The two generally obey different dispersion relations and have different polarizations. When the two interfere, the polarization of the combined wave can vary in ways not observed in isotropic media.

Let us return to Newton's second law, this time including the $\mathbf{J} \times \mathbf{B}$ component of the Lorentz force:

$$m_e \frac{d\mathbf{v}}{dt} = -e(\mathbf{E} + \mathbf{v} \times \mathbf{B}) - m_e \nu \mathbf{v}.$$

In this equation, $\mathbf{B}$ denotes the Earth's background magnetic field and not the magnetic field due to the wave itself, which exerts negligible force on electrons moving at non-relativistic speeds. This equation can be solved for the electron velocity, which is linearly related to the wave electric field, but the relationship is through a matrix, called the mobility matrix.

Its form is simplified by working in Cartesian coordinates and taking the Earth's magnetic field to lie in the $\hat{\mathbf{z}}$ direction:

$$\begin{pmatrix} v_x \\ v_y \\ v_z \end{pmatrix} = \begin{pmatrix} \mu_{11} & \mu_{12} & \\ \mu_{21} & \mu_{22} & \\ & & \mu_{33} \end{pmatrix} \begin{pmatrix} E_x \\ E_y \\ E_z \end{pmatrix}$$

$$\mu_{11} = \mu_{22} = -\left(1 + \frac{(j\omega+\nu)^2}{\Omega_e^2}\right)^{-1} \frac{j\omega+\nu}{\Omega_e B}$$

$$\mu_{12} = -\mu_{21} = \left(1 + \frac{(j\omega+\nu)^2}{\Omega_e^2}\right)^{-1} \frac{1}{B}$$

$$\mu_{33} = -\frac{\Omega_e/B}{j\omega+\nu}.$$

The term $\Omega_e = eB/m_e$ is the angular electron gyrofrequency, which gives the angular frequency of gyration of the electrons around the Earth's magnetic field lines. Anisotropy results from the fact that electrons are bound to orbits around magnetic field lines but are relatively free to move along the field lines. In the ionosphere, $\Omega_e/2\pi \leq 1.4$ MHz and depends on magnetic latitude and altitude. The μ_{33} term is what would be found in an isotropic plasma with no background magnetic field. As the wave frequency increases well beyond the gyrofrequency, μ_{11} and μ_{22} approach μ_{33}, and the off-diagonal terms of the mobility matrix become small. Nevertheless, so-called magneto-ionic effects remain significant even at radio frequencies much higher than $\Omega_e/2\pi$.

The current density is related to the electron drift velocity by $\mathbf{J} = -N_e e \mathbf{v}$, and to the electric field by the constitutive relationship $\mathbf{J} = \sigma \mathbf{E}$, where σ is the conductivity matrix. For wave frequencies much greater than the electron gyrofrequency and collision frequency, the nonzero elements of the conductivity matrix are

$$\sigma_{11} = \sigma_{22} = N_e e \left(1 - \frac{\omega^2}{\Omega_e^2}\right)^{-1} \frac{j\omega}{\Omega_e B}$$

$$\sigma_{12} = -\sigma_{21} = -N_e e \left(1 - \frac{\omega^2}{\Omega_e^2}\right)^{-1} \frac{1}{B}$$

$$\sigma_{33} = -j N_e e \frac{\Omega_e/B}{\omega}.$$

Finally, we can write Ampère's law once again as

$$\nabla \times \mathbf{H} = \mathbf{J} + i\omega\varepsilon_\circ \mathbf{E}$$
$$= (\sigma + i\omega\varepsilon_\circ I)\mathbf{E}$$
$$= i\omega\varepsilon\mathbf{E}$$

$$\varepsilon = \begin{pmatrix} \varepsilon_{11} & \varepsilon_{12} & \\ \varepsilon_{21} & \varepsilon_{22} & \\ & & \varepsilon_{33} \end{pmatrix}$$

$$\varepsilon_{11} = \varepsilon_{22} = \varepsilon_\circ \left(1 + \frac{\omega_p^2}{\Omega_e^2 - \omega^2}\right)$$

$$\varepsilon_{12} = -\varepsilon_{21} = j\varepsilon_\circ \frac{\Omega_e}{\omega} \frac{\omega_p^2}{\Omega_e^2 - \omega^2}$$

$$\varepsilon_{33} = \varepsilon_\circ \left(1 - \frac{\omega_p^2}{\omega^2}\right).$$

The dielectric matrix gives a description of the response of the medium (the ionospheric plasma in this case) to electromagnetic waves. In effect, the dielectric matrix gives the electrical displacement that arises from electric fields applied in different directions. Obviously, the displacement is not parallel to the electric field in general. This has important implications, including the fact that the phase velocity of the waves (which is normal to the electrical displacement – see below) and the group velocity (which is parallel to the Poynting flux and therefore normal to the electric field) need not be parallel. Waves propagating in the same direction but with different polarizations will moreover be governed by different components of the dielectric matrix and so will behave fundamentally differently, having different phase and group velocities, for example, and different reflection conditions.

9.4.1 Fresnel Equation of Wave Normals

To analyze the behavior of electromagnetic waves in the plasma, we combine Faraday's law and Ampère's law in macroscopic form. We seek plane-wave solutions for the field quantities of the form $\exp j(\omega t - \mathbf{k} \cdot \mathbf{r})$. In phasor notation, Faraday's and Ampère's laws in a charge-free, current-free region become

$$\mathbf{k} \times \mathbf{E} = \omega \mu_\circ \mathbf{H}$$
$$\mathbf{k} \times \mathbf{B} = -\omega \mathbf{D}.$$

These equations apply to the ionospheric plasma, which is charge-neutral (ions and electrons everywhere in approximately equal density) so long as the conduction current is included in the definition of the dielectric constant and the electric displacement. From the above equations, we can conclude that the wavevector $\mathbf{k}$ will be normal both to the magnetic field/magnetic induction and also to the electric displacement but not necessarily to the electric field.

Combining Maxwell's equations yields the desired wave equation

$$\mathbf{k}(\mathbf{k} \cdot \mathbf{E}) - k^2 \mathbf{E} + \omega^2 \mu_\circ \boldsymbol{\varepsilon} \cdot \mathbf{E} = 0,$$

where the dielectric constant is the matrix derived just above. In matrix form, the wave equation can be expressed as

$$\begin{pmatrix} \omega^2\mu_\circ\varepsilon_{11} - k^2 + k_x^2 & \omega^2\mu_\circ\varepsilon_{12} + k_x k_y & k_x k_z \\ \omega^2\mu_\circ\varepsilon_{21} + k_x k_y & \omega^2\mu_\circ\varepsilon_{22} - k^2 + k_y^2 & k_y k_z \\ k_x k_z & k_y k_z & \omega^2\mu_\circ\varepsilon_{33} - k^2 + k_z^2 \end{pmatrix} \begin{pmatrix} E_x \\ E_y \\ E_z \end{pmatrix} = 0.$$

For a given wave frequency ω, the objective now is to solve the above for the wavevector $\mathbf{k}(\omega)$, which constitutes a dispersion relation. If a solution other than the trivial solution $(E_x, E_y, E_z = 0)$ is to exist, the 3×3 matrix above must be singular and therefore non-invertible. We therefore find the dispersion relation by setting the determinant of the matrix equal to zero. The resulting equation is called the wave normal equation.

Having found one or more wavevectors that solve the wave normal equation, we can then substitute them back into the 3×3 matrix to see what they imply about $\mathbf{E}$. There is no unique solution for the electric field amplitude corresponding to each wavevector found, since any solution multiplied by a constant remains a solution. However, it is possible to find the ratio of the components $E_x : E_y : E_z$ corresponding to each solution or mode. This ratio determines the polarization of the given mode.

A convenient way to express the dispersion relation is by solving for the wavenumber k as a function of the frequency ω and the prescribed phase propagation direction $\hat{\mathbf{k}} = \mathbf{k}/k$. Since the problem is azimuthally symmetric about the magnetic field, the direction can be expressed wholly in terms of the angle θ between the magnetic field and the wavevector θ. Finally, it is common to write the dispersion relation in terms of the index of refraction, which is related to the wavenumber by $n^2 = c^2 k^2/\omega^2$. Consequently, we will solve the wave normal equation for $n(\omega, \theta)$.

The wave normal equation is quadratic in n^2, and so we expect two independent solutions for any given wave frequency and propagation direction. (The word "birefringence" refers to the dual solutions.) These are the so-called magneto-ionic modes of propagation. The two modes will have distinct, orthogonal polarizations and be governed by different dispersion relations. Radiation launched into the ionosphere at arbitrary direction and polarization will divide into two modes which may then depart from one another. In general, superimposed modes will exhibit a combined polarization that varies in space.

9.4.2 Astrom's Equation

Considerable effort has been invested in finding compact ways of writing the wave normal equation for an ionospheric plasma. One of these is

$$\left(\frac{n_1^2}{n^2 - n_1^2} + \frac{n_{-1}^2}{n^2 - n_{-1}^2} \right) \frac{\sin^2\theta}{2} + \frac{n_\circ^2}{n^2 - n_\circ^2} \cos^2\theta = 0$$

$$n_\alpha^2 \equiv 1 - \frac{X}{1 - \alpha Y}, \quad \alpha = -1, 0, 1$$

$$X \equiv \omega_p^2/\omega^2$$

$$Y \equiv \Omega_e/\omega.$$

Here, ω and θ are independent variables, and n^2 is the dependent variable. The other variables are auxiliary variables; α takes on values from -1 to 1, inclusive.

9.4.3 Booker Quartic

An equivalent formulation is found by multiplying Astrom's equation to clear the fractions:

$$[n_1^2(n^2-n_{-1}^2)+n_{-1}^2(n^2-n_1^2)]\frac{\sin^2\theta}{2}+n_o^2(n^2-n_1^2)(n^2-n_{-1}^2)\cos^2\theta=0.$$

In this formulation, the wave normal equation is clearly quadratic in n^2.

9.4.4 Appleton–Hartree Equation

By factoring the Booker quartic, one can obtain the Appleton–Hartree equation,

$$n^2 = 1 - \frac{X}{1-\frac{Y_T^2}{2(1-X)}\pm\sqrt{\frac{Y_T^4}{4(1-X)^2}+Y_L^2}},$$

where

$$Y_T \equiv Y\sin\theta$$
$$Y_L \equiv Y\cos\theta.$$

The terms Y_T and Y_L refer to transverse and longitudinal propagation with respect to the background magnetic field. In the quasi-transverse and quasi-longitudinal limits, one term is deemed negligible compared to the other.

The three equivalent equations may seem daunting, but a little inspection yields incisive information. The condition for wave propagation is $n^2 > 0$. In the event $n^2 < 0$, the wave is evanescent. Setting $n^2 = 0$ in the Booker quartic yields an equation for the cutoff frequencies

$$n_o^2 n_1^2 n_{-1}^2 = 0$$

for all propagation angles with respect to the magnetic field. Setting $n_o^2 = 0$ gives the condition $\omega^2 = \omega_p^2$. This is the same cutoff condition as for waves in an isotropic plasma (no magnetic field), and so this is referred to as the "ordinary" mode cutoff. Setting $n_{\pm 1}^2 = 0$ meanwhile gives the conditions $X = 1 \mp Y$ or $\omega_p^2 = \omega(\omega \mp \Omega_e)$. This behavior is new, and so the frequencies are known as the "extraordinary" mode cutoffs. The terms "ordinary" and "extraordinary" actually come from crystallography, where similar ideas and terminology apply.

The dispersion relations contained within the three equations can be understood better by examining some illustrative limits. Consider first the case of propagation perpendicular to the magnetic field $\theta = \pi/2$. In this limit, $Y_T = Y$, $Y_L = 0$, and the two solutions to the Appleton–Hartree equation are

$$n^2 = 1 - X$$
$$n^2 = 1 - \frac{X}{1 - Y^2/(1-X)}.$$

The first of these is identical to the dispersion relation in a plasma with no background magnetic field. There is a cutoff at $X = 1$ or $\omega = \omega_p$, and waves with frequencies above (below) the cutoff are propagating (evanescent). Such a wave is called an ordinary or O-mode wave because of its similarity to waves studied in the simpler case of isotropic plasmas. It can be shown to have a characteristic linear polarization with an electric field parallel to the axis of the background magnetic field. Electrons accelerated by the wave electric field will experience no $\mathbf{v} \times \mathbf{B}$ force, and so the magnetic field has no effect on this wave.

The other solution is called the extraordinary or X-mode wave. This wave is also linearly polarized, but with the electric field perpendicular to the background magnetic field. This wave has a resonance ($n^2 \to \infty$) at $X = 1 - Y^2$ which is called the upper-hybrid resonance (hybrid, since the resonant frequency involves both the plasma and the electron cyclotron frequency). The wave propagates at high frequencies and also at frequencies just below the upper-hybrid frequency. It is evanescent at low frequencies and at frequencies just above the upper-hybrid frequency.

Figure 9.5 plots the dispersion relations for the $\theta = \pi/2$ case for both the X- and O-mode waves. It also plots solutions for other propagation angles. Similar features are found at all intermediate angles between $\theta = \pi/2$ and $\theta = 0$. These include an O-mode cutoff at $X = 1$ and an X-mode resonance at the upper-hybrid frequency, given generally by $X = (1 - Y^2)/(1 - Y_L^2)$. We have already determined that the X-mode cutoff frequencies are given by $X = 1 \pm Y$ for all propagation angles. What changes between the plots is the steepness of the curves near the upper-hybrid resonance, where all the waves are clearly very dispersive.

The characteristic polarizations of the X- and O-modes are elliptical. For $\mathbf{k} \cdot \mathbf{B} > 0$, the modes are right and left elliptically polarized, respectively. As the propagation becomes close to parallel, the polarizations become nearly circular. Since electrons orbit the magnetic field line in a right circular sense, the X-mode interacts more strongly with the electrons, explaining the fact that it is more dispersive than the O-mode.

Another special limiting case is that of parallel propagation with $\theta = 0$. Here, the upper-hybrid resonance and the $X = 1$ cutoff vanish, and the X- and O-modes coalesce into two modes with cutoff frequencies still given by $X = 1 \pm Y$. From the Appleton–Hartree equation, we find the dispersion relations for the two modes to be

$$n^2 = 1 - \frac{X}{1 \pm Y}.$$

When $\mathbf{k} \cdot \mathbf{B} > 0$, the $+/-$ solutions correspond, respectively, to left- and right-circularly polarized modes.

The vertical line at $X = 1$ reflects a resonance at the plasma frequency for parallel propagating waves. The waves in question are non-propagating electrostatic plasma oscillations.

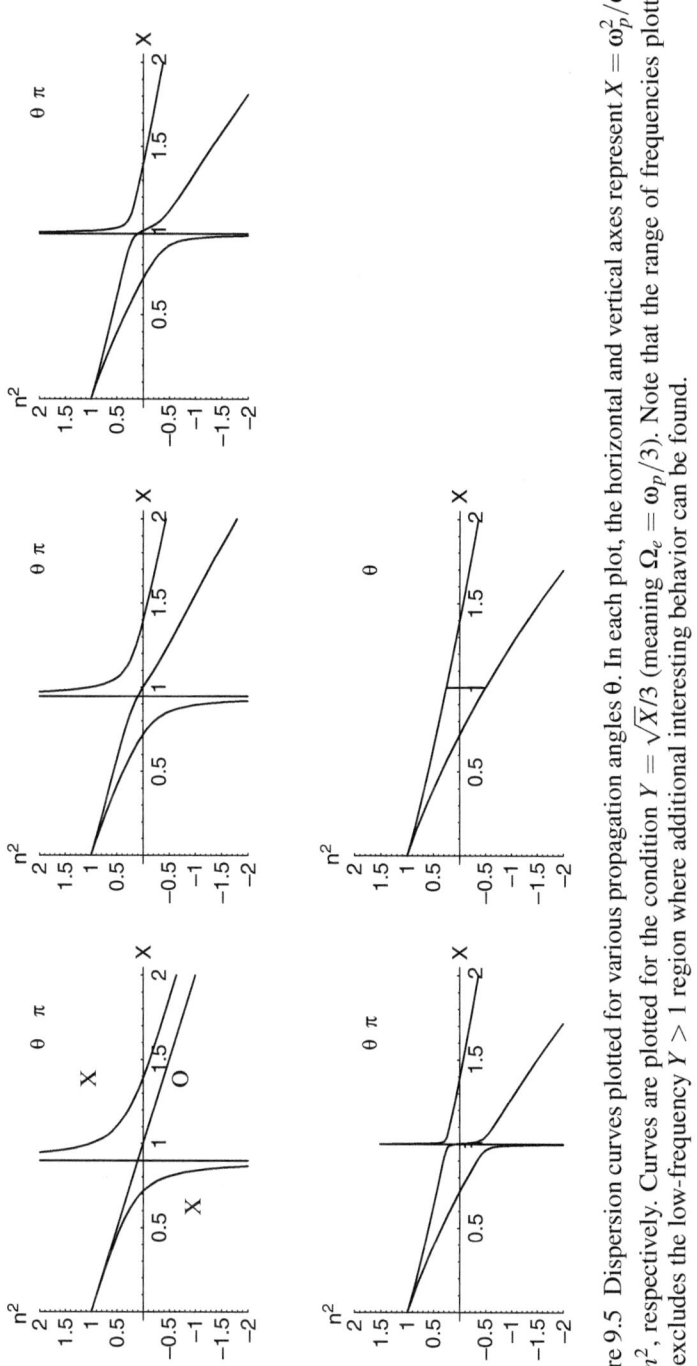

Figure 9.5 Dispersion curves plotted for various propagation angles θ. In each plot, the horizontal and vertical axes represent $X = \omega_p^2/\omega^2$ and n^2, respectively. Curves are plotted for the condition $Y = \sqrt{X}/3$ (meaning $\Omega_e = \omega_p/3$). Note that the range of frequencies plotted here excludes the low-frequency $Y > 1$ region where additional interesting behavior can be found.

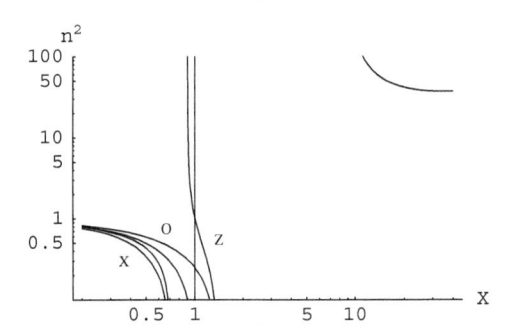

Figure 9.6 Summary of dispersion relations for all modes and propagation angles. Once again, $Y = \sqrt{X}/3$ here. A qualitatively different set of curves results if $Y > \sqrt{X}$ or $\Omega > \omega_p$.

The electric field for these waves is parallel to the wavevector and therefore parallel to the magnetic field, and the electrons it accelerates do not experience a $\mathbf{J} \times \mathbf{B}$ force.

Figure 9.6 summarizes the findings so far. It is similar to Figure 9.5 except that 1) it is plotted on a log-log scale, 2) it extends to larger X (lower frequency) and 3) it only shows propagating modes. Furthermore, curves for both $\theta = 0$ and $\theta = \pi/2$ are plotted simultaneously. Waves propagating at arbitrary, intermediate angles can be thought of as filling in the regions between these two extreme limits. Inspection of the figure reveals that, at any given frequency and angle, there are two, one or no propagating solutions.

At high frequency ($X < 1$), waves can propagate in one of two envelopes collectively referred to as the X and O modes, depending on which of these they approach in the perpendicular limit. Both have indices of refraction that approach unity as $X = \omega_p^2/\omega^2$ approaches zero and the plasma frequency becomes negligible compared to the wave frequency. As the variable X increases, the X-mode cuts off first, followed by the O-mode. At frequencies below the plasma frequency ($X > 1$), wave propagation may take place in the so-called Z-mode. The cutoff frequencies for these three modes are the same three given above and do not depend on propagation angle. The characteristic polarization of the X- and O-mode waves are right- and left-elliptically polarized.

For waves propagating along the magnetic field, another resonance near the electron gyrofrequency ($Y = 1$) occurs at still lower frequencies. This solution is also shown at the right side of Figure 9.6 for $\theta = 0$. Waves propagating in this branch of the dispersion relation are highly dispersive and have a group speed that increases with frequency. Pulses propagating in this mode will disperse in such a way that their high-frequency components arrive at their destination first, followed by the low-frequency components. Radiation from lightning strikes can couple into this mode, giving rise to signals in the audible range with a characteristic whistling noise. Such waves are referred to as "whistlers."

9.4.5 Phase and Group Velocity

As mentioned previously, a remarkable property of birefringent media is that the wave phase and group velocities can differ not just in magnitude but also in direction. The phase velocity is, by definition, in the direction of the wavevector, which is normal to the direction of the electric displacement. The group velocity, meanwhile, is in the direction of the Poynting flux, which is normal to the direction of the electric field. The electric displacement and electric field are related by a matrix rather than a constant and are therefore not necessarily parallel. The phase and group velocities are therefore also not parallel in general. In this case, ray paths will not follow the phase velocity, and propagation must be re-evaluated.

The meaning of phase and group velocity can be examined by considering a wave packet propagating in 3-D space. The packet is composed of Fourier modes (plane waves) with spectral amplitude $A(\mathbf{k})$, which has a finite bandwidth and is centered on the wavevector $\mathbf{k}_\circ$. In physical space, the wave packet is the continuous sum of all the Fourier modes:

$$f(\mathbf{x},t) = \frac{1}{(2\pi)^3} \int d^3k A(\mathbf{k}) e^{i(\mathbf{k}\cdot\mathbf{x}-\omega t)}.$$

We can express the wavenumber $\mathbf{k}$ as the center wavenumber $\mathbf{k}_\circ$ plus the difference $\delta\mathbf{k}$. At $t = 0$, the wave packet looks like

$$f(\mathbf{x},0) = e^{i\mathbf{k}_\circ\cdot\mathbf{x}} \frac{1}{(2\pi)^3} \int d^3k A(\mathbf{k}) e^{i\delta\mathbf{k}\cdot\mathbf{x}}$$
$$\equiv e^{i\mathbf{k}_\circ\cdot\mathbf{x}} F(\mathbf{x}),$$

where $F(\mathbf{x})$ is the initial envelope of the packet and the exponential term is the phase factor. The time evolution of the packet will be governed by the dispersion relation $\omega(\mathbf{k})$ for the wave. This can be expanded about the central wavenumber in a power series

$$\omega(\mathbf{k}) = \omega(\mathbf{k}_\circ) + \delta\mathbf{k} \cdot \left.\frac{\partial\omega}{\partial\mathbf{k}}\right|_{\mathbf{k}_\circ} + O(\delta\mathbf{k}^2),$$

with the first two terms providing a reasonable approximation of the dispersion relation so long as the medium is not too dispersive and the bandwidth of the pulse is not too great. Applying this expansion, the wave packet looks like

$$f(\mathbf{x},t) = e^{i(\mathbf{k}_\circ\cdot\mathbf{x}-\omega(\mathbf{k}_\circ)t)} \frac{1}{(2\pi)^3} \int A(\mathbf{k}) \exp\left[i\delta\mathbf{k}\cdot\left(\mathbf{x}-\left.\frac{\partial\omega}{\partial\mathbf{k}}\right|_{\mathbf{k}_\circ} t\right)\right] d^3k$$
$$= e^{i\mathbf{k}_\circ\cdot(\mathbf{x}-\mathbf{v}_p t)} F(\mathbf{x}-\mathbf{v}_g t).$$

It is evident that the time-dependent description of the wave packet is identical to the initial description, except that both the phase factor and the envelope undergo spatial translation in time. The phase factor translates at a rate given by the phase velocity, $\mathbf{v}_p = \omega(\mathbf{k}_\circ)/|\mathbf{k}_\circ|\hat{\mathbf{k}}_\circ$. The envelope translates at the group velocity, $\mathbf{v}_g = \partial\omega/\partial\mathbf{k}|_{\mathbf{k}_\circ}$. The two velocities need not be the same. If the medium is very dispersive, additional terms in the power series

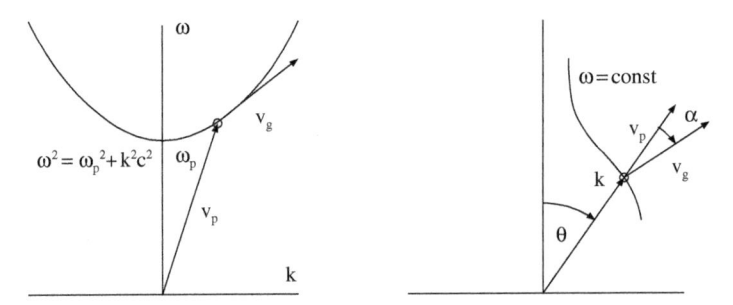

Figure 9.7 Graphical interpretation of phase and group velocities for 1-D isotropic (left) and 2-D anisotropic (right) plasmas.

expansion of ω must be retained. In general, the packet envelope will distort over time as it translates in this case, limiting the range over which information can be communicated reliably.

In the case of an isotropic plasma, we saw that the dispersion relation is simply $n^2 = 1 - \omega_p^2/\omega^2$ or $\omega^2 = \omega_p^2 + k^2 c^2$. (A 1-D treatment is sufficient here.) This dispersion relation forms a parabola in ω-k space. The phase velocity is given by the slope of the line from the origin to any point on the parabola and is always greater than the speed of light c. The group velocity, meanwhile, is given by the slope of the tangent line at that point and is always less than the speed of light. Both velocities share the same direction. Implicit differentiation shows, furthermore, that the product of the phase and group velocities is always the speed of light squared, i.e., $v_p v_g = c^2$.

In the case of an anisotropic plasma, multi-dimensional treatment is necessary in general. Since the problem is symmetric about the axis of the magnetic field, the dispersion relation can be expressed by a wave frequency that is a function of the scalar wavenumber and the angle the wavevector makes with the magnetic field, $\omega(k, \theta)$. The dispersion relation can be plotted readily in polar coordinates in k space as a series of concentric, constant-ω curves or surfaces. The wavevector points from the origin to any point on such a surface. The group velocity at that point can then be found from the gradient in polar coordinates (i.e., $\mathbf{v}_g = \nabla_\mathbf{k} \omega$):

$$v_g = \left.\frac{\partial \omega}{\partial k}\right|_\theta \hat{\mathbf{k}} + \frac{1}{k}\left.\frac{\partial \omega}{\partial \theta}\right|_k \hat{\boldsymbol{\theta}}.$$

By definition, the phase velocity is in the direction $\hat{\mathbf{k}}$. The angle between the group and phase velocity is therefore

$$\alpha = \tan^{-1} \frac{\frac{1}{k}\frac{\partial \omega}{\partial \theta}|_k}{\frac{\partial \omega}{\partial k}|_\theta}$$

$$= \tan^{-1} -\frac{1}{k}\left.\frac{\partial k}{\partial \theta}\right|_\omega,$$

where the last step follows from writing $d\omega(k,\theta) = 0$ on a constant ω surface and considering the dispersion relation in the form $k(\omega, \theta)$. Finally, this angle can be expressed in terms of the phase velocity and the index of refraction:

$$\alpha = \tan^{-1} \frac{1}{v_p} \frac{\partial v_p}{\partial \theta}$$
$$= \tan^{-1} -\frac{1}{n} \frac{\partial n}{\partial \theta}$$
$$= \tan^{-1} -\frac{1}{2n^2} \frac{\partial n^2}{\partial \theta}.$$

9.4.6 Index of Refraction Surfaces

We cannot rely on the equations developed in Section 8.3 for ray tracing in an inhomogeneous medium. That section neglected the effects of birefringence. Its main failing is in equating the direction of the rays with the direction of the phase velocity. As we have argued already, this relationship does not hold for anisotropic media. While a more complete formalism can be developed for ray tracing in inhomogeneous, anisotropic media, it is beyond the scope of this text. Instead, we investigate qualitatively the behavior of the magneto-ionic mode rays with the help of a geometric construction.

The phase and group velocities are said to be in the wave normal and ray normal directions, respectively. The preceding discussion involved calculations of the ray normal direction in terms of dispersion relations stated as $\omega(k, \theta)$. Since the group velocity can be regarded as the gradient of the dispersion relation in wavevector space, it is always normal to surfaces of constant frequency, suggesting a way of studying it graphically.

However, we have seen that the dispersion relations for magneto-ionic modes are often more easily stated in terms of the index of refraction $n(\omega, \theta)$. In order to interpret the phase and ray normal directions graphically, it is useful to define a new vector space with components $n_x = n \sin \theta$ and $n_z = n \cos \theta$, where the direction of the magnetic field continues to be in the $\hat{z}$ direction. (The definition can be generalized to three dimensions using spherical coordinates if desired.) The direction of vectors in this space is understood to be the direction of the wave normal. The length of a vector meanwhile is the index of refraction for a given wave normal direction and frequency. A solution to the Appleton–Hartree equation for some frequency can therefore be represented as a curve or surface in this space. Such surfaces are called refraction index surfaces. There are two at any frequency – the X and O modes for example. Indexes of refraction surfaces for the same mode at different frequencies are concentric.

Since a refractive index surface is obtained for a particular, fixed frequency, the surfaces are also everywhere normal to the ray normal direction. This can be seen by noting the definition of the index of refraction, $n \equiv kc/\omega$, and the consequence of fixing the frequency so that

244   *Propagation*

$$\delta\omega = \frac{\partial \omega}{\partial \mathbf{k}} \cdot \delta\mathbf{k} = 0$$
$$= \mathbf{v}_g \cdot \delta\mathbf{k}.$$

So the group velocity is normal to all allowed displacements in the wavevector at constant frequency, but such displacements all lie along a given refractive index surface calculated for a given frequency. Using this information, it is possible to estimate the shapes of the ray for a magneto-ionic mode using a construction introduced by Poeverlein in the late 1940s. The basic idea is to use the ray tracing equation (or perhaps just Snell's law) for an isotropic medium to estimate the wave-normal direction and then to deduce the corresponding ray-normal directions and ray shapes from the refractive index surface normals. Illustrative examples are given below.

9.4.7 Ionospheric Sounding

We can now begin to address the question of what happens to RF radiation launched vertically into the ionosphere from the ground, this time including the effects of the Earth's magnetic field. An example ionogram is shown in Figure 9.8. Electron density profiles can

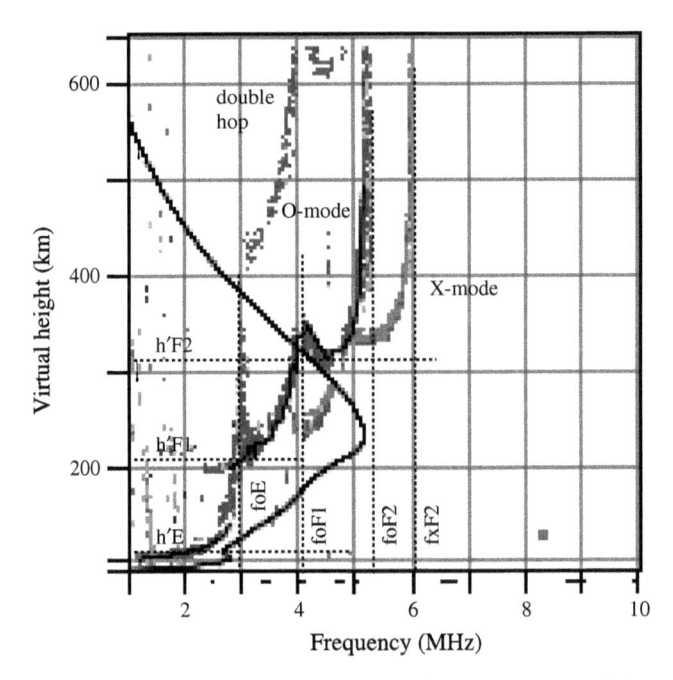

Figure 9.8 Typical vertical incidence ionogram representing winter, solar minimum conditions at high latitudes around noontime. O- and X-mode traces are indicated. Virtual height refers to the echo delay time multiplied by $c/2$. Annotations refer to the virtual heights and critical frequencies of the E, F_1, and F_2 layers. The solid curve is an estimate of the electron density profile based on the traces. The part of the curve above the F_2 layer peak is essentially guesswork.

9.4 Birefringence (Double Refraction)

be deduced, in part, from ionogram traces like those shown in the figure usually with the help of computational inversion algorithms. At ground level, there is no ionization, and the index of refraction of the wave is nearly unity. As the wave ascends into the ionosphere, the plasma frequency increases, X increases, and the wave effectively moves rightward from the far left edge of Figure 9.6. As it does so, the radiation will divide into X and O modes, which will begin to propagate differently. The division depends on the polarization and attack angle of the initial radiation and also on geographic location, since the Earth's magnetic field is oriented differently at different places. It is possible to excite only one of the two magneto-ionic modes by emitting the properly polarized signal from the ground. At high latitudes, the magnetic field is nearly vertical, and a circularly polarized wave launched vertically may excite either the left- or right-circularly polarized mode. Likewise, a linearly polarized signal launched at the magnetic equator can be used to excite either the X or the O mode. In general, however, both modes will be excited by arbitrary radiation sources.

As the waves ascend into the ionosphere, X will continue to increase. At the altitude where $X = 1 - Y$, the X mode will cease to propagate upward. Total reflection will occur, much as happens at the surface of a conductor, and the X-mode signal will be reflected to earth. The O-mode will continue to propagate upward, however, until the condition $X = 1$ is met. Here, the O mode will be reflected and returned to the ground. In Figure 9.8 below about 5.3 MHz, the ionosonde receives X-mode echoes first, followed by O-mode echoes. If the wave frequency is sufficiently high that this condition is never met, the O mode will propagate into space without reflection. In Figure 9.8, only X-mode traces are received between about 5.3 and 6.0 MHz. Likewise, if the $X = 1 - Y$ condition is never met, the X mode too will propagate into space without reflection. No echoes are seen in Figure 9.8 at frequencies above 6.0 MHz.

As was discussed earlier, the altitude at which reflection occurred can be estimated by measuring the travel times for the X- and O-mode waves, which will be detected back at the ground at different times. The travel time signifies the group delay of the signal or the reciprocal of the group speed of the signal, integrated over the propagation path. The virtual height of the echo is defined to be the group delay time multiplied by the speed of light (halved to account for the round trip delay). Virtual heights can be measured as a function of wave frequency in order to gather information about the plasma density as a function of altitude. Inversion algorithms exist for converting the virtual heights into true heights and producing ionospheric density profiles. Radars used for this purpose are called ionosondes. Ionosondes can only measure density profiles below the peak of the F region, since any signals that have not been reflected at this height will never be reflected. Density profiles in the "valley region" between the E and F layers are also difficult or impossible to measure, being obscured by E layer ionization.

The preceding discussion is accurate for vertical incidence but is incomplete and does not address the ray paths of the magneto-ionic modes. For that, we need to analyze the refractive index surface for the two modes. Figure 9.9 (left panel) shows the O-mode refractive index surfaces. Here, the labels indicate the values of X corresponding to the different

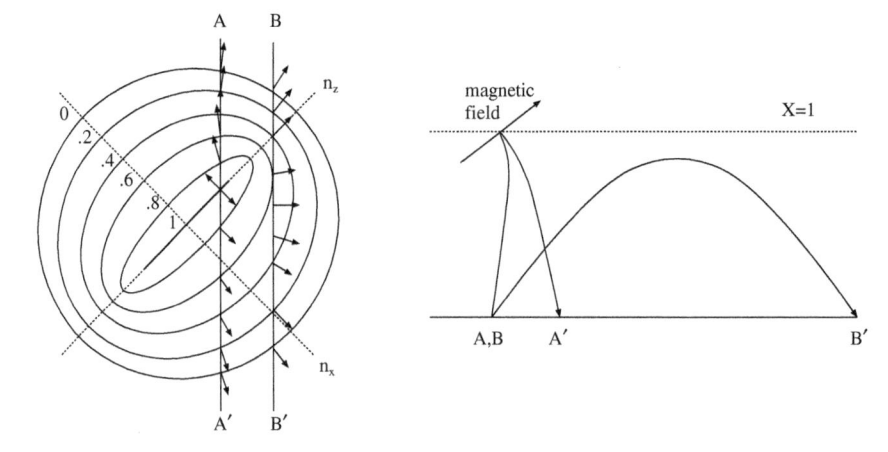

Figure 9.9 Raytracing the O-mode wave in the plane of the magnetic meridian in the southern hemisphere. The magnetic dip angle is set to 45°. (left) Refractive index surfaces for O-mode propagation with $Y < 1$ (oversimplified sketch). The numbers indicate the value of X for which the given surface was drawn. The radius of a point on a surface is the index of refraction $n(\omega, \theta)$ calculated for the given frequency and angle. Vertical lines represent phase normal directions for waves obeying Snell's law. Corresponding ray normal directions are normal to the refractive index surfaces. (right) O-mode rays launched with different zenith angles.

surfaces. The $X = 0$ surface is a circle of radius 1. The $X = 1$ surface collapses to a straight line that passes through the origin. Surfaces at intermediate values of X are ellipsoids. The surfaces have been rotated 45° clockwise so that the direction of the magnetic field is pointed up and to the right. The figure can be viewed as describing wave propagation looking westward in the plane of the magnetic meridian in the southern hemisphere.

The two vertical lines superimposed on the refractive index surfaces represent the wave normals of two waves launched from the ground into a vertically stratified ionosphere. Snell's law demands that the index of refraction times the sine of the zenith angle of the wave normal is a constant, which is the equation for a vertical line in this figure. The greater the initial zenith angle of the wave, the greater the distance of the vertical line from the origin of the figure.

As the waves propagate into and then out of the ionosphere, the ray normal directions are given by the plotted arrows, which are everywhere normal to the refractive index surfaces. Consider first the behavior of the wave $\overline{BB'}$. At the ground, the plasma frequency is zero, and $X = 0$. As the wave enters the ionosphere, X increases, and the ray normal direction rotates clockwise. Eventually, the ray normal direction turns downward, and the wave exits the ionosphere. The turning point occurs in a region where $X < 1$, and so the O-mode reflection height is never reached. The ray path of the wave is illustrated in the right panel of Figure 9.9.

Next, consider what happens to the ray $\overline{AA'}$, which is launched at a much smaller zenith angle. The ray normal direction of this wave initially rotates *counterclockwise*. At the point where it intersects the $X = 1$ refractive index surface, the ray normal direction is normal

9.4 Birefringence (Double Refraction)

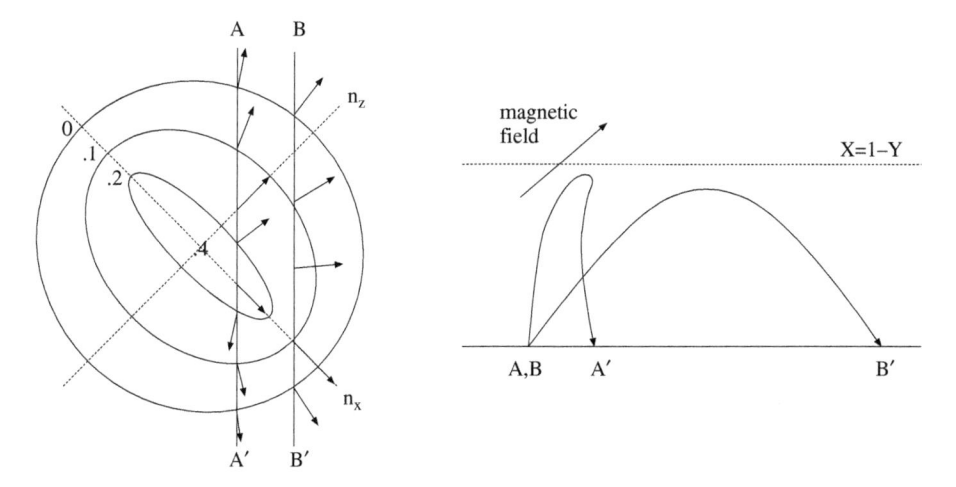

Figure 9.10 Ray tracing the X-mode wave in the plane of the magnetic meridian in the southern hemisphere. (left) Refractive index surfaces for X-mode propagation with $Y < 1$ and $X < 1$ (oversimplified sketch). (right) X-mode rays launched with different zenith angles.

to the direction of the magnetic field. At this point, which is termed the "Spitze," the ray direction reverses, the wave reflects, and it returns through a cusp along the same path through which it came. The ray normal direction rotates clockwise until the wave exits the ionosphere and returns to the ground. In cases where the Spitze exists, reflection occurs at the $X = 1$ height. The range of initial zenith angles for which there is a Spitze is termed the Spitze angle.

Contrast this behavior with that of the X-mode, which is depicted in Figure 9.10. For the X-mode, the refraction index surface collapses to a point at $X = 1 - Y$ rather than to a line; there is consequently no Spitze. Waves launched with finite zenith angles have turning points below the $X = 1 - Y$ altitude. Waves launched with small zenith angles are deflected so as to propagate nearly parallel to the magnetic field near the turning point. For both the X- and O-mode wave solutions, rays return to the ground along the paths by which they entered if the initial incidence is vertical. Note also that the ray paths are reversible.

The X mode is evanescent between $1 - Y < X < X_{uh}$ but propagates again in the regime $X_{uh} < X < 1+Y$, where it is usually referred to as the Z mode. (It is possible to construct refractive index surfaces for the Z mode, just as it was for the X and O modes.) Since the X and O modes are cut off below this range of frequencies, it is difficult to couple radio waves launched from the ground into the Z mode. However, inspection of Figure 9.6 shows that the O and Z modes make contact at $X = 1$. Referring to the left panel of Figure 9.9, the point of contact corresponds to either end of the $X = 1$ line. Waves launched at zenith angles so as to intersect either endpoint can be converted from the O mode to the Z mode at the Spitze. From there, it can be shown (by examining the Z-mode refractive index surfaces – not plotted) that the radiation will propagate nearly horizontally near the $X = 1 - Y^2$

stratum. This radiation will not return to the ground and appear in an ionogram unless additional reflection caused by horizontal inhomogeneity occurs.

9.4.8 Attenuation and Absorption

Collisions mainly between the electrons and neutral atmospheric constituents give rise to Ohmic heating, and the signal is attenuated as a result. The phenomenon is usually termed absorption in the ionosphere. In an anisotropic plasma (no background magnetic field), absorption is governed by (9.5), which retains the electron collision frequency. Defining $Z = \nu/\omega$, that equation can be expressed as

$$n^2 = 1 - X\left(\frac{1+jZ}{1+Z^2}\right),$$

which shows how the index of refraction becomes complex. In the limit that $Z \ll 1$, which holds for most of the HF band and higher frequencies, the real part of the index of refraction is nearly unaffected by collisions. This implies that the ray paths are also unaffected. The imaginary part of the index of refraction, meanwhile, gives rise to attenuation of the signal power, which will be governed by the factor

$$\exp\left(-2k_\circ \int_s \Im(n)ds\right)$$
$$\sim \exp\left(k_\circ \int_s XZ ds\right),$$

where $k_\circ = \omega/c$ and where the integration is over the unperturbed ray path. The factor of 2 comes from the fact that the power is proportional to the square of the signal amplitude. In the case of radar experiments, the path should include the incident and scattered ray paths.

Absorption occurs where the product of the plasma number density and electron collision frequency is high. The collision frequency decreases rapidly with height in the ionosphere, whereas the number density increases rapidly below the F peak. The product is greatest in the D region, which is only present at low and middle latitudes during the day. There, absorption is only important during the day and at frequencies of a few MHz or less. At high latitudes, the D region can be populated by energetic particle precipitation when and where the auroras are active. Very dense D region ionization associated with auroral activity can give rise to significant absorption even at VHF frequencies. Note that when the collision frequency is sufficiently high, the cutoff even vanishes.

The isotropic limit (approximation) holds in a realistic, anisotropic ionospheric plasma so long as the wave frequency is much greater than the cyclotron frequency. At frequencies less than about 30 MHz, however, anisotropy can start to have significant effects. Including the background magnetic field and retaining the electron collision frequency in the derivation, the Appleton–Hartree equation can be shown to assume the form

$$n^2 = 1 - \frac{X}{1 - jZ - \frac{Y_T^2}{2(1-X-jZ)} \pm \sqrt{\frac{Y_T^4}{4(1-X-jZ)^2} + Y_L^2}}$$
$$\equiv (\mu + j\chi)^2,$$

where $\mu = n_r$ and $\chi = n_i$ are referred to as the refractive index and the absorption coefficient in this context. This is a complicated expression, best evaluated computationally. It is illustrative, however, to consider what happens in the quasi-longitudinal ($\theta \sim 0$) and quasi-transverse ($\theta \sim \pi/2$) limits.

The quasi-longitudinal limit applies in situations where $Y_T^2/2Y_L \ll |(1-X) - jZ|$ and as usually quoted as

$$n_{L,R}^2 \approx 1 - \frac{X}{1 - jZ \pm |Y_L|}.$$

Depending on the frequency, the quasi-longitudinal limit applies to angles more than a few degrees from perpendicular propagation. In the opposite limit, which applies only very close to perpendicular, there is the quasi-transverse approximation:

$$n_O^2 \approx 1 - \frac{X \sin^2 \theta}{1 - jZ - X \cos^2 \theta}$$
$$n_X^2 \approx 1 - \frac{X(1 - X - jZ)}{(1 - jZ)(1 - X - jZ) - Y_T^2}.$$

Absorption coefficients are calculated accordingly. The general expressions are quite complicated, but simple results can be obtained in various limiting situations. Information about the collision frequency profile can be derived by comparing the measured O- and X-mode absorption.

9.4.9 Faraday Rotation

Another feature of wave propagation in anisotropic media is the progressive change of polarization along the propagation path. Each magneto-ionic mode has a characteristic polarization, and the polarization of two coincident modes are orthogonal. A signal with arbitrary polarization can therefore be viewed as the superposition of two orthogonal magneto-ionic modes. As these two modes propagate, they obey different dispersion relations, and their phases vary spatially at different rates. The polarization of the combined wave they form is a function of their relative phases. Consequently, polarization can change along the path. One manifestation of this phenomenon is known as Faraday rotation, where the plane of a linearly polarized signal rotates in space.

Consider a linearly polarized wave propagating parallel to the magnetic field in the high frequency limit. This wave is the sum of left- and right-circularly polarized magneto-ionic modes with equal amplitudes. At any point in space, the electric fields of the two modes rotate with the same angular frequency in time but in opposite directions. Measured clockwise from the $\hat{x}$ axis, the angles of the electric fields of the modes are ϕ_L and $-\phi_R$, which are just the phase angles of the modes. The angle of the plane of polarization of the linearly polarized wave formed by the two is then the average, or $(\phi_L - \phi_R)/2$. This angle is constant at any given point in space. Since the wavelengths of the magneto-ionic modes are different, however, the phase difference $\phi_L - \phi_R$ will vary spatially in the direction of propagation, and the plane of the linearly polarized wave will rotate in space.

The fields of the left- and right-circularly polarized modes vary along the path s as

$$E_{L,R}(s) \sim \exp\left[j\left(\omega t - k_\circ \int_s n_\pm(s)ds\right)\right],$$

where the indices of refraction for parallel propagation in the high-frequency, collisionless limit are

$$n_\pm = \left(1 - \frac{X}{1 \pm Y}\right)^{1/2} \sim 1 - \frac{X}{2}(1 \mp Y).$$

Consequently, the accumulated difference in the phase of the left- and right-circularly polarized waves along the path is

$$\Delta\phi = \phi_L - \phi_R = \int_s X(s)Y(s)ds$$

$$= \frac{e^3}{cm_e^2\varepsilon_\circ} \frac{1}{\omega^2} \int_s N(s)B(s)ds.$$

If the propagation is not quite parallel to the magnetic field but falls within the quasi-longitudinal limit, this expression can be written more generally as (in SI units)

$$\Delta\phi = \frac{4.73 \times 10^4}{f^2} \int_s N(s)B(s)\cos\theta \, ds.$$

This formula shows how the plane of polarization of a linearly polarized wave $\Delta\phi/2$ rotates incrementally in a plasma with a background magnetic field. This phenomena can be used to measure the path-integrated plasma number density or magnetic field in a space plasma presuming that the other quantity is somehow known. Differential measurements of the Faraday angle with distance can yield entire profiles of one or the other of these quantities. Differential measurements can be made using spacecraft in relative motion to each other or from radar scatter from different strata in a space plasma. Care must be taken to sample the Faraday angle with sufficient spatial resolution to avoid problems due to aliasing.

Faraday rotation must be taken into account when operating a radio link or radar across the ionosphere. A link made using only linearly polarized antennas will suffer polarization mismatch and fading as the plane of polarization of the received signal changes. Spacecraft communications are often conducted with circularly polarized signals, which approximate a magneto-ionic mode in the quasi-longitudinal limit and so do not undergo Faraday rotation. If linear-polarized signals are used for transmission, both orthogonal polarizations must be received in order to recover all of the power in the link. Circularly polarized receive antennas will not suffer fading but will not recover all the incident power. All of these considerations cease to be important at frequencies well in excess of the plasma frequency (e.g., UHF and above) when the ionospheric plasma is well described by the isotropic approximation.

Note that reciprocity does not hold in this medium. Consider a pair of transceivers using dipole antennas oriented normal to the line separating them and rotated 45° from one another. Because of Faraday rotation, it would be possible for a communication link so constructed to be perfectly efficient for signals going one way and perfectly inefficient for signals going the other. Reciprocity is restored in a broader sense if the direction of the magnetic field is reversed when the roles of the transmitter and receiver are exchanged, however. Note too that, in any event, this argument does not alter the equivalence of antenna effective area and gain.

9.5 Propagation Near Obstacles – Knife-Edge Diffraction

The preceding sections of this chapter considered the effects of refraction on radio- and radar-wave propagation. This section considers the effects of diffraction due to barriers between a transmitter and receiver or between a radar and a scatterer. Scalar diffraction theory was introduced in Chapter 4 in the context of aperture antennas. In that case, aperture antennas were treated as openings in obstacles illuminated from behind by plane waves, i.e., by radiators located infinitely far behind the obstacle. Below, the distances from both the source of radiation and the reception point are allowed to be finite. The tenets of scalar diffraction theory will continue to be observed initially.

Figure 9.11 illustrates the problem of knife-edge diffraction to be considered here, in two dimensions for the sake of simplicity. In between the transmission point T and reception point R is a semi-infinite planar barrier. The distance between T and the edge of the barrier is r_1. We therefore consider the circle of radius r_1 as the aperture. Distances along the

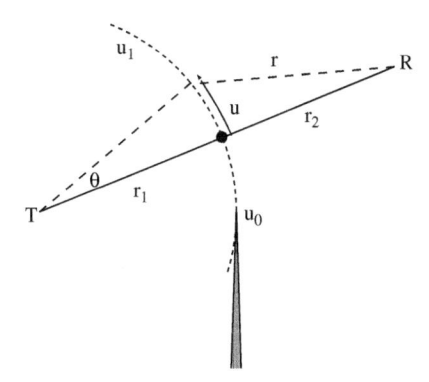

Figure 9.11 Knife-edge diffraction problem geometry. Radiation from the transmission point T passes through an aperture which is a surface of constant range r_1 The coordinate u is measured on the aperture from the origin designated by the filled circle. The aperture is bounded by the points $u_\circ$ and u_1. The distance from a point on the aperture to the reception point R is r.

aperture are marked by the coordinate u, measured from the origin point shown. The aperture is bounded by the obstacle on one side at the point u_o. The point u_1 is the other bound of the aperture as explained below. Since the aperture is a surface of constant distance from T, it can be considered to be uniformly illuminated in $[u_o, u_1]$ in the context of scalar diffraction theory.

Note that we exclude the possibility in this analysis of the transmitter and receiver being situated on the same side of the obstacle such that the condition for reflection from it (angle of incidence equals the angle of reflection) could be satisfied. We consider only the effects of the incident and diffracted fields on the receiver.

Each illuminated point on the aperture will be considered as a Huygens source. The problem is then to calculate the radiation received at the point R. To make this calculation, it will be necessary to determine the distance $r \geq r_2$ to different points on the aperture. Using the law of cosines and taking $\cos(\theta) \approx 1 - \theta^2/2$, $u \sim \theta r_1$ yields

$$\begin{aligned} r^2 &= r_2^2 + u^2 \left(\frac{r_1 + r_2}{r_1} \right) \\ &= (r_2 + \delta)^2 \\ &\sim r_2^2 + 2\delta r_2 \\ \delta &= \frac{u^2}{2} \left(\frac{r_1 + r_2}{r_1 r_2} \right). \end{aligned}$$

The parameter δ is just the difference between r for some point on the aperture and r_2. This parameter is the key to quantifying the effects of diffraction.

The components of the electric field at R can now be estimated by summing the Huygens sources over the aperture (within a constant),

$$E \propto \frac{e^{-jkr_2}}{r_2} \int_{u_o}^{u_1} e^{-jk\delta} du, \qquad (9.16)$$

where the variation in r with u has been neglected in the denominator as usual. By convention, a change of variables from u to a nondimensional parameter v is normally introduced at this point:

$$\begin{aligned} k\delta &= \frac{2\pi}{\lambda} \left(\frac{r_1 + r_2}{r_1 r_2} \right) \frac{u^2}{2} \\ &= \frac{\pi}{2} v^2 \\ v &\equiv \sqrt{\frac{r_1 + r_2}{r_1 r_2} \frac{2}{\lambda}} u \\ &= k_1 u. \end{aligned}$$

9.5 Propagation Near Obstacles – Knife-Edge Diffraction

Substituting into (9.16) then gives

$$E \propto \frac{e^{-jkr_2}}{k_1 r_2} \int_{v_\circ}^{v_1} e^{-j\frac{\pi}{2}v^2} dv,$$

which is very similar to the expression applied to the aperture antenna problem in the Fresnel (near-field) region.

The final step in the derivation of scalar diffraction theory for the knife-edge diffraction problem is the handling of the upper limit of integration v_1. Even though u_1 is finite, v_1 will always be large in the event that the radiation is being observed well within the Fresnel region. With that restriction, we can approximate the electric field as

$$E \propto \frac{e^{-jkr_2}}{k_1 r_2} \int_{v_\circ}^{\infty} e^{-j\frac{\pi}{2}v^2} dv$$

$$= \frac{e^{-jkr_2}}{k_1 r_2} \left(\int_{v_\circ}^{0} e^{-j\frac{\pi}{2}v^2} dv + \int_{0}^{\infty} e^{-j\frac{\pi}{2}v^2} dv \right). \qquad (9.17)$$

The integral is split this way so facilitate the incorporation of the so-called "Fresnel integral" in the calculation. The Fresnel integral is usually defined in terms of its real and imaginary parts

$$C(v) - jS(v) \equiv \int_{0}^{v} e^{-j\frac{\pi}{2}v^2} dv, \qquad (9.18)$$

with the asymptotic limits $C(\pm\infty) = S(\pm\infty) = \pm 0.5$. The Fresnel integral is often depicted graphically by the Euler or Cornu spiral which plots $C(v)$ and $iS(v)$ parametrically in the complex plane (see Figure 9.12). These spirals apply to a number of physical problems and were used by Bernoulli and Euler before being rediscovered by Fresnel. Note that the phase of the Cornu spiral is opposite that of the Fresnel integral by convention.

In the event that there were no screen, $v_\circ \to -\infty$, and the term in the parentheses in (9.17) evaluates to $1 - j$. The ratio of the electric field in the presence of the diffraction screen E to the electric field in the absence of the screen $E_\circ$ may then be written as

$$\frac{E}{E_\circ} = \frac{1}{1-j} \{0.5 - j0.5 - [C(v_\circ) - jS(v_\circ)]\}, \qquad (9.19)$$

which is the Kirchhoff–Fresnel formula for scalar knife-edge diffraction. Note that the boundary $v_\circ$ may be positive or negative. In terms of the Cornu spiral, the term in the angle brackets in (9.19) is the distance on the curve from the point $(0.5 + j0.5)$ to the point $C(v_\circ) + jS(v_\circ)$.

9.5.1 Geometric Interpretation

The modulus of (9.19) is plotted in Figure 9.13 as a function of the parameter v. For positive values of v, the receiver is in the shadow of the obstacle. Here, the radiation amplitude trails

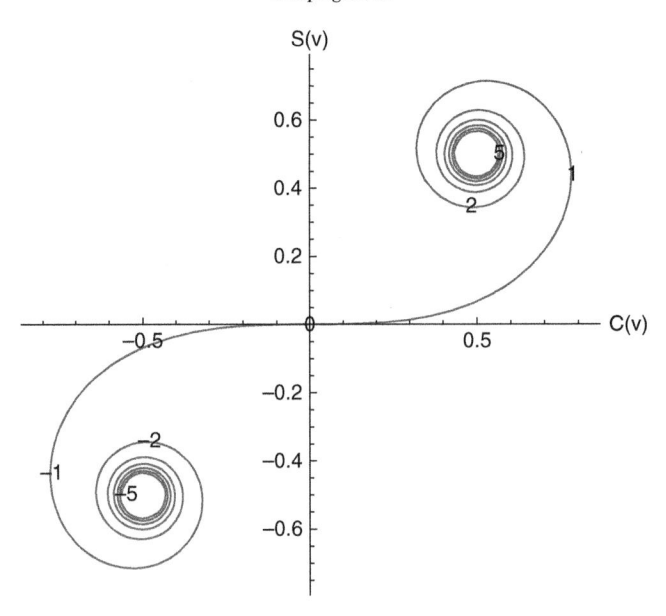

Figure 9.12 Euler or Cornu spiral. The curve offers a graphical means of calculating Fresnel integrals (see text).

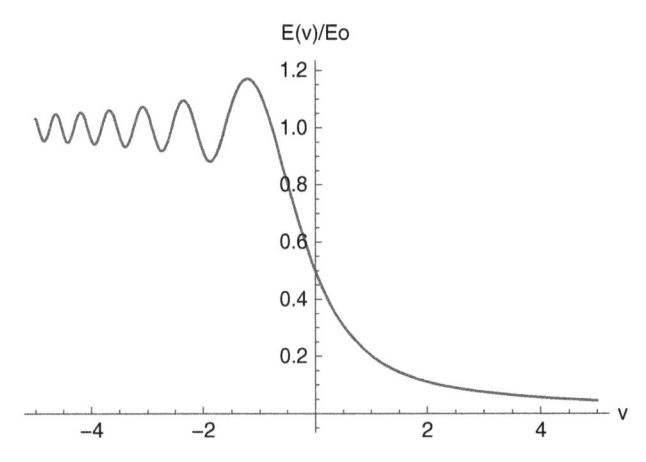

Figure 9.13 Graphical representation of the knife-edge diffraction factor $|E/E_\circ|$ versus the parameter v. Positive values of v place the receiver in the shadow of the obstacle, whereas negative values place it in the direct path of the radiation from the transmitter. The power pattern is the square of the diffraction factor.

off gradually with increasing v from a value of 0.5 at $v = 0$. In the negative-v regime, the receiver is in the direct line of sight of the transmitter. That there is an interference pattern in this regime suggests that knife-edge diffraction can be likened to the presence of an additional radiation source. This virtual source alone remains visible to a receiver in the shadow region.

The parameter δ was derived so that the distance from a point on the aperture to the receiver is $r = r_2 + \delta$. The sum of $r_1 + r + 2 + \delta$ is therefore the geometric length of a path from the transmitter to the aperture to the receiver. When δ is evaluated on the tip of the obstruction at u_o (or equivalently v_o), then $r_1 + r + 2 + \delta$ is the distance from the transmitter to the tip to the receiver. Moreover, δ is the difference between this length and the direct-path length $r_1 + r_2$. Since the interference pattern is defined entirely in terms of this length, we can conclude that the virtual source lies at the tip of the obstruction.

The different peaks and valleys in the interference pattern in Figure 9.13 are the "Fresnel zones" and represent the geometric conditions where the received signal can expect to be stronger and weaker. The crests and troughs are designated by odd and even integers, respectively, starting from the first and largest crest or the first Fresnel zone. The zones depend on the parameter v. Given fixed transmission and reception points and fixed wavelength, v depends only on δ or, equivalently, on the total distance from the transmitter to the tip of the obstruction to the receiver. Surfaces of constant δ are ellipsoids with foci at R and T. These ellipsoids can also be labeled with the appropriate Fresnel-zone numbers. The effect of an obstacle depends on the ellipsoid on which its tip falls.

9.5.2 Other Diffraction Models

The section concludes with an outline of more accurate and comprehensive theoretical treatments of diffraction. The first of these is Sommerfeld's theory which considers the demands of Maxwell's equations both on the relationship between different components of the incident and diffracted vector fields and on the boundary conditions which may be assumed near obstacles. The boundary condition issue will be addressed first.

Rayleigh–Sommerfeld Model

The mathematical basis for diffraction theory is Green's theorem, which is introduced in the Appendix B. The theorem states that two scalar functions U and G are related by

$$\oint_S \left(U \frac{\partial G}{\partial n} - G \frac{\partial U}{\partial n} \right) ds = \int_v \left(U \nabla^2 G - G \nabla^2 U \right) dv,$$

which involves surface and volume integrals over a region of space. We take $U(\mathbf{x})$ to be the solution to a time-independent wave equation (the Helmholtz equation) for an arbitrary source $V(\mathbf{x})$ and $G(\mathbf{x}, \mathbf{x}')$ to be the solution to the same wave equation for a point source, i.e., an impulse response:

$$\nabla^2 U(\mathbf{x}) + k^2 U(\mathbf{x}) = V(\mathbf{x})$$
$$\nabla^2 G(\mathbf{x} - \mathbf{x}') + k^2 G(\mathbf{x} - \mathbf{x}') = \delta(\mathbf{x} - \mathbf{x}').$$

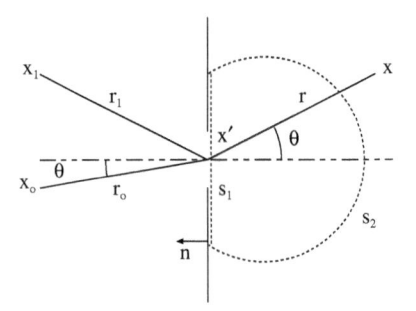

Figure 9.14 Geometry for Sommerfeld's vector diffraction theory. Waves are incident on the aperture plane from the left. A surface encloses an infinite volume on the right side of the aperture plane. The surface has two parts. S_1 is coincident with the aperture plane.

If the source region is infinitesimally small, Green's theorem gives the solution for $U(\mathbf{x})$ in terms of field quantities evaluated on the infinite surface illustrated in Figure 9.14:

$$U(\mathbf{x}) = \oint_{s_1+s_2} \left(U(\mathbf{x}') \frac{\partial G(\mathbf{x}-\mathbf{x}')}{\partial n'} - G(\mathbf{x}-\mathbf{x}') \frac{\partial U(\mathbf{x}')}{\partial n'} \right) ds'. \qquad (9.20)$$

The domain of integration can be divided into two surfaces, one of which (s_1) being coincident with the aperture plane. So long as U vanishes at least as rapidly as a diverging spherical wavefront, the contribution from the other part of the surface (s_2) vanishes. This is the Sommerfeld radiation condition.

Kirchhoff–Fresnel diffraction theory utilizes the Green's function $G(r) = \exp(-jkr)/r$, $r = |\mathbf{x}-\mathbf{x}'|$, which was shown to be a solution of the Helmholtz equation. In order that the domain of integration be further limited to the part of s_1 that is coincident with the aperture, the field quantity U and its normal derivative are taken to vanish on the obstacle. If the source of U is a point source at a distance $r_\circ$ to the left of the aperture plane, then the normal derivatives of G and U on the aperture plane can be expressed as

$$\frac{\partial G}{\partial n'} = \left(-jk - \frac{1}{r} \right) G \cos\theta$$

$$\frac{\partial U}{\partial n'} = -\left(-jk - \frac{1}{r_\circ} \right) U \cos\theta_\circ,$$

where the r^{-1} and $r_\circ^{-1}$ terms can be discarded with the assumption that the source and observation points are located many wavelengths away from the aperture. Finally, the Kirchhoff–Fresnel solution can be expressed as

$$U(\mathbf{x}) = \int_{\text{aper}} U(\mathbf{x}') \frac{e^{-jkr}}{r} jk \left[\cos(\theta_\circ) + \cos(\theta) \right] ds', \qquad (9.21)$$

where the cosine terms are called obliquity factors, heretofore neglected in the analysis. Otherwise, this is the same formula for scalar diffraction theory (to within a multiplicative constant) used throughout the text to this point.

The Rayleigh–Sommerfeld theory differs from the Fresnel–Kirchhoff theory in the choice of the Green's function. The Green's function is an auxiliary function that can be chosen to suit the demands of the problem. The function is required only to be a solution to the Helmholtz equation and to satisfy certain boundary conditions. With the appropriate choice of G, the requirements on the field U on the obstacle can be relaxed. In particular, if either G or its normal derivative can be made to vanish on the entire surface s_1, then it will no longer be necessary to require that both U and its normal derivative vanish on the obstruction in order to limit the region of integration to the aperture region (see (9.20)).

The Green's function that accomplishes this is

$$G_\pm(\mathbf{x},\mathbf{x}') = \frac{e^{-ikr}}{r} \pm \frac{e^{-ikr_1}}{r_1},$$

where $r_1 = |\mathbf{x}' - \mathbf{x}_1|$ and where $\mathbf{x}_1$ is a point at the image of the observation point (see Figure 9.14). The introduction of the image point in this problem is completely analogous to image theory in the context of reflection. On the surface s_1, the function G_- and the normal derivative of the function G_+ both vanish. The function G_- is used in problems where U can be taken to vanish on the diffraction obstacle, and G_+ where $\partial U/\partial n'$ can be taken to vanish on the obstacle. Consequently, the two Rayleigh–Sommerfeld diffraction solutions are

$$U_-(\mathbf{x}) = \int_{\text{aper}} U(\mathbf{x}') \frac{\partial G_-(\mathbf{x}-\mathbf{x}')}{\partial n'} ds'$$

$$U_+(\mathbf{x}) = -\int_{\text{aper}} G_+(\mathbf{x}-\mathbf{x}') \frac{\partial U(\mathbf{x}')}{\partial n'} ds',$$

where which solution to use depends on what boundary condition can reasonably be applied on the obstacle. In terms of the previous Green's function definition, $G_+ = 2G$ and $\partial G_-/\partial n' = 2\partial G/\partial n'$ on s_1. In the limit in which (9.21) applies, the solutions can be written

$$U_-(\mathbf{x}) = \int_{\text{aper}} U(\mathbf{x}') \frac{e^{-jkr}}{r} 2jk\cos(\theta)\, ds'$$

$$U_+(\mathbf{x}) = \int_{\text{aper}} U(\mathbf{x}') \frac{e^{-jkr}}{r} 2jk\cos(\theta_\circ)\, ds'.$$

Finally, the Kirchhoff–Fresnel diffraction solution can be seen to be the arithmetic average of the two Rayleigh–Sommerfeld solutions.

In vector diffraction theory, the vector components of the electromagnetic field are calculated individually to be mutually consistent with Maxwell's equations and the boundary conditions at the obstacle. The solution to the problem depends on the polarization of the incident radiation. Perpendicular (or sometimes "soft") polarization refers to an incident wave with an electric field normal to the plane of Figure 9.14. Parallel (or sometimes "hard") polarization refers to an incident wave with a magnetic field normal to this plane.

Adjacent to a perfectly conducting obstacle, the tangential component of the electric field and the normal derivative of the tangential component of the magnetic field must both vanish. In the case of perpendicular polarization, the Rayleigh–Sommerfeld solution is applied to the incident electric field using G_-. For parallel polarization, the solution is applied to the incident magnetic field using G_+. The other field components downstream of the aperture are then determined through the application of Maxwell's equations. An arbitrarily polarized incident wave can be analyzed using the principle of superposition.

Geometric Theory of Diffraction

In the second half of the twentieth century, the geometric theory of diffraction or GTD was introduced to allow the expedient solution of complicated diffraction problems in three dimensions. GTD combines Rayleigh–Sommerfeld theory with geometric optics, considering rays propagating from a source to a target and allowing for reflection, refraction and diffraction. The net downstream field found through GTD is continuous across the geometric optics boundaries that determine where the direct, reflected and diffracted components can be seen. Diffraction in GTD is seen as occurring at the obstacle edges. Diffraction coefficients, which are calculated from the Rayleigh–Sommerfeld formulas and function like reflection coefficients, are used to determine the transmissivity of the given ray path. The ray paths themselves are calculated using Fermat's principle, i.e., so as to minimize the propagation time. Approximate series representations of the Fresnel integrals are used to speed computations. The uniform theory of diffraction or UTD generalizes the diffraction coefficients of GTD to allow for radiation sources other than point sources. GTD and its successors have proven effective tools for analyzing realistic propagation scenarios.

Spectral Diffraction Theory

The last topic in this section concerns the treatment of diffraction in three dimensions using the theory of linear invariant systems. The treatment is based on Fourier transform theory and is at once powerful and straightforward. The idea is to consider a wave field specified on a plane, e.g., $U(x,y,z=0)$, and to infer the field everywhere downstream of the plane, i.e., $U(x,y,z)$. Some of the complications arising from the physical-space geometry of the problem disappear when the problem is worked in Fourier space.

The analysis begins with the consideration of the 2-D Fourier transform of the field in the plane

$$A(k_x,k_y;0) = \iint_{-\infty}^{\infty} U(x,y,0) e^{-j(k_x x + k_y y)} \, dx dy$$

9.5 Propagation Near Obstacles – Knife-Edge Diffraction

$$U(x,y,0) = \frac{1}{(2\pi)^2} \iint_{-\infty}^{\infty} A(k_x, k_y; 0) e^{j(k_x x + k_y y)} dk_x dk_y,$$

where A is called the spatial spectrum of the field. To simplify the notation, define α, β and γ as the direction cosines of the wavevector **k** such that $k_x = k\alpha$, $k_y = k\beta$ and $k_z = k\gamma = k\sqrt{1-\alpha^2-\beta^2}$. This makes the angular spectrum

$$A(k\alpha, k\beta; 0) = \iint_{-\infty}^{\infty} U(x,y,0) e^{-jk(\alpha x + \beta y)} dx dy.$$

Now, at a point z downstream, the relationship between the field and the angular spectrum is evidently given by

$$A(k\alpha, k\beta; z) = \iint_{-\infty}^{\infty} U(x,y,z) e^{-jk(\alpha x + \beta y)} dx dy$$

$$U(x,y,z) = \frac{1}{(2\pi)^2} \iint_{-\infty}^{\infty} A(k\alpha, k\beta; z) e^{-jk(\alpha x + \beta y)} dk\alpha dk\beta.$$

A constraint on the field quantity U is that it must satisfy the Helmholtz equation downstream of the aperture, i.e., $\nabla^2 U + k^2 U = 0$. Applying this to the formulation of U given immediately above gives an ordinary differential equation for the spatial spectrum,

$$\frac{d}{dz^2} A(k\alpha, k\beta; z) + k^2 \left(1 - \alpha^2 - \beta^2\right) A(\alpha, \beta; z) = 0,$$

which has a solution of the form

$$A(k\alpha, k\beta; z) = A(k\alpha, k\beta; 0) e^{jk\sqrt{1-\alpha^2-\beta^2} z}. \tag{9.22}$$

This solution has two regimes of interest. In the first, $\alpha^2 + \beta^2 < 1$, and the progression in z implies a phase shift for each component of the spectrum. The components are propagating. This is the case wherein α and β correspond to actual, realizable direction cosines. However, the Fourier transform of the angular spectrum involves integrating over all α and β including values that do not correspond to actual direction cosines. In this regime, $\alpha^2 + \beta^2 > 1$, the exponential term in (9.22) is real and negative, and progression in z implies attenuation of the given spectral component. Such components represent evanescent waves that do not support the flow of power away from the aperture.

It is now possible to calculate the field quantity downstream of the aperture by applying the inverse 2-D transform to the angular spectrum found above. In doing so, we apply a weight of unity to the propagating spectral components, zero to the evanescent components, and one-half to components on the boundary,

$$U(x,y,z)$$
$$= \iint_{-\infty}^{\infty} A(k\alpha, k\beta; 0) e^{jk\sqrt{1-\alpha^2-\beta^2} z} H(1-\alpha^2-\beta^2) e^{jk(\alpha x + \beta y)} dk\alpha dk\beta, \tag{9.23}$$

where H is the Heaviside step function. In this formulation, diffraction can be viewed as a linear, invariant system in which the transfer function is the product of the new exponential term and the Heaviside step function. The transfer function is a linear, dispersive spatial filter with limited bandwidth. The Heaviside function prevents angular spectral components beyond the evanescent cutoff $\sqrt{\alpha^2 + \beta^2} = 1$ from contributing to the downstream field and imposes the bandwidth limit, which is the diffraction limit for downstream imaging.

The advantage of this formulation becomes clear when one considers the effects of an obstacle in the $z = 0$ plane on downstream propagation. Define a transmission function $t(x,y)$ as a function which, when multiplied by $U(x,y,0)$, gives $U_t(x,y,0)$, the field quantity immediately behind the obstacle. The 2-D Fourier transform of $t(x,y)$ is $t(k\alpha, k\beta)$, and the corresponding angular spectrum is

$$A_t(k\alpha, k\beta; 0) = A(k\alpha, k\beta; 0) \otimes t(k\alpha, k\beta; 0),$$

i.e., the convolution of the incident angular spectrum and the angular spectrum of the obstruction. This function, replacing A in (9.23), gives $U_t(x,y,z)$, the downstream field quantity after diffraction by the obstacle. In the event that the original illumination is a plane wave, its angular spectrum is a Dirac delta function, and the net angular spectrum is the angular spectrum of the obstacle. In general, the convolution will produce a broadening of the angular spectrum and a diffraction pattern in the downstream field quantity.

9.6 Notes and Further Reading

The propagation of radio waves through inhomogeneous, anisotropic, lossy, random media and/or near obstacles is an enormous topic which has hardly been resolved. R. E. Collin's text treats problems in radio propagation generally. This chapter focused on ionospheric radio propagation because the ionosphere exhibits most every complication that can possibly occur. Four texts that cover this material well are the classics by K. G. Budden, K. Davies and R. D. Hunsucker.

A definitive report explicitly showing how to implement a ray tracing algorithm in the ionosphere was written by R. M. Jones and J. J. Stephenson. The interpretation of the Doppler shift in the ray-theory context was explained by J. A. Bennett. An exhaustive theory of geometric optics in space- and time-varying plasmas formulated in terms of energy density was presented by I. B. Bernstein.

Historical references related to diffraction theory were given already at the end of Chapter 4. A more thorough treatment of the various diffraction theories summarized in this chapter is given in J. W. Goodman's text. The geometric theory of diffraction was developed by J. B. Keller and colleagues and is reviewed in W. L. Studzman's and G. A. Thiele's antenna textbook. The spectral theory of diffraction is reviewed by R. Mittra et al.

E. V. Appleton. Wireless studies of the ionosphere. *J. Inst. Elec. Engrs.*, 71:642, 1932.

J. A. Bennet. The ray theory of Doppler frequency shifts. *Aust. J. Phys.*, 21:259–272, 1968.

I. B. Bernstein. Geometric optics in space- and time-varying plasmas. *Phys. Fluids*, 18(3):320–324, 1974.

H. G. Booker. Oblique propagation of electromagnetic waves in a slowly varying non-isotropic medium. *Proc. Roy. Soc. A*, 237:411, 1936.

M. Born and E. Wolf. *Principles of Optics*. Pergamon Press, New York, 1970.

K. G. Budden. *The Propagation of Radio Waves*. Cambridge University Press, New York, 1985.

R. E. Collin. *Antennas and Radiowave Propagation*. McGraw-Hill, New York, 1985.

K. Davies. *Ionospheric Radio*. Peter Peregrinus Ltd., London, 1990.

J. W. Goodman. *Introduction to Fourier Optica*, 3rd. ed. Roberts & Company, Englewood, CO, 2005.

M. P. M. Hall, L. W. Barclay and M. T. Hewitt, editors. *Propagation of Radiowaves*. The Institute of Electronic Engineers, London, 1996.

R. D. Hunsucker. *Radio Techniques for Probing the Terrestrial Ionosphere*. Springer Verlag, New York, 1991.

R. M. Jones and J. J. Stephenson. A versatile three-dimensional ray tracing computer program for radio waves in the ionosphere. Technical Report 75–76, US Department of Commerce, 1975.

J. B. Keller. Geometrical theory of diffraction. *J. Opt. Soc. of Am.*, 52:116–130, 1962.

R. Mittra, Y. Rahmat-Samii and W. L. Ko. Spectral theory of diffraction. *Appl. Phys.*, 10:1–13, 1976.

W. L. Stutzman and G. A. Thiele. *Antenna Theory and Design*, 3rd ed. John Wiley & Sons, New York, 2012.

Exercises

9.1 Verify the following relationship given in the text to help evaluate long-distance ionospheric propagation paths. Here, h' is the virtual reflection height (see Fig. 9.15).

$$\left(1 + \frac{h'}{R_e} - \cos\theta\right)\csc\theta = \cot\psi_i$$

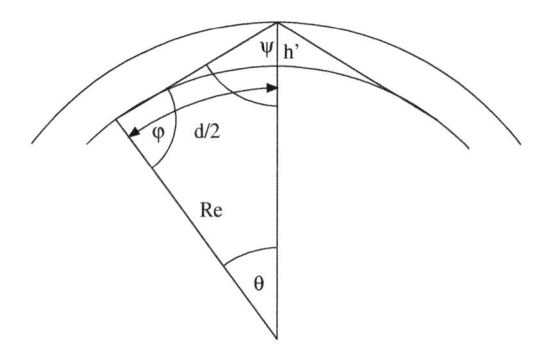

Figure 9.15 Geometry for problem 9.1.

262   *Propagation*

9.2 Suppose a shortwave station in Ohio wants to broadcast to central Europe, a great-circle distance of $d = 6760$ km away. We know that $d = 2\sqrt{2R_e h'}$ for radiation beamed horizontally from the transmitter, implying that the virtual reflection height h' for a single-hop link would have to be 900 km, an altitude which is above the peak of the ionosphere. Hence, the required link must involve two hops or skips, each a distance of only 3380 km.

Assuming a nominal virtual reflection height of 300 km and a peak F-region electron density of 10^{12} m^{-3}, determine the elevation angle of the transmitted radiation and the maximum usable frequency. Repeat the calculation for a nighttime peak F-region density of 10^{11} m^{-3}.

The station would like to operate somewhere within the 31 m shortwave band which occupies 9.2–9.7 MHz. Comment on whether this will be possible during the day and at night.

9.3 Consider a ray propagating from an air-filled region ($y < 0$) into a semi-infinite region ($y > 0$) with an index of refraction given by $n^2(y) = 1 - \varepsilon y^2$. The initial attack angle of the ray at ($x = 0, y = 0$) is given by θ. Find the formula that describes the ray curve, expressed as $y(x)$. Does the ray turn around? If so, at what value of x does it emerge from the medium? Under what conditions does the ray penetrate to the $n = 0$ reflection point?

9.4 Consider the general case of a wave propagating in a plasma parallel to the background magnetic field with Y allowed to be greater or less than one. (You can assume $\omega_p > \Omega_e$ however.) (a) Study the Appleton Hartree equation, and show that one of the solutions can propagate even at frequencies much lower than the plasma frequency. This is the so-called "whistler wave," which has an interesting history related to the discovery of the magnetosphere, the outer part of the ionosphere that extends outward several Earth radii. (b) Calculate the group velocity of this wave and sketch a curve of the delay time for the propagation of a pulse vs. frequency. Hint: for the whistler wave, take $n^2 \gg 1$.

Pulses can be launched by lightning and propagate distances of thousands of kilometers. The sketch should reveal why those pulses are known as whistlers. (c) Find the "nose frequency" of the whistler, the frequency for which the delay time is a minimum. (d) Suppose the path length is 10,000 km, the gyro-frequency is 100 kHz, and the plasma frequency is 2 MHz. Fine the "nose" delay time.

9.5 A satellite emits a linearly polarized wave at a frequency of 200 MHz, and another satellite a distance 1000 km away receives it. Take the propagation path to be parallel to the Earth's magnetic field, which is uniform and has a strength of 0.2 Gauss near the satellites. The electron number density is also constant in this region of space.

a) The plane of polarization is found to differ in rotation by $\pi/2$ radians between the transmission and reception points. On the basis of this information alone, is it possible to determine the electron number density unambiguously?

b) Give a formula for the possible electron number densities.
c) When the distance between the satellites increases to 1100 km, the plane of polarization is found to differ by $3\pi/4$ radians between the transmission and reception points. Estimate the electron number density.

10
Overspread Targets

The signal processing techniques emphasized so far can be classified as "pulse-to-pulse" techniques, deriving spectral information from changes in the scattered signal detected over the course of several interpulse periods. Such techniques are suitable for underspread targets only. Given a low enough radar frequency, a target can often be made to be underspread, and radar engineers may choose the frequency so that simple and effective pulse-to-pulse techniques can be used, even at the expense of other considerations like antenna gain and system noise temperature. However, there are situations where targets cannot be made to be underspread, either because of practical limitations in radar frequency or because of other more fundamental limitations. We saw in Chapter 7 that planetary radar echoes are necessarily overspread in the case of some planets but that pulse-to-pulse techniques can still be applied by ignoring echoes coming from beyond a certain range. This technique is only possible because of the rapid decrease of echo power with range and is imperfect in that it effectively limits the regions of the planets that can be imaged. For most overspread targets, no comparable mitigation strategy exists.

For example, consider radar echoes from meteor heads, or more precisely from the envelopes of plasma surrounding micrometeoroids passing downward through the Earth's upper atmosphere. The echoes are weak and can only be detected with high-power, large-aperture radars that operate, for all practical purposes, at VHF frequencies and above. The meteor heads are detected at altitudes between about 80 and 110 km, implying a minimum interpulse period of $200\mu s$ for unambiguous range detection. For a 50 MHz radar operating at this IPP, Doppler velocities between ± 7.5 km/s are unambiguously detectable with pulse-to-pulse methods. (Since the meteor heads only travel downward, one could argue that descent speeds limited to 15 km/s would be uniquely resolvable.) However, the velocity distribution for meteor heads is broad, and peaks in the vicinity of 50 km/s. Meteor heads are therefore decidedly overspread.

The ionosphere provides another example of an overspread target in the form of modified Thomson scatter (termed "incoherent scatter" by the community that studies it) from free thermal electrons at altitudes above about 70 km. Since the radar-scattering cross-section of an electron is small ($\sim 10^{-28}$ m^2), incoherent scatter is very weak, and its detection is again limited to high-power, large-aperture radars operating at VHF frequencies and above. Incoherent scatter is strongest at altitudes near the F region ionization peak. The effective thickness of the ionosphere as a radar target is at least several hundred km. Since the strongest spectral line in the incoherent scatter spectrum has a width related to the ion acoustic speed, which is typically of the order of a few km/s, this target, too, is necessarily overspread under most circumstances.

These two examples differ in some important respects. The meteor head echo, coming from a hard target, has a narrow spectrum and a long correlation time compared to the shortest IPP that can be employed. Incoherent scatter, meanwhile, comes from a beam-filling target and has a broad spectrum and a short correlation time compared to a practical IPP. Whereas the meteor head problem is one of determining the range and Doppler shift of a point target simultaneously and unambiguously, the incoherent scatter problem is the more difficult one of reproducing spectral shape as a function of range.

Consider once more the meteor head echo problem. The range of the target can be determined using an IPP of 200 μs. Pulse-to-pulse analysis offers a multi-valued solution for the Doppler velocity, with solutions differing by integer multiples of 15 km/s. What is required is a coarse Doppler velocity estimator to help resolve the ambiguity. Such an estimator is available in the form of range-time-intensity (RTI) data. The time rate of change in the echo range, or its range rate, should be evident in an RTI plot. Given a pulse rate of 5000 s^{-1}, even a 50 km/s meteor head will only move 10 m between successive pulses. Allowing, for example, for 10 incoherent integrations to improve detectability, this translates to 100 m displacements between successive power estimates. Incoherent integration is only helpful so long as the target remains within a given range gate, suggesting that the range gates need to be at least 100 m wide in this scenario. In any event, it should clearly be possible to make a coarse estimate of the meteor head velocity on the basis of its range rate. Subsequent spectral analysis could then provide finer estimates of the Doppler shift, which might vary over time as the micrometeorite decelerates due to atmospheric drag.

The forgoing analysis was only possible because of the simplicity of a hard target which presents as a Dirac delta function in range and Doppler frequency. Overspread soft targets are more complicated and difficult to analyze. Soft targets do not exhibit unitary, clearly identifiable range rates as a rule, and their spectra are likely to be distorted by frequency aliasing upon pulse-to-pulse analysis. Fundamentally, it is misguided to study overspread targets with pulse-to-pulse processing. Instead, one must infer their properties by interrogating samples from within individual pulses. We refer to this kind of analysis as "long-pulse" methodology since the length of the pulse now will ultimately determine the frequency resolution of the resulting Doppler spectra.

10.1 Conventional Long-Pulse Analysis

The simplest technique for probing overspread targets involves transmitting long, uncoded pulses with pulse lengths T comparable to or longer than the correlation time of the target. This situation is depicted in the range-time diagram in Figure 10.1. The pulse is oversampled, meaning that sampling takes place at subintervals of the pulse length. The

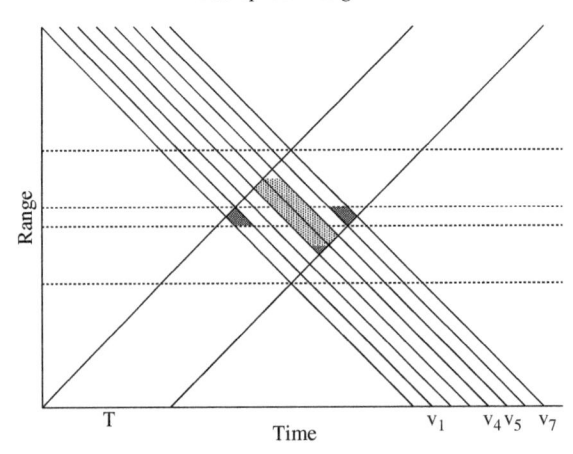

Figure 10.1 Range-time diagram for a conventional long-pulse experiment. In this example, seven samples are taken within the pulse length T.

receiver filter is taken here to have a boxcar-shaped impulse response function with a width equal to the sampling interval τ. The sampling interval itself is chosen to be short compared to the correlation time of the target. The pulse rate is not important for this analysis except insofar as faster pulses mean more statistically independent samples in time and better statistics in the final analysis.

In Figure 10.1, samples $v_1, v_2, \cdots, v_7$ represent the temporal variation in the scattering volume over a period of time T and can form the basis of a spectral estimate. The overall depth of the scattering volume in question, which is bounded above and below by widely spaced, horizontal, dashed lines in the figure, is cT. Clearly, the voltage samples will be related to the spectrum of the signals scattered within the volume. Because of the conflation of range and time, however, contributions to the spectrum from different parts of the scattering volume are unequal. Earlier samples arise from shorter ranges than later samples, for example, and long-period spectral information is only available from intermediate ranges, since only those ranges contribute to all of the samples in the sequence. The root of the problem lies in the attempt to derive both spatial and temporal information from a single independent variable, which is time. This is the central dilemma of long-pulse analysis.

It is often more expedient to analyze long-pulse experiments in terms of autocorrelation functions than spectra, and we will begin to do so here. Autocorrelation functions in long-pulse radar applications are nonstationary as a rule because of variations in the signal with range. Autocorrelation functions can be estimated directly from so-called "lag products," or averaged products of pairs of voltage samples, i.e., $A_{ij} \equiv \langle v_i v_j^* \rangle$. The complete set of lag products for a long-pulse experiment is related to the autocorrelation functions at all ranges by linear transformation, as will be seen later. Considerable further simplification of the problem occurs if the scatter from different parts of the scattering volume can be considered to be statistically independent. This approximation holds in a great many circumstances and rests upon the assumption that the scattering processes in different parts

10.1 Conventional Long-Pulse Analysis

of the volume are not in close communication with one another. Soft targets very often fulfill this requirement.

Suppose we calculated the lag product A_{17} as shown in Figure 10.1. The shaded regions associated with samples 1 and 7 represent the scattering volume that is common to both. Only scatter from this part of the volume will be correlated and contribute to the expectation of the averaged cross product. This volume is smaller than the common volume for the lag product A_{45}, which is depicted by a the lightly shaded region in the figure. This, in turn, is somewhat smaller than the volume for the lag product A_{44}. When computing autocorrelation functions vs. range, we must take into account the fact that different lag products have different effective scattering volume sizes and are consequently weighted differently. For a long-pulse experiment with n times oversampling such that $T = n\tau$, the relative size of the scattering volume of a lag product decreases with the sample offset (lag) according to the sequence $n, n-1/2, n-3/2, \cdots, 1/2$, with only zeros after that.

Aside from weighting issues, the varying range resolution of different lag products poses problems for estimating range-sorted autocorrelation functions in an inhomogeneous medium. One way of dealing with this problem is to degrade the resolution of the lag products through averaging. In Figure 10.1, the range resolution of the zero-lag estimator A_{44} is roughly comparable to first-lag estimator formed from the average of A_{34} and A_{45}. This is roughly comparable to the second-lag estimator formed from the average of A_{24}, A_{35} and A_{46}, etc. Continuing this way, it is possible to form estimators of all the lags of the autocorrelation function for any range, representing roughly the same scattering volume and with a common range resolution of $\sim cT/2$. This is the best range resolution possible for an unmodulated long-pulse experiment. The prescription for estimating the autocorrelation function for range gate r_i and lag j from the lag products is then given by

$$\rho(r_i, j\tau) \propto \sum_{k=i-j}^{i} A_{k,k+j} w_j^{-1}, \qquad (10.1)$$

where the weight w_j is the product of the correct term from the series given above and the number of terms in the sum in (10.1) being added, viz. $w_0 = n, w_j = (n-j+1/2)(j+1), n \geq j > 0$. It should be noted that (10.1) is only an approximation at best and breaks down in the event the filtering is not ideal boxcar. Furthermore, the range resolution of the technique may be quite poor, depending on the pulse length T. If the scattering medium is inhomogeneous and varies appreciably in range, the autocorrelation estimate given by (10.1) may cease to be physically representative or meaningful.

A helpful means of analyzing autocorrelation function estimators is the lag profile matrix. This is a graphical representation of the relationship between lag products and autocorrelation function estimates that may be constructed from them. Figure 10.2 shows representative lag profile matrices for the long-pulse experiment described above and also the coded long-pulse experiment described later in the chapter. The pulse in either case is sampled four times. The points in the matrices represent $A(i, j)$. Diagonals correspond to different lags. The cluster of points highlighted represent lag-products used in estimating the autocorrelation function for a given radar range.

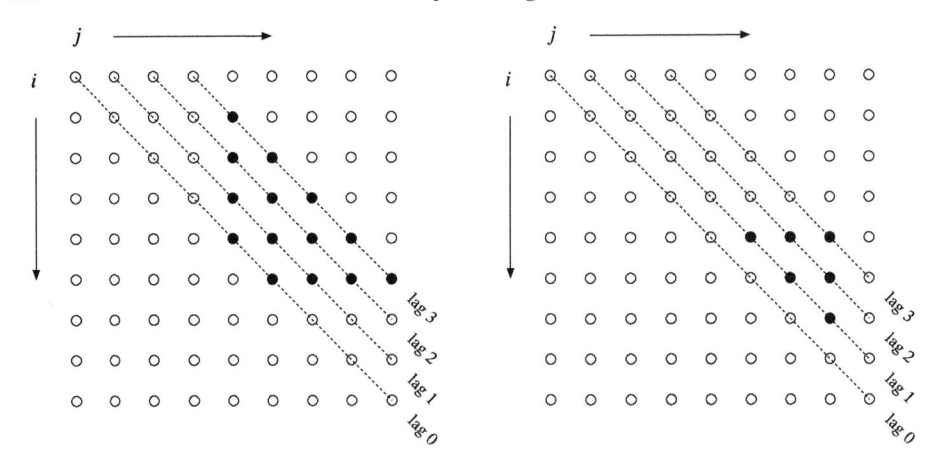

Figure 10.2 Lag profile matrices for long-pulse (left) and coded long-pulse (right) experiments.

Finally, the performance of the long pulse described here is degraded by self clutter. Throughout this text, the term "clutter" refers to radar echoes from a target other than the target of interest. Such echoes behave like interference. Self clutter, meanwhile, refers to echoes from the target of interest but from ranges other than the given range of interest. Recall the discussion pertaining to the calculation of A_{17}, above. Only echoes from the scattering volume in the shaded region highlighted in Figure 10.1 were said to contribute to the expectation of the cross product $\langle v_1 v_7^* \rangle$. However, all of the echos contained within the samples v_1 and v_7 contribute to the variance. The clutter power for the lag product is actually much larger than the signal power in this case, and degrades the experiment in much the same way as noise. In cases where the signal-to-noise ratio is large, self clutter can dominate noise. In such cases, increasing the signal power has little effect on the accuracy of the experiment, since the clutter power is also increased. Note that in the long-pulse experiment, there is no clutter power in the zero lag, and the signal-to-clutter ratio decreases with increasing lag number. We will return to the concept of self clutter in the error analysis section of the chapter.

10.2 Amplitude Modulation and the Multipulse

Both the poor range resolution and the high vulnerability to self clutter inherent in the long-pulse experiment can be mitigated to some extent with a multipulse experiment. A multipulse experiment can be regarded as a long-pulse experiment where the transmitter is inactive part or most of the time. The simplest example is a double-pulse experiment like the one depicted in the left panel of Figure 10.3. Here, a pair of pulses of width τ is transmitted with a delay T and received using a filter with a boxcar impulse response function matched to the pulse width. Two echoes will be received from each range, separated by T. Samples taken at these times form the basis for an estimate of the autocorrelation function for the lag time T. By varying T from pulse to pulse, it is possible to measure multiple lags

10.2 Amplitude Modulation and the Multipulse

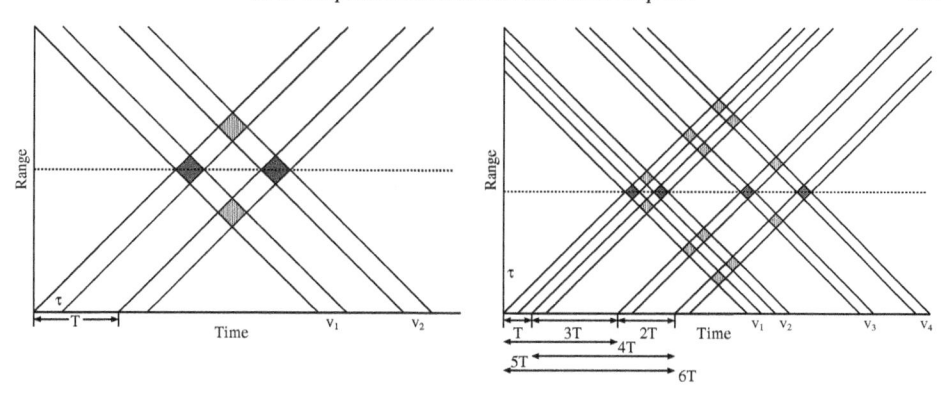

Figure 10.3 Range-time diagram for a double-pulse (left) and multipulse (right) radar experiment. The horizontal dashed lines represent the given range gate under investigation.

of the autocorrelation function, including the zero lag. Furthermore, additional improvements in range resolution can be realized by using pulse compression techniques on the sub pulses.

Self clutter continues to be an issue in the double-pulse experiment, although to a much lesser extent than in the long pulse. In Figure 10.3, self clutter is represented by the light gray diamonds. The clutter does not contribute to the expectation of the cross product, but it will increase the variance. In a double-pulse experiment, each sample receives clutter from a single range in addition to the signal in the range of interest. Roughly speaking, we might expect the signal-to-clutter power ratio to be of order unity in typical double-pulse experiments.

The main limitation of a double-pulse experiment is that it acquires data very slowly – one lag at a time. Multiple lags of the autocorrelation function can be measured using multiple pulses. A four-pulse experiment is illustrated in the right panel of Figure 10.3. Note that the pulse spacings have been chosen judiciously in this example. From the four samples with signals corresponding to a given range gate of interest, estimates of six distinct, nonzero lags of the autocorrelation function (i.e., $nT, n = 1, \cdots, 6$) can be formed. Given a sequence of m strategically placed pulses, it is possible to recover up to $m(m-1)/2$ non-redundant lags of the autocorrelation function, although the lag sequence will necessarily have gaps if $m > 4$. In order to recover the zero lag, a single-pulse experiment must be interleaved with the multipulse.

In the examples shown in Figure 10.3, the minimum lag spacing T is greater than or equal to twice the pulse width τ. This ensures that the clutter is uncorrelated with the main echoes of interest and does not contribute to the lag product estimators. More generally, if the pulse width τ and the filter impulse response time/sample interval τ' are permitted to differ, then the condition becomes $T > \tau + \tau'$. In this case, the range resolution of the experiment is determined by the longer of τ and τ', but in no case by T. Another condition that must be satisfied by the multipulse is that there be no redundant lag spacings. This prevents clutter from any given range from appearing in more than one receiver sample, which would also lead to contamination of the lag product estimates. Procedures exist for generating arbitrarily long-pulse sequences that satisfy this condition. Long sequences may also be found using stochastic (trial-and-error) searches. Very long "aperiodic" sequences reminiscent of the aperiodic pulses employed by some MTI radars are sometimes used.

With additional pulses and lags comes additional self clutter. In the case of the four-pulse experiment, each receiver sample receives clutter from three range gates in addition to the signal from the range gate of interest. We might therefore expect a signal-to-clutter power ratio of roughly 1/3 in a typical four-pulse multipulse experiment. As the multipulse pattern becomes denser, the clutter performance of the experiment becomes poorer.

Spectral analysis of multipulse data can be carried out in one of two ways. First, lag products can be estimated and spectra computed through Fourier transformation of the associated autocorrelation functions. Second, the receiver samples (e.g., v_1 through v_4 in the right panel of Figure 10.3) can undergo Fourier transformation and detection directly. In either case, it will generally be necessary to calculate the Fourier transforms of nonuniformly sampled data. Strategies based on linear constrained minimum variance (LCMV), Kalman filtering and Bayesian inversion have been developed to address this problem. In some applications, the lag products provide sufficient information and spectral analysis need not be undertaken.

It is important to note that the effects of clutter in a multipulse experiment cannot be removed through coding. Code sequences that eliminate the clutter also eliminate the signal. An amazing proof of this exists, but it is too long to be contained in the margins of this text. Clutter can be mitigated through the introduction of polarization and frequency diversity.

10.3 Pulse Compression and Binary Phase Coding

Incorporating binary phase codes in a long-pulse experiment is another way to improve the range resolution of and avoid distortion inherent in the "sum rule" long-pulse analysis represented by (10.1). Coded pulses offer multiple estimates of each lag of the autocorrelation function from each pulse. Furthermore, coded long pulses produce autocorrelation function estimates with uniformly spaced lags, sidestepping the aforementioned complication inherent in multipulse data. Finally, coded long pulses can offer optimal utilization of the duty-cycle and bandwidth capabilities of a transmitter. Of all of the pulsing methods described in this section, however, coded long pulses generally suffer the most from self clutter.

Suppose (case I) we tried to observe meteor head echoes using Barker coded pulses. If the resulting echoes were processed in the conventional way outlined in Chapter 8 using a baseband filter matched to the coded transmitted waveform, we could expect to see very little filter output in general. Only the rare echoes with small Doppler shifts and long periods compared to the pulse length would emerge from the filter. Other echoes would add destructively under filtering and be attenuated. This behavior could be predicted with the help of the range-Doppler ambiguity function for the experiment, which falls off sharply with increasing Doppler frequency.

10.3 Pulse Compression and Binary Phase Coding

Suppose next (case II) that we somehow knew the Doppler shift of the echo in advance. We could then incorporate this information into the matched filter, which would now be matched to a Doppler-shifted copy of the transmitted waveform. The resulting range-Doppler ambiguity function would look like the original, only shifted in the frequency axis such that its peak coincided with the Doppler-shifted frequency of the echo (recall (8.3)). The filter output would be large for the anticipated echo and small for echoes with different Doppler frequencies. In the former case, the echo signal would add constructively throughout filtering, while noise would not, and pulse compression and sidelobe suppression would occur, albeit still with limited effectiveness for echoes with short correlation times.

Suppose finally (case III) that, not knowing the Doppler shift of the echo *a priori*, we simply processed the signal in the manner described above for every possible Doppler shift, one by one. The process is analogous to ordinary coherent processing as described in Chapter 7, only for samples taken within the pulse rather than pulse-to-pulse. After conventional detection and incoherent integration, we would be left with power estimators ordered by frequency, which is a Doppler spectrum. The data would also be ordered by range, as usual. The lowest frequency in the spectrum would be set by the pulse length and the highest by the bandwidth and sample rate, which also set the range resolution. Since the sample rate can be much faster than the pulse rate, the upper frequency bound is much higher than what could be obtained in a pulse-to-pulse experiment. For the 50 MHz meteor radar example in question, given a range resolution of 1 km, the corresponding maximum Doppler velocity would be 75 km/s.

We can explore the simple pulse-code example more closely through numerical simulation (see Figure 10.4). Here, we consider a 5-bit Barker code. The radar scatter comes entirely from a single range and is represented by a random process with a finite correlation time. The received signal is a delayed copy of the random process, multiplied by the transmitted pulse code. Signal processing can be performed either in the method described above (i.e., matched filtering incorporating DFTs) or, equivalently, using long-pulse analysis, incorporating the pulse code in the lag-product computations. Since the latter option is new to this chapter, it is instructive to examine it in more detail.

Lag products for lags between 0 and 4 and for 11 ranges centered on the range containing the echo were computed by summing results from a large number of statistically independent trials. The lag products for each range formed the basis for estimating the autocorrelation function for that range. The pulse code was incorporated into the lag product computations. Since there are 5, 4, ..., 2, 1 lag products contributing to the 0, ,1, ..., 3, 4 lags of the autocorrelation function estimate for each range gate, the autocorrelation function thus computed has an implicit triangular taper. We retain this taper since it facilitates spectral computation with the knowledge that the resulting spectra are artificially broadened.

The left panel of Figure 10.4 shows a spectrogram (range-sorted Doppler spectra) for a simulated echo with a small Doppler shift and a long correlation time (narrow spectral width). The spectrum for the range containing the target is highlighted in bold. The spectra in the four ranges above and the four below represent the range sidelobes. White noise

272   *Overspread Targets*

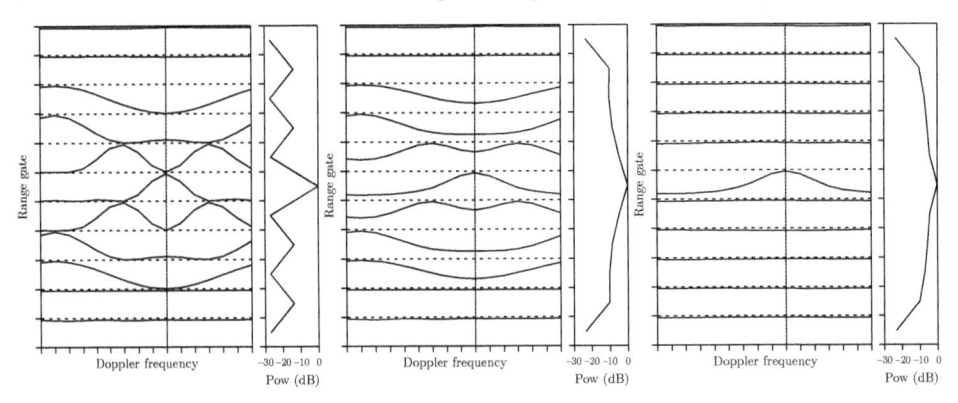

Figure 10.4 Long-pulse spectrograms (range-sorted Doppler spectra) for three cases: (a) 5-bit Barker code analysis of an echo with a long correlation time; (b) same as above, only for an echo with a shorter correlation time; (c) same as above, only for a random coded long pulse. Beside each spectrogram is a profile representing the relative power of the signal in the Doppler frequency bin indicated by the dashed vertical line.

has been added to the simulated signal, and this is all that is present in the upper- and lower-most spectra in the spectrogram. The individual spectra are all self-normalized for plotting. The profile to the right of the spectrogram shows the relative power level versus range computed for the Doppler frequency bin where the actual echo power is a maximum. This bin is designated by the dashed vertical line in the spectrogram.

We can see pulse compression at work in the left panel of Figure 10.4, where the noise is distributed over all Doppler frequency bins, but the signal is concentrated in just a subset of them. The signal-to-noise ratio in those bins is therefore effectively improved. Whereas the total signal-to-noise ratio is 20 dB in this example, the ratio in the frequency bin where the echo strength is maximum is about 28 dB here. Note that in ordinary pulse compression (case I), only the zero frequency bin is retained by matched filtering, and the echo spectrum must therefore be concentrated near zero frequency for pulse compression to work. In case II, any single frequency bin may be retained. In long-pulse analysis (case III), all bins are retained, and the benefits of pulse compression are realized in whatever frequency bins are occupied by signal.

The left panel of Figure 10.4 also demonstrates some sidelobe suppression. With long-pulse analysis, we see that the spectra of the sidelobes are actually distorted and broadened copies of the main-lobe spectrum. What the figure reveals is how sidelobe suppression really involves the shifting of the sidelobes to frequency bins away from the frequencies where the echo signal resides. In the left panel of the figure, the sidelobe power profile for the frequency bin through which the dashed vertical lines passes follows the familiar (0, 1/25, 0, 1/25, 0, 1, 0, 1/25, 0, 1/25, 0) pattern. (In the gaps between sidelobes, the power is reduced to the noise power level rather than to zero.) Considering all frequency bins, however, we see that the total power in each sidelobe is actually comparable to the power in the main lobe.

The middle panel of Figure 10.4 demonstrates how both pulse compression and sidelobe suppression break down when the correlation time of the echo is decreased. Since the echo spectrum is less concentrated in any one frequency bin, the effective signal-to-noise ratio in any frequency bin is improved less. Moreover, as the sidelobe spectra become broader, they can no longer be excluded from the frequencies occupied by the main-lobe spectrum. As the width of the echo spectrum increases, the merits of the coding scheme vanish.

10.3.1 Coded Long Pulse

Consider now how things change when the Barker code is replaced by a sequence of random or pseudorandom binary phase codes that change from pulse to pulse. The right panel in Figure 10.4 illustrates this case, which is termed a "coded long pulse." Here, a different, random, 5-bit binary phase code is transmitted for each pulse, and the lag products representing a large number of pulses are summed and Fourier analyzed to produce the spectrogram. In this case, the sidelobe spectra become flat (uniform) even as the main-lobe spectrum remains accurately rendered. What has happened is that the random pulse code has turned the radar clutter, after summation over the entire pulse sequence, into white noise. The total power appearing in the sidelobes obeys the ratio 1:2:3:4:5:4:3:2:1, with the maximum power in the main lobe. The sharp peak in the power profile in the right panel of Figure 10.4 reflects the concentration of the main-lobe spectrum near the frequency bin of interest there (i.e., pulse compression).

The noise-like nature of the range sidelobes results from the randomness of the binary phase code, which is decoded optimally only in the case of the main lobe. Equivalently, the random code has caused the lag products associated with the sidelobes to add destructively. This is true for all the lag products except the zero lags, which cannot be coded and which always add constructively. Consequently, range-sidelobes add a pedestal to the spectra they contaminate.

It is instructive to explore lag product cancellation more explicitly. Figure 10.5 illustrates the measurement of the decoded lag product $\langle a_\circ a_1 v_\circ v_1^* \rangle = \langle a_\circ a_1 v(t_\circ) v^*(t_1) \rangle$ corresponding to range $r_\circ$ in a coded long-pulse experiment where the transmitted code for the ith pulse is represented by the real sequence $a_{\circ i}, a_{1i}, a_{2i}, a_{3i}$. Explicitly, we have (neglecting numerous constants of proportionality)

$$\langle a_\circ a_1 v_\circ v_1^* \rangle = \langle a_\circ [a_\circ s_\circ(t_\circ) + a_1 s_{-1}(t_\circ) + a_2 s_{-2}(t_\circ) + a_3 s_{-3}(t_\circ)]$$
$$\cdot a_1 [a_\circ s_1(t_1) + a_1 s_\circ(t_1) + a_2 s_{-1}(t_1) + a_3 s_{-2}(t_1)]^* \rangle$$
$$= \langle s_\circ(t_\circ) s_\circ^*(t_1) \rangle$$

$$\overbrace{+ \langle a_\circ a_1 \rangle \langle s_\circ(t_\circ) s_1^*(t_1) \rangle + \langle a_1 a_2 \rangle \langle s_\circ(t_\circ) s_{-1}^*(t_1) \rangle}^{\text{clutter}}$$
$$+ \langle a_\circ a_1 \rangle \langle s_{-1}(t_\circ) s_\circ^*(t_1) \rangle + \langle a_\circ a_2 \rangle \langle s_{-1}(t_\circ) s_{-1}^*(t_1) \rangle$$
$$+ \langle a_\circ a_3 \rangle \langle s_{-1}(t_\circ) s_{-2}^*(t_1) \rangle + \langle a_\circ a_1 \rangle \langle s_{-2}(t_\circ) s_{-1}^*(t_1) \rangle$$
$$+ \langle a_\circ a_1 a_2 a_3 \rangle \langle s_{-2}(t_\circ) s_{-2}^*(t_1) \rangle + \langle a_\circ a_1 \rangle \langle s_{-3}(t_\circ) s_{-2}^*(t_1) \rangle,$$

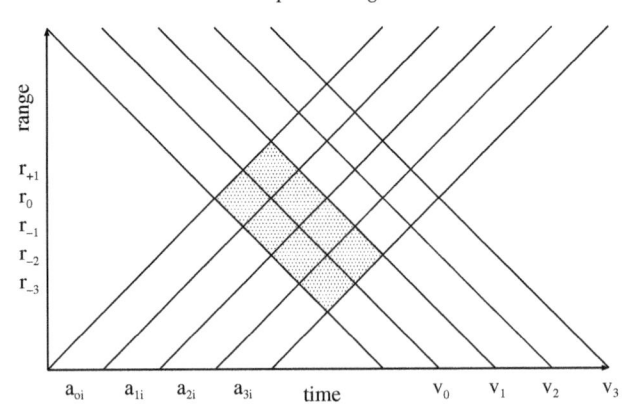

Figure 10.5 Range time intensity representation showing the four range gates involved in the computation of the lag product involving samples $v_\circ = v(t_\circ)$ and $v_1 = v(t_1)$.

where $s_j(t)$ refers to the echo signal from range j received at time t. The result shows that the lag product in question gives the desired lag of the autocorrelation function for the desired range plus, in this case, eight self clutter terms. Each of these is associated either with repeated range gates or with adjacent range gates that, as shown in Figure 10.5, actually overlap halfway in range. (By sub-coding the bits of the code, for example by using a Barker code, these half-overlap terms could be made to vanish, leaving only two clutter terms. We will not explore this strategy in detail, however.) Seven other potential self clutter terms have already vanished from the result by virtue of having no range overlap. (Recall that backscatter from different volumes is statistically uncorrelated.)

While the remaining self clutter terms do not vanish for any single pulse code, they can be made to vanish upon summation given an appropriate code sequence. Random coding accomplishes this for all but the zero lags. Self clutter will subsequently be absent from autocorrelation functions constructed from the lag-products. More specifically, it will be absent from the mean of the autocorrelation function estimates. As self clutter remains present in the individual lag products, it still contributes to the variance of the autocorrelation function estimates and to the statistical errors.

The utility of the coded long pulse depends on the fact that the zero-lag range sidelobes can either removed or neglected. Removal can be accomplished in much the same way as noise removal by estimating and subtracting the spectral pedestal. Alternatively, if the shape of the spectrum is mainly of interest, the analyst may be able to simply ignore the spectral pedestal (or the zero-lag of the corresponding ACF). Note, however, that the self clutter will still increase the variance of the ACF estimate, whether it is subtracted from the mean or not.

10.3.2 Alternating Code

With the coded long pulse, the lag product sidelobes for the nonzero lags are made to cancel asymptotically. However, there are deterministic, repeating coded pulse sequences

10.3 Pulse Compression and Binary Phase Coding

that cause the sidelobes to cancel arithmetically. These sets are referred to as "alternating codes." Two types of codes, weak and strong, have been found. Both have bit lengths that are a power of 2. Strong codes are necessary when of the bits of the code are transmitted continuously and without subcoding. For strong codes, the number of pulses in a complete sequence is twice the number of bits. Codes with lengths longer than 32 bits remained unknown for some time, but practical means of generating very long strong codes now exist.

The alternating codes are derived from the so-called Walsh matrices, which form the basis of an alternative form of fast Fourier transform. The basic Walsh matrix is $[[1,1],[1,-1]]$. Higher-order Walsh matrices are derived by successive multiplications by this matrix. For example, the matrix of order 8 is

$$\begin{bmatrix} + & + & + & + & + & + & + & + \\ + & - & + & - & + & - & + & - \\ + & + & - & - & + & + & - & - \\ + & - & - & + & + & - & - & + \\ + & + & + & + & - & - & - & - \\ + & - & + & - & - & + & - & + \\ + & + & - & - & - & - & + & + \\ + & - & - & + & - & + & + & - \end{bmatrix},$$

where the plus and minus signs denote $+1$ and -1, respectively. The elements of a Walsh matrix can be computed explicitly using

$$\text{Wsh}(i,j) = \text{par}(i \wedge j),$$

where i and j are the matrix indices. Here, the $\wedge$ operator is the parity of a binary number (the bitwise logical *and* of i and j), which is 1 (-1) if that number has an even (odd) number of 1s in its binary representation.

Alternating codes are composed from the columns of Walsh matrices. In order to construct a 4-bit strong alternating code, for example, utilize columns 0, 1, 2 and 4 of the 8th-order Walsh matrix.

```
    0   1   2   4
    +   +   +   +
    +   -   +   +
    +   +   -   +
    +   -   -   +
    +   +   +   -
    +   -   +   -
    +   +   -   -
    +   -   -   -
```

The code to transmit is a sequence of eight 4-bit codes, each code in the sequence being given by one of the rows of the Walsh matrix shown above.

Table 10.1 *Strong alternating code sequences. The number of distinct pulses in a sequence, NP, is twice the number of bits in each coded pulse. Values are represented here in octal.*

NP	b_0	b_1	b_2	b_3	b_4	b_5	b_6	b_7	b_8	b_9	b_{10}	b_{11}	b_{12}	b_{13}	b_{14}	b_{15}
8	00	01	02	04												
16	00	01	02	04	10	03	07	16								
32	00	01	03	04	10	20	17	37	21	14	31	35	24	06	15	32

Finding alternating codes amounts to selecting the right columns of the Walsh matrix and putting them in the right order. The derivation of the method for doing so is beyond the scope of this text. Instead, we merely give the generating functions for a few strong code sequences. It is a straightforward matter to verify that the codes have the required properties. Table 10.1 specifies alternating codes of length 4, and 8, and 16 bits, where NP, the number of distinct pulses in the code sequence, is twice the number of bits. The various b_i refer to columns of the Walsh matrix so that the value of the ith bit of the jth code in the sequence is given by

$$a_{ij} = \text{Wsh}(j, b_i).$$

Note that any alternating code multiplied by a constant code remains an alternating code. Note also that the first n bits of an alternating code is also an alternating code.

One or two differences between the alternating codes and the coded long pulse bear mentioning. Systematic clutter cancellation in the former is only guaranteed to the extent that the filter impulse response function is ideal and exact. In the event the filter is imperfect, exact cancellation may not occur, and the residual error will become systematic. Furthermore, in the event that the echoes from different ranges are not statistically uncorrelated, the seven resurgent clutter terms will also make systematic contributions to the lag products, and these might easily be mistaken for signals of interest. By virtue of its randomness, the coded long pulse is potentially more robust and resistant to these two failure modes. In practice, the alternating code can be randomized through multiplication with codes that change randomly or pseudorandomly at the completion of every alternating code sequence. Such a strategy possesses the strengths of both methods.

Note finally that another family of alternating codes has been developed that is not restricted to power-of-2 lengths. Sidelobe cancellation is achieved with these codes only with the summing of all the lag product measurements that correspond to a given lag of the ACF. (With the alternating code, self clutter is removed from every distinct lag product.) Such an approach becomes impractical if the transmitter power varies over the length of the pulse. This code family is not considered in the text that follows.

10.4 Range-Lag Ambiguity Functions

Additional mathematical machinery is necessary at this point to evaluate and compare the efficacy of the various long-pulse methodologies. Just as the range-Doppler ambiguity

10.4 Range-Lag Ambiguity Functions

function was helpful for assessing the performance of pulse-to-pulse experiments, the so-called "range-lag ambiguity function" can be useful for evaluating long-pulse experiments. The derivation of the function is given below. Some illustrative examples are presented afterward. The radar target is taken to be a soft target with echoes arising from index of refraction fluctuations ΔN distributed throughout the scattering volume. The scattering will be regarded as a Gaussian random process in light of the central limit theorem.

A radar signal radiated at time $t - 2r/c$ and scattered off index of refraction fluctuations at time $t - r/c$ and radar range r will be received at the radar at time t. The relative phase of the transmitted and received signals will be $\mathbf{k} \cdot \mathbf{r}$ radians, where $\mathbf{k}$ is the scattering wavevector (the difference of the incident and scattered wavevectors) and $\mathbf{r}$ is the scattering coordinate. In light of the first Born approximation, summing over the radar-scattering volume gives the total signal applied to the receiver

$$x(t) = \int d^3\mathbf{r}\, e^{i\mathbf{k}\cdot\mathbf{r}} s(t - 2r/c) \Delta N(\mathbf{r}, t - r/c).$$

We exclude constants associated with the transmitter power, antenna gain, electron scattering cross-section and target range that are not relevant to the current analysis. Here, $s(t)$ is the transmitted pulse shape that defines the range extent of the scattering volume. If the impulse response function of the receiver is $h(t)$, then the corresponding receiver output at the sample time t_s will be

$$y(t_s) = \int dt\, x(t) h^*(t_s - t)$$
$$= \int dt\, d^3\mathbf{r}\, e^{i\mathbf{k}\cdot\mathbf{r}}$$
$$\cdot s(t - 2r/c, \sigma) \Delta N(\mathbf{r}, t - r/c) h^*(t_s - t).$$

The receiver output can be regarded as a zero-mean Gaussian random variable, with information contained in the second-order statistics:

$$\langle y(t_{s2}) y^*(t_{s1}) \rangle =$$
$$\int dt_1 dt_2 d^3\mathbf{r}_1 d^3\mathbf{r}_2\, e^{i\mathbf{k}\cdot(\mathbf{r}_2 - \mathbf{r}_1)}$$
$$\cdot \langle \Delta N(\mathbf{r}_2, t_2 - r_2/c) \Delta N^*(\mathbf{r}_1, t_1 - r_1/c) \rangle$$
$$\cdot s(t_2 - 2r_2/c) h^*(t_{s2} - t_2) s^*(t_1 - 2r_1/c) h(t_{s1} - t_1).$$

The following variable substitutions will be expedient:

$$\mathbf{r}_1 \equiv \mathbf{r}$$
$$\mathbf{r}_2 \equiv \mathbf{r} + \mathbf{r}'.$$

Taking the index of refraction fluctuations to have a physically short correlation length, we can take $\mathbf{r}_2 - \mathbf{r}_1 = \mathbf{r}' \sim 0$ everywhere in this expression except in the exponent. This yields

$$\langle y(t_{s2})y^*(t_{s1})\rangle =$$
$$\int dt_1 dt_2 d^3\mathbf{r} \int d^3\mathbf{r}' e^{i\mathbf{k}\cdot\mathbf{r}'}$$
$$\cdot \langle \Delta N(\mathbf{r}+\mathbf{r}', t_2 - r/c)\Delta N^*(\mathbf{r}, t_1 - r/c)\rangle$$
$$\cdot s(t_2 - 2r/c)h^*(t_{s2} - t_2)s^*(t_1 - 2r/c)h(t_{s1} - t_1).$$

We can now perform the $\int d^3\mathbf{r}'$ integral, noting that the term in the angle brackets is the spatio-temporal autocorrelation function of the density irregularities and that the integral in question is then its spatial Fourier transform. The result is the temporal autocorrelation function for electron density irregularities with a wavevector $\mathbf{k}$ or $\rho(\mathbf{k}, t2, t1; \mathbf{r})$. This term is stationary and depends only on $\tau \equiv t_2 - t_1$. Consequently, another change of variables can now be made:

$$t_1 \equiv t$$
$$t_2 \equiv t + \tau,$$

resulting in

$$\langle y(t_{s2})y^*(t_{s1})\rangle =$$
$$\int d\tau d^3\mathbf{r} \rho(\mathbf{k}, \tau; \mathbf{r}) \int dt\, s(t + \tau - 2r/c)$$
$$\cdot h^*(t_{s2} - t - \tau)s^*(t - 2r/c)h(t_{s1} - t).$$

Finally, defining the 1-D and 2-D radar ambiguity functions

$$W_{ts}(t, r) \equiv s(t - 2r/c)h^*(t_s - t)$$
$$W_{ts1,ts2}(\tau, r) \equiv \int dt\, W_{ts2}(t + \tau, r) W_{ts1}^*(t, r)$$

leads to the following compact result:

$$\langle y(t_{s2})y^*(t_{s1})\rangle$$
$$= \int d\tau dr \rho(\mathbf{k}, \tau; \mathbf{r}) W_{ts1,ts2}(\tau, r). \qquad (10.2)$$

The quantity on the left is the lag product matrix, which constitutes the measurements for the experiment with associated noise and statistical uncertainty. The ambiguity functions are known for any lag product measurement in a given radar experiment. The autocorrelation function is related to the lag products through a 2-D convolution with the 2-D ambiguity function. The autocorrelation function itself is related to the state parameters of the scattering medium through whatever theory is appropriate.

Figure 10.6 shows five ambiguity functions plotted side-by-side for a long-pulse experiment with a pulse width 5 times the receiver sampling interval. What is plotted here is $W_{ts1,ts2}(\tau, r)$ for $ts2 - ts1 = 0$, 1, 2, 3 and 4 times the receiver sample interval, i.e., for lag products corresponding to the 0, 1, 2, 3 and 4 lags. The curves show the ranges and autocorrelation function lag times included in or represented by the given lag product.

10.4 Range-Lag Ambiguity Functions

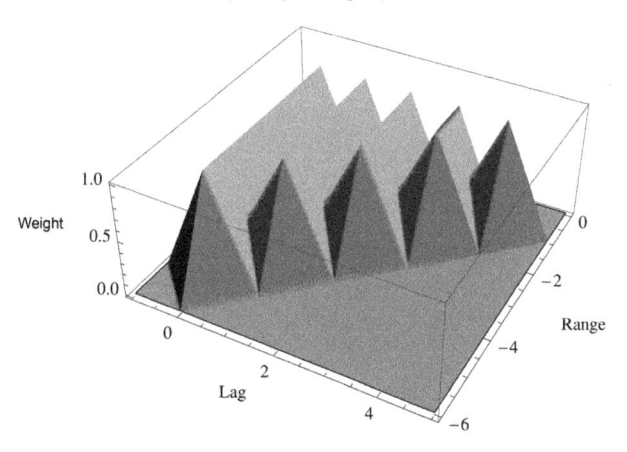

Figure 10.6 Ambiguity functions $W_{ts1,ts2}(\tau, r)$ for a long-pulse experiment with a pulse 5 times the width of the sampling interval. Five different ambiguity functions are plotted here, representing the lags from 0 to 4.

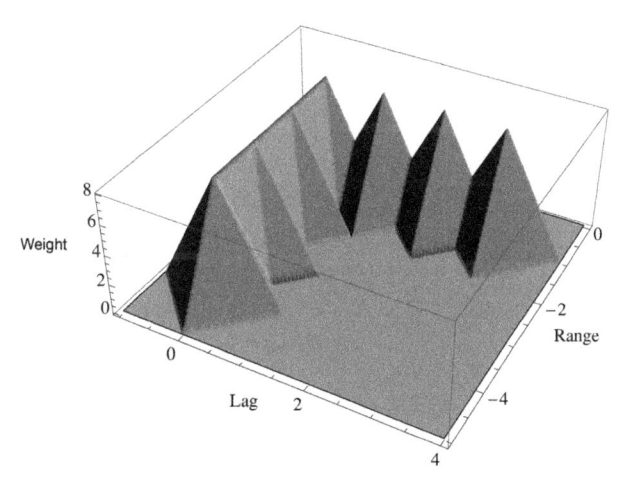

Figure 10.7 Ambiguity functions $W_{ts1,ts2}(\tau, r)$ for a 4-bit alternating code experiment. Four different ambiguity functions are plotted here, representing the lags from 0 to 3 decoded for range gate -1. Eight different ambiguity functions representing eight different pulse sequences have been added in constructing this figure.

They illustrate how the different lag products have different range centers and range resolutions. Furthermore, they show how the lag products actually span overlapping ranges and times. This information is necessary, for example, for predicting measured lag products on the basis of model autocorrelation functions and power profiles. Inverting the problem and predicting the latter from the former is challenging, and the summation rule given by (10.1) gives only a crude, approximate solution to the problem.

Figure 10.7 shows four superimposed ambiguity functions for a 4-bit alternating code experiment, i.e., $W_{ts1,ts2}(\tau, r)$ for $ts2-ts1 = 0, 1, 2, 3$ times the sampling interval. The coded pulses were decoded for range gate -1, and the ambiguity functions are tightly concentrated around the desired range and lag except for the zero lag, for which coding cannot be effective. Except for the zero lag, the correlated clutter has been made to vanish in the course of summing the eight alternating pulse codes in the sequence. Note that correlated clutter is present to varying degrees in each of the pulse codes taken separately. This correlated clutter, along with the clutter represented by the uncoded zero lag, can contribute significantly to the variance of the lag product estimators, particularly in the high SNR limit. In order to understand why and to formulate mitigation strategies, we much undertake a statistical error analysis.

10.5 Error Analysis

Range-lag ambiguity functions permit the assessment of different coding strategies for overspread targets on the basis of the range and time resolution of lag product measurements. A complete assessment of the merits of the strategies also requires comprehensive error analysis. We are interested in calculating the variances and covariances of lag product estimators and the degree to which they are aggravated by self clutter. Much of the methodology outlined below and some of the resulting formulas can also be applied to pulse-to-pulse techniques.

Back in Chapter 7, it was argued that the variance of the estimate of the power relative to the mean of the estimate is reduced through incoherent integration. We begin this section of the text by reproducing that result. Let $V_i + V_\circ$ stand for the ith sample of the receiver output voltage, where V_i can be regarded as a Gaussian random variable and where $V_\circ$ represents a constant (DC) bias in the voltage due, for example, to an finite amplifier offset. An obvious estimator for the receiver output power based on N statistically independent samples is

$$\hat{\mathbf{P}} \equiv \frac{1}{N} \sum_{i=1}^{N} (V_i + V_\circ)(V_i + V_\circ)^*.$$

Taking the expectation of this estimator reveals it to be biased:

$$\langle \hat{\mathbf{P}} \rangle = P + P_\circ,$$

where $P \equiv \langle V_i V_i^* \rangle$ and $P_\circ \equiv \langle V_\circ V_\circ^* \rangle$ define the signal power and DC offset power, respectively. Since $\langle V_i + V_\circ \rangle = V_\circ$, it would be a straightforward matter to estimate the DC bias voltage from the raw receiver samples and remove it, either at the voltage level or at the power level. The two procedures might appear to be equivalent since they both lead to power estimates with the same expectation. The former procedure is performed routinely with analog receivers, which are prone to DC bias problems. The latter procedure turns out to be decidedly suboptimal for reasons made clear below, and is not performed in practice.

Let us estimate the mean squared error or variance σ^2 of the power estimator $\hat{\mathbf{P}}$. This is the expectation of the square of the difference of the estimator and its expectation:

10.5 Error Analysis

$$\sigma_v^2 \equiv \langle (\hat{\mathbf{P}} - (P+P_\circ))^2 \rangle$$

$$= \left\langle \left(\frac{1}{N} \sum_{i=1}^{N} (V_i + V_\circ)(V_i + V_\circ)^* - (P+P_\circ) \right)^2 \right\rangle$$

$$= \left\langle \frac{1}{N^2} \sum_{i,j=1}^{N} (V_i + V_\circ)(V_i + V_\circ)^*(V_j + V_\circ)(V_j + V_\circ)^* \right.$$

$$\left. - (P+P_\circ)\frac{2}{N} \sum_{i=1}^{N} (V_i + V_\circ)(V_i + V_\circ)^* + (P+P_\circ)^2 \right\rangle.$$

With the help of the even-moment theorem for Gaussian random variables, and being sure to distinguish between products of voltages with like and unlike indices, it is possible to evaluate this expression and arrive at

$$\sigma_v^2 = \frac{1}{N} \left(P^2 + 2PP_\circ \right),$$

making the relative root mean squared error of the power estimator

$$\sigma_v/P = \frac{1}{\sqrt{N}} \sqrt{1 + 2P_\circ/P}. \qquad (10.3)$$

In the event that there is no DC bias in the receiver voltage samples at the time of detection, this reverts to the earlier result that the relative root mean squared (RMS) error of the power estimator is given by the number of statistically independent samples to the minus one-half power.

In the event of a voltage bias $V_\circ$, the variance of the power estimator increases, even if the bias $P_\circ$ is itself subtracted from the power estimator. That correction has no effect on the estimator variance, which can be thought of as the discrepancy between an estimator and its expectation due to the finiteness of the sample size. This example therefore illustrates how certain aspects of a signal can affect the estimator variances even if they do not affect the mean, something that is prone to happen when bias corrections are made at the power rather than the voltage level.

Let us now turn to the more general problem of variances and covariances in correlation function estimators. Suppose there are four complex data streams labeled V_{1i} through V_{4i}, where i is a counting index. Take the data to be zero-mean Gaussian random variables. Let A represent the inner product estimator based on k statistically independent samples, i.e.,

$$A_{12} \equiv \frac{1}{k} \sum_{i=1}^{k} V_{1i} V_{2i}^* \qquad (10.4)$$

$$= \langle A_{12} \rangle (1 + \varepsilon_{12}), \qquad (10.5)$$

where A_{12} has been represented in terms of its expectation and a departure arising from the finite number of samples used to form the estimate. The meaningful statistic characterizing this departure is the error covariance,

$$\langle (A_{12} - \langle A_{12}\rangle)(A_{34} - \langle A_{34}\rangle)^*\rangle = \langle A_{12}\rangle \langle A_{34}^*\rangle \langle \varepsilon_{12}\varepsilon_{34}^*\rangle. \tag{10.6}$$

By definition, we may write $\langle A_{12}A_{34}^*\rangle = \langle A_{12}\rangle\langle A_{34}^*\rangle(1 + \langle \varepsilon_{12}\varepsilon_{34}^*\rangle)$. The term on the left of this equation can be estimated from

$$\langle A_{12}A_{34}^*\rangle = \left\langle \frac{1}{k^2}\sum_{i,j}(V_{1i}V_{2i}^*)(V_{3j}V_{4j}^*)^*\right\rangle$$
$$= \frac{1}{k}\langle V_{1i}V_{2i}^*V_{3i}^*V_{4i}\rangle + \frac{k(k-1)}{k^2}\langle V_{1i}V_{2i}^*V_{3j}^*V_{4j}\rangle_{i\neq j},$$

where products involving like and unlike indices have been segregated. With the help of a theorem for even moments of Gaussian random variables, we then obtain

$$\langle A_{12}A_{34}^*\rangle = \langle A_{12}\rangle\langle A_{34}^*\rangle + \frac{1}{k}\langle A_{13}\rangle\langle A_{24}^*\rangle.$$

Substituting for $\langle \varepsilon_{12}\varepsilon_{34}^*\rangle$ and solving for the corresponding error covariance term in (10.6) gives

$$\langle \delta_{12}\delta_{34}^*\rangle \equiv \langle (A_{12} - \langle A_{12}\rangle)(A_{34} - \langle A_{34}\rangle)^*\rangle$$
$$= \frac{1}{k}\langle A_{13}\rangle\langle A_{24}^*\rangle, \tag{10.7}$$

which demonstrates how the error covariance of a two cross-product estimators depends on the values of other cross products that can be formed from the available samples.

Following the same methodology, we can equally well determine that

$$\langle \delta_{12}\delta_{34}\rangle \equiv \langle (A_{12} - \langle A_{12}\rangle)(A_{34} - \langle A_{34}\rangle)\rangle$$
$$= \frac{1}{k}\langle A_{14}\rangle\langle A_{23}^*\rangle. \tag{10.8}$$

The purpose of the dual calculation is that it provides a means of separating the covariances of the real and imaginary parts of the cross products, which can be expressed in terms of the real and imaginary parts of the average and difference of (10.7) and (10.8). Expanding the δs in terms of their real and imaginary parts gives

$$\langle \Re\delta_{12}\Re\delta_{34}\rangle = \frac{1}{2}\Re(\langle \delta_{12}\delta_{34}^*\rangle + \langle \delta_{12}\delta_{34}\rangle) \tag{10.9}$$

$$\langle \Im\delta_{12}\Im\delta_{34}\rangle = \frac{1}{2}\Re(\langle \delta_{12}\delta_{34}^*\rangle - \langle \delta_{12}\delta_{34}\rangle) \tag{10.10}$$

$$\langle \Im\delta_{12}\Re\delta_{34}\rangle = \frac{1}{2}\Im(\langle \delta_{12}\delta_{34}\rangle + \langle \delta_{12}\delta_{34}^*\rangle) \tag{10.11}$$

$$\langle \Re\delta_{12}\Im\delta_{34}\rangle = \frac{1}{2}\Im(\langle \delta_{12}\delta_{34}\rangle - \langle \delta_{12}\delta_{34}^*\rangle), \tag{10.12}$$

which together completely specify the error covariance for the cross-product estimators for the four data streams. Note that the autocovariances can be found by replacing the 3–4 indices with 1–2 indices in the formulas.

The error covariance matrix for a standard long-pulse experiment can be calculated by identifying A_{ij} with the long-pulse lag products. For illustrative purposes, consider the covariance of the real parts of the lag products, which may be sufficient to specify the statistical errors for radar targets with small Doppler shifts or with Doppler shifts that have been estimated from the phase angles of the lag products and removed. Rewriting (10.9) gives

$$\langle \Re(A_{ij} - \langle A_{ij} \rangle) \Re(A_{kl} - \langle A_{kl} \rangle) \rangle = \frac{1}{2k} \Re \left(\langle A_{ik} \rangle \langle A_{jl}^* \rangle + \langle A_{il} \rangle \langle A_{jk}^* \rangle \right), \quad (10.13)$$

where the indices i, j, k, l need not be distinct. To the extent that the various lag products on the right side of (10.13) are large, the errors are likely to be significant and highly correlated, i.e., the error covariance matrix will not be diagonally dominant. There may be important consequences for theoretical analysis of the lag products using least-squares fitting, for example. Here, diagonalization through similarity transformation is an expedient way to handle error analysis properly.

10.5.1 Noise and Self Clutter

Equation (10.13) also illustrates how noise and clutter influence and degrade long-pulse experiments. Noise and clutter do not affect the lag products means (after summing over pulses in a sequence) except for the zero lags. It may seem as if those effects could be negated by either correcting the zero lags for noise and clutter (through estimation and subtraction) or by ignoring the zero lags in subsequent analyses, if possible. However, note that the zero lag expectations appear on the right side of (10.13) whenever repeated indices appear on the left side, i.e., whenever i or j equals k or l. Repeated indices denote shared samples. Noise and clutter consequently can contribute to the covariances of lag products even when they do not contribute to the means. In the case of the autocovariances, for which $i = k$ and $j = l$, zero lag expectations necessarily appear in the right side of (10.13), and noise and self clutter absolutely contribute.

Just as lag product measurements generally have to be manipulated to produce range-sorted autocorrelation function estimates, so the error estimators described in the previous section of the text also have to be combined and manipulated to predict ACF covariances. The manipulation amounts to combining different lag products that relate to the same component of the ACF for a given range gate or range gates. If the manipulation can be cast in the form of a linear transformation $\rho = BA$, then the error covariance matrices propagate according to the standard formula

$$C_\rho = B C_A B^T,$$

which can be applied directly, for example, to the summation rule (10.1) for long-pulse analysis. Note that this formulation assumes that the estimators are unbiased and that the errors are normally distributed, which is a good approximation, given sufficient incoherent integration.

Equation (10.13) can be applied to coded long-pulse and alternating-code experiments if the voltage samples are interpreted as having been decoded for the range gate of interest. However, error propagation is complicated by the fact that the error covariance matrix C_A is different for different pulses in the pulse sequence. This implies an extra layer of averaging over the pulses in the sequence, i.e.,

$$\overline{C_\rho} = B\overline{C_A}B^T,$$

where the overbar represents an average over the pulses in a sequence and where the target is assumed to be statistically stationary over the time it takes to complete the sequence.

Pulse coding has two major effects on statistical errors. First, there is an overall reduction in the magnitude of the error terms, especially the off-diagonal terms of the error covariance matrix, due to the fact that the lag product expectations on the right side of (10.13) are generally smaller than in a long-pulse experiment. After summing over all pulses in the sequence, we know that the nonzero lag products will be weighted by the power in a single range gate rather than multiple range gates. For any given pulse, the lag products reflect contributions from adjacent range gates also, but these contributions are as likely to have negative as positive weights, and so considerable cancellation is possible for the terms on the right side of (10.13). The diagonal terms, meanwhile, are not drastically reduced by pulse coding because the zero-lag lag products that dominate them are unaffected by coding.

The deterministic nature of the alternating code can be a liability if any of the pulse codes in the sequence contributes disproportionately to $\overline{C_A}$. Including the summation over all the pulses in a sequence, the right side of (10.13) contains terms of the form $\overline{\langle A_{ij}\rangle\langle A_{kl}\rangle}$. In the event that the clutter terms in the individual lag products are uncorrelated, such terms can be approximated by $\overline{\langle A_{ij}\rangle}\,\overline{\langle A_{kl}\rangle}$, i.e., the averaged lag products in which self clutter does not appear. Such is the case for the coded long pulse. It is not the case for the alternating code for which correlated clutter is an issue and the corresponding terms can assume much larger values. A remedy for this, called the "randomized alternating code," combines the best characteristics of both methods. Here, the alternating code pulse sequences are multiplied by random pulse codes that vary from one sequence to the next. This prevents codes with particularly unfortunate properties from being transmitted. An alternating code multiplied by a constant is still an alternating code, and so arithmetic rather than asymptotic cancellation of the range sidelobes is still afforded by the method. This is a significant simplification, since the quantities in the latter expression can be estimated empirically.

Second, the signal-to-clutter power ratio tends to be much greater in coded long-pulse and alternating-code experiments than in long-pulse experiments, as was discussed earlier in the chapter. Consequently, although the error covariance matrix in coded pulse experiments tends to be diagonally dominant, the diagonal terms (autocovariances) tend to be larger than they are in long-pulse experiments in the high signal-to-noise ratio limit, where clutter is most important. Even in the very low SNR limit, where the effects of self clutter become negligible, the SNR will be higher in a long-pulse experiment than in a comparable coded-pulse experiment. The price paid for the former is increased ambiguity, since long-pulse lag products are much more difficult to relate directly to autocorrelation functions.

In summary, (10.13) can be used to evaluate the error covariances of autocorrelation function estimates from coded long-pulse and randomized alternating-code experiments through the use of the following associations:

$$\overline{\langle \Re(A_{ij} - \langle A_{ij}\rangle)\Re(A_{kl} - \langle A_{kl}\rangle)\rangle} \longrightarrow S^2 \langle \Re(\rho_{ij} - \langle \rho_{ij}\rangle)\Re(\rho_{kl} - \langle \rho_{kl}\rangle)\rangle$$

$$\overline{\langle A_{ij}\rangle} \longrightarrow \begin{cases} S\rho(\tau_{ij}), i \neq j \\ S+N+C, i = j, \end{cases}$$

where τ_{ij} is the lag time associated with the given lag product, and S, N and C refer to the signal power in the range gate of interest, the clutter power (power scattered from other ranges), and the noise power, respectively.

10.6 Notes and Further Reading

The literature concerning radar probing of overspread targets is application specific and not very extensive. The references cited below come from the area of upper-atmospheric physics. Multipulse experiments were introduced into this field by D. T. Farley. The coded long pulse was the invention of M. P. Sulzer, working at Arecibo, while the alternating code, discovered by M. S. Lehtinen, was adopted first among the EISCAT radars in Europe. The alternating code has undergone steady refinement, as documented below.

A topic that was not discussed in this chapter is the idea of inferring parameters from lag profile matrices using statistical inverse methods and so-called "full profile analysis." This method was discussed first by M. S. Lehtinen and developed in more detail by J. M. Holt. Another overlooked but promising approach is the use of aperiodic pulse sequences.

D. T. Farley. Multiple-pulse incoherent-scatter correlation function measurements. *Radio Sci.*, 7:661, 1972.

B. Gustavsson and T. Grydeland. Orthogonal-polarization alternating codes. *Radio Sci.*, 44(RS6005):doi:10.1029/2008RS004132, 2009.

Orthogonal-polarization multipulse sequences. *Radio Sci.*, 46(RS1003):doi:10.1029/2010RS004425, 2009.

J. M. Holt, D. A. Rhoda, D. Tetenbaum and A. P. van Eyken. Optimal analysis of incoherent scatter radar data. *Radio Sci.*, 27:435–447, 1992.

A. Huuskonen, M. S. Lehtinen and J. Pirttilä. Fractional lags in alternating codes: Improving incoherent scatter measurements by using lag estimates at noninteger multiples of baud length. *Radio Sci.*, 31:245, 1996.

D. L. Hysell, F. S. Rodrigues, J. L. Chau and J. D. Huba. Full profile incoherent scatter analysis at Jicamarca. *Ann. Geophys.*, 26:59–75, 2008.

M. S. Lehtinen. Statistical theory of incoherent scatter radar measurements. Technical Report 86/45, Eur. Incoherent Scatter Sci. Assoc., Kiruna, Sweden, 1986.

M. S. Lehtinen and I Häggström. A new modulation principle for incoherent scatter measurements. *Radio Sci.*, 22:625–634, 1987.

M. S. Lehtinen, A. Huuskonen and M. Markkanen. Randomization of alternating codes: Improving incoherent scatter measurements by reducing correlations of gated ACF estimates,. *Radio Sci.*, 1997.

M. S. Lehtinen, A. Huuskonen and J. Pirttilä. First experiences of full-profile analysis with GUISDAP. *Ann. Geophys.*, 14(12):1487–1495, 1997.

M. Markkanen and T. Nygrén. A 64-bit strong alternating code discovered. *Radio Sci.*, 31(2):241–243, 1996.

Long alternating codes, 2, Practical search method. *Radio Sci.*, 32(1):9–18, 1997.

M. Markkanen, J. Vierinen and J. Markkanen. Polyphase alternating codes. *Ann. Geophys.*, 26(9):2237–2243, 2008.

T. Nygrén. *Introduction to Incoherent Scatter Measurements*. Invers OY, Sodankylä, Finland, 1996.

M. P. Sulzer. A radar technique for high range resolution incoherent scatter autocorrelation function measurements utilizing the full average power of klystron radars. *Radio Sci.*, 21:1033–1040, 1986.

A new type of alternating code for incoherent scatter measurements. *Radio Sci.*, 28:995, 1993.

S. V. Uppala and J. D. Sahr. Spectrum estimation moderately overspread radar targets using aperiodic transmitter coding. *Radio Sci.*, 29:611, 1994.

J. Vierinen. Fractional baud-length coding. *Ann. Geophys.*, 29(6):1189–1196, 2011.

I. I. Virtanen, M. S. Lehtinen, T. Nygrén, M. Orispää and J. Vierinen. Lag profile inversion method for EISCAT data analysis. *Ann. Geophys.*, 26:571–581, 2008.

I. I. Virtanen, J. Vierinen and M. S. Lehtinen. Phase-coded pulse aperiodic transmitter coding. *Ann. Geophys.*, 27:2799–2811, 2009.

J. L. Walsh. A closed set of normal orthogonal functions. *Amer. J. Math*, 45:5–24, 1923.

Exercises

10.1 Write a program to calculate and plot the range-lag ambiguity functions for a long pulse with a width five times that of the sampling interval. Take the filter impulse response function to be rectangular with a width equal to the sample interval. Plot functions for lags 0–4. The results should look like Figure 10.6.

10.2 Write a program to generate 4-bit strong alternating codes. The program will need a routine to calculate the Walsh sign sequences on the basis of parity. Verify that the codes have the properties advertised for them by calculating and plotting the corresponding range-lag ambiguity functions. Make use of the work done for Exercise 10.1. The results should look like Figure 10.7.

11
Weather Radar

Meteorological phenomena were vividly apparent to radar operators during the Second World War, and while the main concern lay in mitigating the effects of the radar clutter the weather produced, monitoring and forecasting the weather were also recognized as being important to the war effort. Precipitation was first observed by radar operators in Britain in 1940 and at the MIT Radiation Laboratory in 1941, and the first publication on meteorological echoes appeared in 1943. Also in 1943, Project Stormy Weather in Canada began, fostering an era of pioneering work in precipitation and cloud physics. Academic symposia and textbooks appeared immediately after the conclusion of the war. The availability of surplus radars spurred rapid development in the emerging science of radar meteorology. In the 1950s, operational networks of weather radar began to proliferate. Radar networks expanded throughout the succeeding decades, culminating in the United States with the deployment of the NEXRAD network of WSR-88D Doppler weather radars in the 1990s and the Terminal Doppler Warning Radar (TDWR) network, completed in 2011.

The principles of radar covered throughout this text apply directly to the study of weather. As in any subdiscipline, however, weather radar is accompanied by its own vernacular and specialized formalisms which have evolved historically to address the most pressing problems in the field. This chapter presents an overview of those problems and the attendant language and formalism.

When precipitation is present, backscatter from water or ice droplets or *hydrometeors* dominates all other forms of backscatter. Backscatter from hydrometeors is governed by the principles of Lorenz–Mie theory as outlined in Chapter 6. It is usually desirable to observe precipitation at wavelengths approximately an order of magnitude larger than the diameter of the precipitation drop size of interest. This maximizes scattering efficiency while preserving the applicability of Rayleigh scattering theory, simplifying the interpretation of the observations greatly. Table 11.1 lists representative drop sizes for different hydrometeor types. The NEXRAD radars operate at S-band frequencies between 2.7 and 3 GHz or wavelengths close to 10 cm. This is a good choice for observing most hydrometeors associated with severe weather. For observing clouds or clear air, C- and X-band radars are more commonly used. These radars present a much smaller physical footprint and can operate at lower power levels.

Table 11.1 *Representative diameters of hydrometeors detected by radar.*

Hydrometeor	diameter (mm)
Fog	0.01
Mist	0.1
Drizzle	1.0
Light rain	1.5
Heavy rain	2.0
Graupel	2–5
Hail	5–10
Sleet	5–50
Snow	1–100

The subsequent sections of this chapter concern the interpretation of radar backscatter from weather systems, severe weather, and clouds and clear air. One of the main goals of radar meteorology is the estimation of rainfall rates from radar echoes. Another goal is the estimation of wind speed and direction. This information is available to meteorologists operationally now, although the methodology is undergoing improvement. Finally, the microphysics of clouds and of turbulence constitute frontier areas of research.

11.1 Radar Cross-Section and Radar Equation

The radar-scattering cross-section from precipitation can be understood on the basis of concepts already developed in this text. In the Rayleigh scattering limit, the scattering cross-section of a hyrdometeor can be written as

$$\sigma = \frac{\pi^5}{\lambda^4} |K|^2 D^6$$

$$K \equiv \frac{\varepsilon_r - 1}{\varepsilon_r + 2},$$

where D is its diameter and ε_r is its relative permittivity or dielectric constant. The radar meteorology problem is necessarily one of volume scattering. Summing the contributions from n scatterers in the scattering volume V and assuming the Born scattering limit allows the expression of the scattering cross-section in terms of two useful auxiliary variables, the radar reflectivity Z and the radar reflectivity per unit volume η,

$$Z \equiv \frac{1}{V} \sum_{i=1}^{n} D_i^6 \qquad (11.1)$$

$$= N \langle D_i^6 \rangle \qquad (11.2)$$

$$\eta \equiv \frac{\pi^5}{\lambda^4} |K|^2 Z,$$

where N is the number of scatterers per unit volume.

For quantitative analysis of volume scatter from precipitation, a specification for the antenna backscatter gain is also required. An approximation commonly employed for weather-radar applications is that the antenna radiation pattern has a Gaussian shape, i.e., $P_\Omega = \exp(-\theta^2/2\sigma_b^2)$, where θ is the polar angle measured from the antenna boresight and σ_b describes the width of the pattern. In terms of the σ parameter, the half-power full beamwidth of the pattern is readily found to be $\Theta = \sqrt{8\ln 2}\sigma_b \approx 2.35\sigma_b$. The solid angle is $\Omega = \int P_\Omega d\Omega \approx 2\pi\sigma_b^2$ in the limit of small σ_b in which the $\sin\theta$ term in $d\Omega$ can be replaced with θ to simplify the integral. The corresponding gain is $G_m = 2/\sigma_b^2$ or $G(\theta) = G_m \exp(-\theta^2/2\sigma_b^2)$. Finally, the backscatter gain for a Guassian beam shape is

$$\begin{aligned} G_{bs} &= \frac{1}{4\pi} \int G^2(\theta) d\Omega \\ &= \frac{G_m^2}{2} \int_0^\pi e^{-\theta^2/\sigma_b^2} \theta d\theta \\ &= \frac{G_m^2}{4} \sigma_b^2 \\ &= \frac{G_m^2 \Theta^2}{32\ln 2}, \end{aligned} \quad (11.3)$$

where the last result has been stated in terms of the half-power full beamwidth.

Incorporating the scattering cross-section and the backscatter gain above into the radar equation for volume scatter given in Section 8.3.2 yields an expression suitable for weather-radar applications:

$$\begin{aligned} P_r &= \frac{P_t \sigma L_r c\tau \lambda^2}{32\pi^2 R^2} G_{bs} \\ &= \frac{\pi^3 P_t G_m^2 L_r}{1024\ln 2 R^2 \lambda^2}, \Theta^2 c\tau |K|^2 Z, \end{aligned} \quad (11.4)$$

for a volume at range R, illuminated by rectangular pulses of width τ and incorporating matched filtering in the processing. (Recall from Chapter 8 that the scattering volume width correction associated with square pulses and matched filtering is $L_r = -1.76$ dB.) In the event the beam shape is ellipsoidal rather than circular, the factor Θ^2 in (11.4) can be replaced with $\Theta\Phi$, the product of the two half-power beamwidths that describe such a beam.

The radar equation (11.4) neglects factors for losses in the antenna feedlines, the antenna coupling network, and the receiver. These factors need to be measured and introduced as multiplicative constants if calibrated reflectivity measurements are to be made. Depending on the system and the application, cross-polarization effects may have to be considered as well.

As is generally the case with volume scatter, the received power is inversely proportional to the square of the distance to the scattering volume rather than the fourth power. It is also proportional to the pulse width. All of the factors other than $|K|^2 Z$ reflect characteristics of the radar system. The $|K|^2 Z$ factor, meanwhile, reflects characteristics of the precipitation.

It is an interesting historical note that a more naïve treatment of the problem ignoring the volume-scatter aspect of the backscatter results in a formula similar to (11.4), only missing a factor of $2 \ln 2$ in the denominator. This factor was identified empirically in early weather radar analyses before the theoretical reasons for it were fully understood.

11.1.1 Reflectivity

The reflectivity Z is one of the main observables of weather radar. Its theoretical definition is given by (11.1) and its experimental or effective definition in terms of power measurements is given by solving the radar Equation (11.4) for Z. A discrepancy between the theoretical and measured definitions will arise if the Rayleigh scattering limit or another simplifying assumption leading to (11.4) is violated. In that case, the measured value of Z is known as the effective reflectivity, sometimes written Z_e.

While the natural units for Z would seem to be m^6/m^3, the convention is to refer to the reflectivity in decibels compared to a reference of $1 \text{ mm}^6/m^3$, i.e.,

$$\text{dBZ} \equiv 10 \log \left(\frac{Z}{1 \text{ mm}^6/m^3} \right).$$

The decibel scaling is necessary because of the large dynamic range of the scattering cross-section in practice. Large hailstones may produce backscatter at the level of 75 dBZ, whereas scatter from fog might be detected at the level of -25 dBZ, for example. The dynamic range of a radar system, the ratio of the largest to the smallest discernible received power level, is typically about 90 dB in conventional systems. This precludes the possibility of reliably observing both kinds of precipitation at the same range at once.

Bright bands One of the most prominent features encountered with precipitation radars is the appearance of bright bands of backscatter in annular regions surrounding the radar in plan-position indicator (PPI) displays. Bright bands were seen in early radar observations of weather, but the correct interpretation came only after reflectivity was better understood. It is now known that the echoes that form the bright bands come from the melting layer between snow and rain, something that is most apparent in range-height indicator (RHI) displays.

Hydrometeors made of snow can be very large, but their reflectivity is limited by the fact that $|K|^2$ for ice is relatively small, as will be seen in the section on relative permittivity, below. As the snow melts from the outside in, the hydrometeors become covered with a layer of liquid water, and $|K|^2$ and the reflectivity increase dramatically as a result. This is the source of the bright bands. As the snow continues to melt, however, diameter and reflectivity decrease. In addition, with decreasing diameter comes increasing terminal velocity. The faster-falling raindrops occupy the radar-scattering volume less densely, and the reflectivity decreases even further.

Drop-size distribution A tacit assumption in the foregoing analysis has been that the hydrometeors responsible for the radar scatter are uniform in size. In fact, different distributions of sizes will be present depending on the prevailing meteorological conditions. Stratiform clouds produce rain with a different distribution of sizes than convective clouds, for example. The distribution can be measured using particle probes on aircraft as well as ground-based distrometers. An ongoing area of study is the identification of parametrized statistical drop-size distributions which can accurately capture both the radar reflectivity and other less-accessible weather parameters, the goal being to make inferences about the latter from measurements of the former.

For a drop-size distribution $N(D)$ that gives the number of drops with diameters in the interval between D and $D + dD$ per unit volume and that carries units of number per cubic meter per mm, the reflectivity will be

$$Z = \int_0^\infty N(D) D^6 \, dD. \tag{11.5}$$

A reasonable drop-size distribution should resemble measurements while being analytically tractable. *A priori*, it is known that water drops larger than about 5 mm in diameter are unstable, implying a large-diameter limit on the distribution. Additional information has historically come from observing campaigns conducted under different weather conditions. In 1948, J. S. Marshall and W. M. K. Palmer promoted an exponential drop-size distribution that is still widely used and referenced today:

$$N(D) = N_\circ \exp(-\lambda D), \tag{11.6}$$

where λ is the slope parameter and $N_\circ$ is the intercept. Although this exponential function has two adjustable parameters, Marshall and Palmer fixed the value of $N_\circ$ at 8000 m^{-3}mm^{-1} (and also tied the slope parameter to the rainfall rate R through the formula $\lambda = 4.1 R^{-0.21}$ – see below). Their model consequently had but a single free parameter. The parametrization was developed on the basis of observations of stratiform rain at middle latitudes. Other studies conducted under different conditions have pointed to somewhat different parameter sets.

Another candidate distribution in widespread use is the gamma distribution, a generalization of the exponential function,

$$N(D) = N_\circ (D/D_\circ)^\mu \exp(-\lambda D), \tag{11.7}$$

which combines an additional shape parameter μ for a total of three. (The $D_\circ$ figure here is not a true free parameter and has been introduced to preserve the appropriate units. Its value can be set to unity without loss of generality.) The shape parameter controls the curvature of the distribution. While it can take on a negative sign, this causes the function to diverge at small diameters, something that is not physical. In some studies, the shape parameter is either held fixed or constrained while the others are fit to data.

Still another candidate is the three-parameter log-normal distribution,

$$N(D) = \frac{N_d}{\sqrt{2\pi}\sigma D} \exp\left(-\frac{(\ln D - \mu)^2}{2\sigma^2}\right), \tag{11.8}$$

where N_d is the drop count with units of m^{-3} and where μ and σ^2 are the logarithmized mean and variance of D, respectively. The log-normal model has the strength that the three parameters are related to the first three moments of the distribution and so have intuitive meanings. Another feature of the distribution is that it tends to zero at small diameters, a characteristic of rain seen in distrometer data.

Note that the aforementioned distributions are empirical and do not arise from theory, modeling, or simulation of precipitation per se. They are, however, able to capture the main features of distrometer measurements to varying degrees. Naturally, the more parameters a model has, the smaller the fitting residuals will be. Future studies may uncover candidate functions inspired by theory that are even better able to reproduce experimental data using just a few parameters.

With a specification of the drop-size distribution, other meteorological parameters can be obtained. These include the rainfall rate R which is given by

$$R = \int_0^\infty N(D)V(D)v(D)\,dD \tag{11.9}$$

$$= 6\pi \times 10^{-4} \int_0^\infty N(D)D^3 v(D)\,dD, \tag{11.10}$$

where $V(D)$ is the volume of a hydrometeor in mm^3 and $v(D)$ is its terminal fall speed in m/s. Semi-empirical formulas for the terminal fall speed of different types of hydrometeors have been found in different studies, as will be discussed below. The units of rainfall rate are mm/hr.

Another derivable parameter is the mass-weighted drop diameter D_m:

$$D_m = \frac{\int_0^\infty N(D)D^4\,dD}{\int_0^\infty N(D)D^3\,dD}. \tag{11.11}$$

A third parameter of interest is the liquid water content W,

$$W = \int_0^\infty m(D)N(D)\,dD,$$

in which $m(D) = \rho(\pi/6)D^3$ is the hydrometeor mass, ρ being the mass density. The units of W are km/m^3. Together, the parameters Z, R, D_m and W constrain the shape of the drop-size distribution as well as depend upon it. Reflectivity measurements therefore constrain both the drop-size distribution and the other parameters that depend upon it. The relationships between these quantities will be discussed further below in the section of the text that relates reflectivity and rainfall rates.

11.1.2 Relative Permittivity of Hydrometeors

Another factor influencing the power scattered by hydrometeors is their relative permittivity. This factor must be known in order to calculate the reflectivity from radar data. Water is a dipolar fluid composed of bent molecules with permanent electric dipole moments. Thermal agitation produces random orientations in these dipoles. The application of an electric field tends to orient the dipoles together. The competing factors determine the overall polarization or dipole moment per unit volume in the fluid. Scattering is radiation by the time-varying polarization, and the dielectric function captures the tendency to radiate. Unsurprisingly, the relative dielectric constant of hydrometeors depends on their physical state.

To model the dielectric constant of hydrometeors, a few simplifying assumptions can be made. The first is that the individual dipole moments **p** of the water molecules are independent of temperature and electric field. The second is that the density of the fluid is low enough that the energy of the dipolar interactions is small compared to the thermal energy $K_B T$. This means that Maxwell–Boltzmann statistics apply. The third is that the dipoles can assume any orientation relative to the applied electric field.

The potential energy u of a dipole **p** in a field **E** is $u = -\mathbf{p} \cdot \mathbf{E} = -pE\cos\theta$. The number of dipoles per unit volume within the differential solid angle $d\Omega$ is $dn = A\exp(-u/K_B T)d\Omega$, where A is a constant. Finally, the average dipole moment in the direction of the applied electric field is given by

$$\begin{aligned}
\langle p_e \rangle &= \frac{\int p\cos\theta \, dn}{\int dn} \\
&= \frac{p\int_0^{2\pi}\int_0^{\pi} \cos\theta \, e^{\frac{pE\cos\theta}{K_B T}} \sin\theta \, d\theta \, d\phi}{\int_0^{2\pi}\int_0^{\pi} e^{\frac{pE\cos\theta}{K_B T}} \sin\theta \, d\theta \, d\phi} \\
&= p\frac{\int_{-1}^{1} e^{\lambda u} u \, du}{\int_{-1}^{1} e^{\lambda u} du} \\
&= p\left[\coth(\lambda) - \lambda^{-1}\right],
\end{aligned}$$

where $\lambda \equiv pE/K_B T$ and where the function in the square brackets is called the Langevin function $L(\lambda)$. For realistic parameter values pursuant to weather radar applications, $\lambda \ll 1$, and the Langevin function is well approximated by the first term in its Taylor-series expansion, which is $\lambda/3$. This gives an average dipole moment in the direction of the applied electric field $\langle p_e \rangle = p^2 E / 3K_B T$. The corresponding polarization or dipole moment per unit volume is

$$\mathbf{P} = \frac{Np^2}{3K_B T} \mathbf{E}$$
$$= \varepsilon_\circ \chi_\circ \mathbf{E},$$

where N is the number density of molecules. This relationship defines the static susceptibility of the medium $\chi_\circ$. Note that the applied electric field is assumed to dominate the electric field due to the dipole moments in the above analysis.

Missing in this analysis but crucial to scattering is the dynamic behavior of polarization under a time-varying electric field. A reasonable phenomenological model for this behavior is one of simple relaxation, i.e., the response of the polarization in time to step-wise changes in the applied electric field is exponential growth and decay:

$$P(t) = \varepsilon_\circ \chi_\circ E \left(1 - e^{-t/\tau}\right) \quad t \geq 0 \quad \text{(turn on)}$$
$$= \varepsilon_\circ \chi_\circ E e^{-t/\tau} \quad t \geq 0 \quad \text{(turn off)},$$

where τ is the characteristic relaxation time which is determined by the mobility of the dipoles. In the frequency domain, this behavior is consistent with a dynamic susceptibility of the form

$$\chi(\omega) = \frac{\chi_\circ}{1 + j\omega\tau}.$$

The dynamic dielectric function is then finally related to the dynamic susceptibility by the relationship $\varepsilon(\omega)/\varepsilon_\circ = 1 + \chi(\omega)$, which can be decomposed into its real and imaginary parts $\varepsilon = \varepsilon' - j\varepsilon''$ as

$$\frac{\varepsilon(\omega)}{\varepsilon_\circ} = 1 + \frac{\chi_\circ}{1 + \omega^2 \tau^2} - j\omega\tau \frac{\chi_\circ}{1 + \omega^2 \tau^2}. \quad (11.12)$$

One last modification to the model accounts for the fact that the relative permittivity of the dielectric material $\varepsilon_r(\omega) = \varepsilon/\varepsilon_\circ$ generally assumes a value other than unity at high frequencies. Calling this value $\varepsilon_{r\infty}$ and shifting the formula accordingly yields the desired result,

$$\varepsilon_r(\omega) = \varepsilon_{r\infty} + \frac{\varepsilon_{r\circ} - \varepsilon_{r\infty}}{1 + \omega^2 \tau^2} - j\omega\tau \frac{\varepsilon_{r\circ} - \varepsilon_{r\infty}}{1 + \omega^2 \tau^2}, \quad (11.13)$$

where $\varepsilon_{r\circ}$ is the static relative permittivity of the dielectric. This is the Debye relaxation model for dipolar polarization. In addition, the molecules can exhibit ionic and electronic polarization as well, albeit typically at infrared and optical rather than microwave and millimeter-wave frequencies.

Figure 11.1 shows plots of the real and imaginary parts of the relative permittivity for ice and liquid water at 0°C. To generate these curves, representative values for water (ice) were taken to be $\tau = 1.7 \times 10^{-11}$ s (2.2×10^{-5} s), $\varepsilon_{r\infty} = 5$ (3.3), and $\varepsilon_{r\circ} = 88$ (96), respectively. (These parameters are approximate empirical matches to laboratory data.)

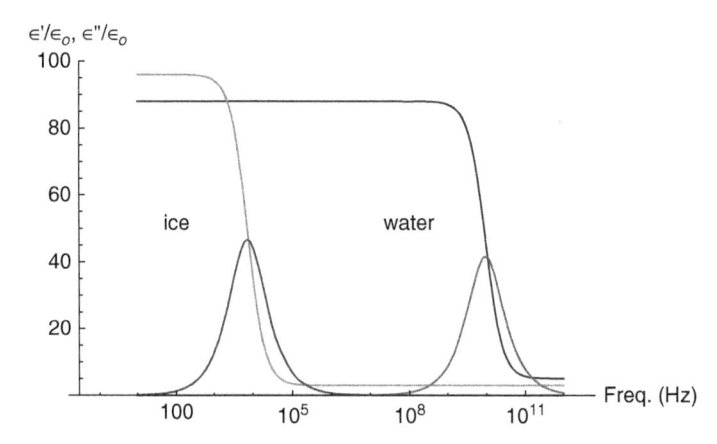

Figure 11.1 Relative permittivity for ice and water as a function of frequency in Hz according to the Debye polarization model.

Broad resonances for ice and water occur at frequencies centered near 10 kHz and 10 GHz, respectively. Near the resonances, the imaginary part of the dielectric function is large, and the medium is highly absorbing. At frequencies just below or above the resonances, the medium is highly dispersive. The difference in the resonance frequencies for ice and water arises from the fact that the dipole mobility is restricted in the solid phase. At microwave frequencies, $K \sim 0.44$ for ice whereas $K \sim 0.96$ for liquid water. Consequently, the scattering cross-section of the former is about 7 dB lower than the latter, all other factors being equal.

11.1.3 Mixed-Phase Hydrometeors and Effective Permittivity

The preceding treatment considered simple, homogeneous dielectrics. In general, however, hydrometeors are a mixture of air, liquid water, and ice. The conventional way to treat mixed-phase dielectrics is to combine the permittivities of the component materials into a single, effective permittivity parameter based on the material mixing ratios. The question is, how to weight the component terms – by mass, volume or something else?

Dielectrics are composed of atoms and molecules from which dipole moments are induced by the application of an electric field. The ratio of the dipole moment to the electric field is the polarizability, a microscopic quantity that depends on the properties of the atom or molecule. Finding the net electric field resulting from the applied field and the induced dipoles would be very complicated in practice in a sample with Avogadro's number of atoms or molecules, and the field thus found would exhibit singularities at the charge sites. The alternative and usually more practical approach is to regard the sample as a continuum in which the spatially averaged (over many charge sites) or macroscopic fields vary gradually. The relative permittivity is a parameter that applies in the more tractable macroscopic domain.

Whereas microscopic parameters for mixtures are straightforward, macroscopic parameters are not. Some way of relating the latter to the former is therefore required.

Consider a dielectric slab sandwiched between conducting parallel plates across which a potential difference is established. The macroscopic electric field in the dielectric will be the sum of the external electric field $\mathbf{E}_\circ$ in which the dielectric is immersed and a depolarization electric field $\mathbf{E}_1$ associated with polarization charge density at the dielectric boundaries. Let the sum $\mathbf{E} = \mathbf{E}_\circ + \mathbf{E}_1$ be the net electric field inside the dielectric.

Consider next a small sample of dielectric a few atoms or molecules in size in each dimension somewhere within the slab. We can regard this sample as inhabiting a small sphere separating it from the remainder of the dielectric. This is called a Lorentz sphere. If the sphere were the boundary of an empty cavity, the polarization (dipole moment per unit volume) inside would be zero, there would be a discontinuity at the boundary, and additional polarization charge $\sigma = \mathbf{P} \cdot \hat{\mathbf{n}}$ would exist, where $\mathbf{P}$ is the polarization outside the sphere, which is uniform, and $\hat{\mathbf{n}}$ is the outward unit normal at the spherical boundary. The potential at the center of the cavity due to this polarization charge would be

$$\phi(\mathbf{x}) = \frac{1}{4\pi\varepsilon_\circ} \oint \frac{\mathbf{P}(\mathbf{x}') \cdot \hat{\mathbf{n}}(\mathbf{x}')}{|\mathbf{x} - \mathbf{x}'|} dS', \qquad (11.14)$$

where the primed coordinates are at the spherical surface. The corresponding electric field within the cavity can be found from

$$-\nabla_x \phi(\mathbf{x}) = \frac{1}{4\pi\varepsilon_\circ} \oint \mathbf{P}(\mathbf{x}') \cdot \hat{\mathbf{n}}(\mathbf{x}') \frac{\mathbf{x} - \mathbf{x}'}{|\mathbf{x} - \mathbf{x}'|^3} dS'. \qquad (11.15)$$

This is an elementary integral which yields the result $\mathbf{E}_2 = (1/3\varepsilon_\circ)\mathbf{P}$, which is known as the Lorentz field. The sum of the external, depolarization, and Lorentz fields is termed the local field. This local field $\mathbf{E}_L = \mathbf{E} + \mathbf{E}_2 = \mathbf{E} + (1/3\varepsilon_\circ)\mathbf{P}$ is the field seen by the sample within the Lorentz sphere.

Meanwhile, the sample itself is small enough to be described by microscopic parameters. The local field will induce a dipole moment in the sample. Shifting to the microscopic point of view, the polarization in the sample is given by

$$\begin{aligned}\mathbf{P} &= N\mathbf{p} \\ &= N\alpha \mathbf{E}_L \\ &= N\alpha \left(\mathbf{E} + \frac{1}{3\varepsilon_\circ}\mathbf{P}\right),\end{aligned}$$

in which N is the number density of molecules in the sample and α is the polarizability. Now, the polarization inside and outside the Lorentz sphere should be the same. Solving for the polarization in the sample gives

$$\mathbf{P} = \frac{N\alpha}{1 - N\alpha/3\varepsilon_\circ} \mathbf{E} \;=\; \varepsilon_\circ \chi \mathbf{E},$$

where χ is defined as the susceptibility. With $\varepsilon_r = \varepsilon/\varepsilon_\circ = 1 + \chi$, we find (after some rearranging)

$$\frac{\varepsilon_r - 1}{\varepsilon_r + 2} = \frac{N\alpha}{3\varepsilon_\circ}, \qquad (11.16)$$

which is the Clausius–Mossotti relation or the Lorentz Lorenz equation. The relation bridges the gap between the macroscopic and microscopic properties of a dielectric.

Now, if the dielectric and the sample were composed of a mixture of n different materials with volume fractions f_i and polarizabilities α_i, the Clausius–Mossotti equation would read

$$\frac{\varepsilon_{rm} - 1}{\varepsilon_{rm} + 2} = \sum_{i=1}^{n} f_i \frac{N\alpha_i}{3\varepsilon_\circ}, \qquad (11.17)$$

where the relative permittivity ε_{rm} is the effective value for the mixture. This expression is accurate but less useful than one stated entirely in terms of macroscopic parameters. However, by (11.16), note that each quotient term on the right side of (11.17) can be replaced with a term like the one on the left side of (11.16), leading to the formula

$$\frac{\varepsilon_{rm} - 1}{\varepsilon_{rm} + 2} = \sum_{i=1}^{n} f_i \frac{\varepsilon_{ri} - 1}{\varepsilon_{ri} + 2}, \qquad (11.18)$$

where ε_{ri} is the relative permittivity of the ith component in the mixture. This is the Maxwell Garnett equation and the recipe for calculating the complex permittivity of a hydrometeor as a weighted average. The effective relative permeability can be viewed as the ratio of the average electric displacement in a compound dielectric to the average electric field.

Another form of effective medium approximation is given by Bruggeman's mixing formula (or the Polder–van Santen formula),

$$0 = \sum_{i=1}^{n} f_i \frac{\varepsilon_{ri} - \varepsilon_{rm}}{\varepsilon_{ri} + 2\varepsilon_{rm}},$$

which can be derived by repeating the derivation for the Maxwell Garnett equation, only taking the electric field seen by the sample to be $\mathbf{E}$ rather than $\mathbf{E}_L$. Many other variations on the theme have been examined in the literature and compared with observations and with simulations. These simple formulas often do a satisfactory job of representing mixed-phase hydrometeors, so long as the intrusions are approximately isotropic in shape.

11.1.4 Attenuation

So far in the analysis, losses due to attenuation of the radar signal between the radar and the target have been neglected. In fact, attenuation due to scattering and absorption affect the measured reflectivity and need to be taken into account. In the Rayleigh long-wavelength limit, the formulas given in Chapter 6 for the total scattering cross-section σ_t

11.1 Radar Cross-Section and Radar Equation

and the absorption cross section σ_a can be used to analyze attenuation. For reference, those formulas are repeated here:

$$\sigma_t = \frac{128}{3}\left(\pi a^2\right)\pi^4|K|^2\left(\frac{a}{\lambda}\right)^4 \tag{11.19}$$

$$\sigma_a = \frac{8\pi^2}{\lambda}a^3\Im(-K). \tag{11.20}$$

These expressions apply to a single target. The extension to volume scatter follows the same prescription as the radar-scattering cross-section, viz.

$$K_t = N\langle\sigma_t\rangle$$
$$K_a = N\langle\sigma_a\rangle$$
$$K_L = K_t + K_a,$$

where K_t, K_a and K_L are the total, absorption and attenuation cross-sections for volume scatter with units of inverse length. The factor N is the number of particles per unit volume. The angle brackets denote an average over the volume. Since σ_t and σ_a depend on the particle radius a, the averages should be expressed as integrals over the drop-size distribution

$$N\langle\sigma_{t,a}\rangle = \int_0^\infty N(D)\sigma_{t,a}(D)\,dD.$$

It is noteworthy that both the absorption cross-section σ_a and the mass of a hydrometeor are proportional to D^3, signifying that K_a and the water content W are closely related. Likewise, both the total scattering cross-section σ_t and the radar-scattering cross-section σ are proportional to D^6 so that K_t and and the reflectivity Z are also closely related.

At microwave frequencies, the Rayleigh limit is appropriate for all except the largest hydrometeors (hailstones, water-coated or "spongy" hail and melting snow). When considering attenuation from large hydrometeors or at millimeter wavelengths, full Mie scattering theory and the associated extinction coefficient may be necessary.

As discussed in Chapter 5, path loss is described by the Beer–Lambert law,

$$dP_r = -2K_L P_r dr,$$

where P is the returned power, K_L is the attenuation coefficient, r is range, and where the factor of two accounts for two-way losses in the radar context. The solution is

$$P_r = P_{r\circ}e^{-2\int K_L dr}$$
$$10\log(P_r/P_{r\circ}) = -2\int \kappa_L dr,$$

where the integral covers the path between the radar and the scattering volume at range r. The second variant of the formula is in decibels and makes use of attenuation coefficients

κ_L expressed in terms of dB/km. In the event that more than one type of material causes attenuation, the appropriate attenuation coefficients can be added.

The radiative balance equations that govern absorption were treated in the text in Chapter 5, and the physics behind absorption in lossy dielectrics was addressed just above. We have only to apply the aforementioned tools to some specific meteorological examples. Below, the main sources of attenuation in weather radar are considered.

Atmospheric Gasses At radio frequencies, scattering due to atmospheric gasses is negligible, and attenuation is due only to absorption. (This is not true at optical frequencies, where the effects of scattering by oxygen and nitrogen molecules, water vapor and aerosols are readily apparent.) At radio frequencies, water vapor is a strong absorber at wavelengths less than about 10 cm and presents a narrow absorption line at about 3 cm wavelength at which point the attenuation is greater than 0.1 dB/km. Oxygen is a strong absorber at wavelengths less than 100 cm and has a sharp absorption line at 0.5 cm wavelength, where the attenuation is greater than 10 dB/km. Such frequencies should be avoided in radar design.

Note that attenuation is a function not only of frequency and distance but also of elevation, since low elevations imply longer paths through the densest part of the atmosphere. A number of empirical formulas giving absorption coefficients for different radar wavelengths and elevation angles have been published. The reader is referred to Battan's encyclopedic text.

Clouds and Fog Attenuation due to hydrometeors is in general due to a combination of scattering and absorption, depending on composition, shape and size. The larger the hydrometeor (or the shorter the wavelength), the more important the role of scattering. Recall that in the Rayleigh limit, the total scattering cross-section σ_t is proportional to a^6/λ^4, whereas the absorption cross-section σ_a is proportional to a^3/λ. The attenuation cross-section combines both factors.

The attenuation coefficient κ_L is often specified in units of dB/km/g/M^3. Since the absorption cross-section K_a is related to the drop-size distribution in the same way as the water content W, the ratio of the former to the latter does not depend on the drop-size distribution, and so can be readily estimated. The attenuation constant is also a function of temperature. Tables listing representative values of κ_L for clouds and fog at different temperatures are also given in Battan's text.

Rain Raindrops can still be analyzed using Rayleigh's formulas at S-band frequencies. At higher frequencies, at least in principle, complete Mie scattering theory and the extinction coefficients referred to in Chapter 6 must be used. In practice, however, attenuation in rain can often be calculated using empirical relationships with other parameters.

Shortly after the end of WWII, it was found that the attenuation κ in units of dB/km could be related to the rainfall rate through a power-law relationship, i.e., $\kappa = aR^b$, with a and b being functions of the radar wavelength and R having units of mm/hr. Since rainfall rates can themselves be related to the reflectivity through a power law with some degree of accuracy, as discussed below, the attenuation can equally well be expressed as $\kappa = cZ^d$, where c and d are wavelength-dependent and where the units of Z are mm^6/m^3. The values of a, b, c and d can be calculated if the distribution of rain drop sizes along with their terminal fall speeds is known. Once again, tables for these coefficients can be found in Battan's text. Somewhat better estimates of the attenuation can be made if the rainfall rate and reflectivity are known independently.

Snow The situation for ordinary snowfall is similar to that for rainfall. The Rayleigh limit generally applies, and both scattering and absorption contribute to attenuation. The empirical Z-R relationship is different for rain and snow. More importantly, the terminal fall speed of snow is relatively slow and can be taken to be approximately $v = 1$ m/s independent of diameter.

It was already noted that K_t is proportional to the reflectivity Z. If the reflectivity is taken to be proportional to R^2 for snow, then $\kappa_t \propto R^2/\lambda^4$. Likewise, K_a is proportional to the water content W, which is proportional to the rainfall rate R in the case of uniform terminal fall speeds. Consequently, $\kappa_a \propto R/\lambda$. As the constants of proportionality can be shown to be comparable, we can surmise that scattering from snowfall dominates absorption at short radar wavelengths and/or during intervals of very strong precipitation. Absorption dominates otherwise.

Hail Precipitation by hail generally falls within the domain of Mie theory at microwave frequencies, and the attenuation rate is governed by the extinction efficiency developed in Chapter 6. The extinction cross-section quantifies the power that is prevented from propagating through a medium, and the total scattering cross-section quantifies the total power scattered away from the forward direction. The difference between these implies absorption. The extinction and scattering efficiencies, denoted Q_t and Q_s, are the respective cross-sections divided by the physical cross-section of the target. Both are usually calculated as a function of the circumference of the target measured in wavelengths, C_λ.

The extinction and total scattering cross-sections calculated in Chapter 6 were for individual targets. To generalize the results for volume scatter, the procedure outlined at the start of this section of the text, which considered at first the Rayleigh limit, can be followed.

Consider now a radar operating at 10 GHz (3 cm wavelength). The Debye relaxation model using the coefficients given in Section 11.1.2 yields estimates for the complex relative permittivity for ice ($\varepsilon_r = 3.3$) and liquid water ($\varepsilon_r = 43.8 + j41$) at 0° C at this frequency. Assuming a uniform mixture of both phases in the hail, the Maxwell Garnett relation then gives the effective relative permittivity ε_{rm} for different mixtures.

Three representative cases corresponding to 10, 50 and 90 percent liquid water mixtures are analyzed in Figure 11.2. For each case, the extinction and scattering efficiency are plotted. The absorption efficiency is the difference between the two curves. The curves illustrate how absorption becomes increasingly significant with increased liquid water content, eventually matching scattering at about the 50 percent level for large targets and at about the 90 percent level for small targets.

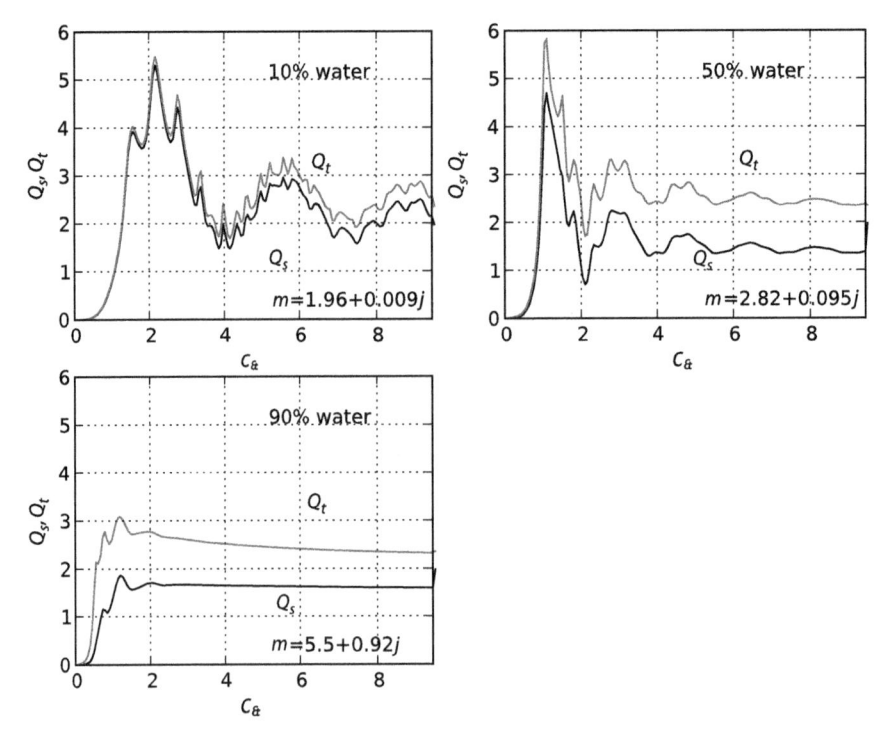

Figure 11.2 Mie extinction and scattering efficiencies for hydrometeors composed of ice/water mixtures. The liquid water fraction is indicated in each case, as is the complex index of refraction $m = \sqrt{\varepsilon_{rm}}$.

Note that the situation is entirely different for inhomogeneous hydrometeors with liquid water coatings surrounding ice cores. Analyzing this kind of "spongy ice" requires a reformulation of Mie theory that takes into account layered composition. While the calculations required are beyond the scope of this text, the problem has been solved in closed form. To a reasonable approximation, hail stones covered in liquid water behave like liquid water droplets of the same size.

11.2 Z-R Relationships

The most sought-after relationship in radar meteorology is the one between reflectivity and rainfall rate. A reliable, universal relationship would confer the ability to estimate rainfall rates in expansive, sparsely populated, and otherwise inaccessible regions on the basis of remote sensing information alone. It is hard to imagine a radar application with greater practical utility. Completely satisfactory Z-R relationships have yet to be formulated, however. The field remains in some sense immature, relying mainly on empirical rather than theoretical or simulation-based methods to the present.

11.2 Z-R Relationships

The most common relationships in operational use have the form $Z = AR^B$, where A and B are constants. This form is not arbitrary but can be seen as a consequence of the drop-size distributions examined above along with other conventional parametrizations. If the gamma distribution is used, then the reflectivity takes the form

$$Z = \int_0^\infty N_\circ (D/D_\circ)^\mu D^6 e^{-\lambda D} \, dD$$
$$= N_\circ \frac{\Gamma(\mu+7)}{D_\circ^\mu \lambda^{\mu+7}}, \qquad (11.21)$$

where Γ is the Gamma function. Recall that the Marshall–Palmer distribution is a special case of the Gamma distribution with $\mu = 0$.

In order to calculate the corresponding rainfall rate, a specification of the terminal fall speed of hydrometeors is required. Wind-tunnel and other laboratory and environmental tests offer different options for this specification, some with the elementary form $v(D) = a(D/D_\circ)^b$, where a and b are constants. This parametrization yields rainfall-rate estimates of the form

$$R = 6\pi \times 10^{-4} \int_0^\infty N_\circ a (D/D_\circ)^{\mu+b} D^3 e^{-\lambda D} \, dD$$
$$= 6\pi \times 10^{-4} N_\circ a \frac{\Gamma(\mu+4+b)}{D_\circ^{\mu+b} \lambda^{\mu+4+b}}. \qquad (11.22)$$

Combining (11.21) and (11.22) gives the formal Z-R relationship

$$Z = \frac{N_\circ^{1-\gamma} \Gamma(\mu+7) D_\circ^{(\gamma-1)\mu + b\mu}}{((6\pi \times 10^{-4}) a \Gamma(\mu+4+b))^\gamma} R^\gamma, \qquad (11.23)$$

where $\gamma \equiv (\mu+7)/(\mu+4+b)$. Note that the $D_\circ$ term is immaterial and may be set to unity in the appropriate units.

Equation (11.23) is clearly of the form $Z = AR^B$. For raindrops with diameters between 0.5 and 5 mm, Atlas and Ulbrich's work showed that the terminal fall speeds determined from earlier experiments followed $v(D) = a(D/D_\circ)^b$ with $a = 3.78$ m/s and $b = 0.67$ (and $D_\circ = 1$ mm). Incorporating these constants along with $N_\circ = 8000$, $\mu = 0$ gives the Z-R relationship

$$Z = 240 R^{1.5}. \qquad (11.24)$$

In fact, hundreds of Z-R relationships have been found in different studies published in the literature over the years. Five of these recommended by the National Weather Service are listed in Table 11.2. Equation (11.24) is close to the Marshall–Palmer Z-R relationship, which is therefore roughly consistent with the Marshall–Palmer drop-size distribution and the Atlas–Ulbrich terminal-speed formula. Still other formulas have been found to be accurate in different circumstances.

Table 11.2 *Z-R relationships recommended by the Radar Operations Center of the National Weather Service.*

name	Z-R	used for
Marshall–Palmer	$Z = 200R^{1.6}$	general stratiform precip.
Rosenfeld Tropical	$Z = 250R^{1.2}$	tropical convective systems
WSR-88D Convective	$Z = 300R^{1.4}$	summer deep convection
East Cool Stratiform	$Z = 130R^2$	winter stratiform east of continental divide
West Cool Stratiform	$Z = 75R^2$	winter stratiform west of continental divide

11.3 Doppler Radar

A breakthrough occurred in radar meteorology with the introduction of Doppler radar. As in most every radar application, Doppler signal processing adds another dimension from which to extract information about the target. In weather radar, the Doppler spectrum gives information about wind speed and direction, material and momentum flux, severe-weather indicators, turbulence and other critical factors. Wind measurements from radars form the basis for short-term persistence forecasting.

Considerations for possible range and frequency aliasing, spectral resolution and statistical accuracy must be made, as in all radar applications. The spectral widths seen by radars looking vertically may be small enough to warrant coherent integration. Echoes from severe weather observed via tropospheric ducts may meanwhile be overspread. Complex spectra may appear in echoes from severe storms, tornadoes and hurricanes.

Radar echoes from weather can be regarded as a random processes. For different realizations of the random signal, we require a function that gives consistent estimates of the distribution of power versus frequency. The power spectral density does this. If the random process is wide-sense stationary, its autocorrelation function should be invariant over time intervals that are long compared to the estimation time. The power spectral density should therefore be derivable from the autocorrelation function, which can be estimated unambiguously on the basis of a finite number of samples.

Chapter 1 introduced the continuous and discrete Fourier transforms and their inverses and also described the power spectral density of continuous random processes. Here, we consider the power spectral density for discretely sampled random processes. If the discrete samples are denoted f_n, the coefficients of discrete Fourier transform f_m are

$$f_m = \sum_{n=0}^{N-1} f_n e^{-j2\pi nm/N}$$

for N samples. We define the power spectral density estimator to be $\tilde{S}_m \equiv |f_m|^2/N$. The spectral density is then the expectation of this estimator in the limit $N \to \infty$:

11.3 Doppler Radar

$$S_m \equiv \lim_{N \to \infty} E \left\{ \frac{1}{N} |f_m|^2 \right\}$$

$$\equiv \lim_{N \to \infty} E \left\{ \frac{1}{N} \sum_{n=0}^{N-1} f_n e^{-j2\pi nm/N} \sum_{n'=0}^{N-1} f_{n'}^* e^{j2\pi n'm/N} \right\}.$$

Evaluation of this expression is facilitated by the double-summation identity:

$$\sum_{i=0}^{N-1} \sum_{j=0}^{N-1} g(i-j) = \sum_{k=-N+1}^{N-1} (N - |k|) g(k).$$

Substitution into the expression for the discrete power spectrum yields

$$E\{|f_m|^2\} = \sum_{n=0}^{N-1} \sum_{n'=0}^{N-1} E\left(f_n f_{n'}^*\right) e^{-j2\pi(n-n')m/N}$$

$$= \sum_{n=0}^{N-1} \sum_{n'=0}^{N-1} \rho_{n-n'} e^{-j2\pi(n-n')m/N}$$

$$= \sum_{k=-N+1}^{N-1} (N - |k|) \rho_k e^{-j2\pi km/N},$$

which has the form of the inverse Fourier transform of the discrete autocorrelation function ρ_k multiplied by a weight function or window with a triangular shape. This window is implicit in finite-interval samples and implies a limit on spectral resolution that is stricter than the sampling theorem enforces.

To complete the derivation, we apply the limit

$$\lim_{N \to \infty} E\left\{ \frac{1}{N} |f_m|^2 \right\} = \lim_{N \to \infty} \frac{1}{N} \sum_{k=-N+1}^{N-1} (N - |k|) \rho_k e^{-j2\pi km/N}$$

$$= \sum_{k=-\infty}^{\infty} \rho_k e^{-j2\pi km/N}$$

$$= S_m,$$

proving that the power spectral density estimator for N discrete samples has the required limiting behavior. The only assumption required to arrive at the last result is that the autocorrelation function decays at long lags for rapidly than $|k|^{-1}$.

11.3.1 Spectral Moments and Errors

Often times, the salient features of the Doppler spectrum and the features of interest for weather reader are the Doppler shift V_d and the spectral width σ_d. The former can be indicative of bulk motion of material through the scattering volume, and the latter of variations in the motion across the volume. These features are related to the first three frequency moments of the Doppler-frequency spectrum:

$$P = \int S(f)\,df$$

$$\langle f \rangle = \frac{1}{P}\int fS(f)\,df$$

$$\langle f^2 \rangle = \frac{1}{P}\int f^2 S(f)\,df,$$

i.e., $V_d = \lambda_s \langle f \rangle$ and $\sigma_d = \lambda_s \sqrt{\langle f^2 \rangle - \langle f \rangle^2}$, where $\lambda_s = \lambda/2$ is the scattering wavelength. The spectral moments themselves can be computed from the discrete Doppler spectrum using conventional numerical quadrature methods, as can even higher-order moments, if desired.

If only Doppler-shift and spectral-width estimates are required, it may be expedient to replace frequency-domain processing with time-domain processing. This is because the complex autocorrelation function measured at a single short-lag τ also contains the first three moments of the Doppler spectrum. This can be seen by examining the relationship between the normalized autocorrelation function $\rho(\tau)$ and the Doppler spectrum:

$$\begin{aligned}
\rho(\tau) &= \frac{\int S(f) e^{i2\pi f\tau}\,df}{\int S(f)\,df} \\
&= \langle e^{i2\pi f\tau} \rangle \\
&= e^{i2\pi \langle f \rangle \tau} \langle e^{i2\pi (f - \langle f \rangle)\tau} \rangle \\
&\approx e^{i2\pi \langle f \rangle \tau} \left(1 - \frac{\tau^2}{2}(2\pi)^2 \langle (f - \langle f \rangle)^2 \rangle \right).
\end{aligned}$$

This shows how $\rho(\tau)$ can be viewed as the spectral moment of the characteristic function $\exp(i2\pi f\tau)$. So long as τ is sufficiently short, the characteristic function can be represented accurately by the first three terms in its Taylor-series expansion. The linear term, $\langle i2\pi\tau(f - \langle f \rangle) \rangle$, is identically zero, leaving the constant and quadratic terms written above. This calculation shows how the modulus of the normalized autocorrelation function, called the coherence, gives an estimate of the spectral width, whereas the phase angle gives an estimate of the Doppler shift.

In practice, $\rho(\tau)$ can be measured at one lag simply by correlating signals received in a given range gate from adjacent pulses. Normalization is through division by the zero-lag estimate. Since noise power contributes to the denominator but not to the numerator for finite τ, the noise power must be subtracted from the zero-lag estimate prior to normalization. Overall, time-domain processing can be much less computationally intensive than spectral analysis followed by numerical quadrature. Naturally, cautions pertaining to spectral aliasing still apply.

Without proof, we write the statistical errors expected for the estimate of the normalized complex cross-correlation function thus formed,

$$\langle \delta |\rho|^2 \rangle = \frac{1}{2n} \left(1 - |\rho|^2\right)^2$$
$$\langle \delta \phi^2 \rangle = \frac{1}{2n} \left(1 - |\rho|^2\right) / |\rho|^2,$$

in which n is the number of statistically independent samples. The top expression gives the statistical variance of estimates of the square modulus of the normalized correlation function, and the bottom the statistical variance of estimates of the phase angle. These formulas are appropriate for the noise-free case. To include the effects of noise in error analysis, the following substitution should be made when using the above formulas:

$$|\rho|^2 \leftarrow \left(\frac{S}{S+N}\right)^2 |\rho|^2.$$

The resulting expression forms the basis for calculating the error covariance matrix for Doppler-shift measurements, for example. As the sampling errors for different range gates, pointing positions, and times are uncorrelated in simple pulse-to-pulse processing, the error covariance matrix C_d can usually be considered to be diagonal.

11.3.2 Spectral Noise Estimates

Estimating and removing the noise from spectra is essential if spectral moments and other observables are to be measured accurately. Noise sources, systemic and natural, vary in time as well as with changing experimental conditions, and so the noise level needs to be measured dynamically. The noise level can be estimated from transmitter-off data collection intervals or from receiver samples from ranges or antenna pointing positions where no signal is expected. This necessitates dedicating experiment time to noise-sample collection, and may be impractical, depending on the radar and the measurement in question. Most desirable are noise estimates that can be derived from the spectra that also contain signals.

If the radar measurement can be designed such that the spectrum of the signal is confined to a fraction of the frequency bins, then it should be possible to estimate the background noise from the remaining bins. The problem then is to determine which spectral bins are signal free. A common approach is to employ rank-ordered statistics, where the spectral bins are reordered by power spectral density, from lowest to highest. Spectral bins containing signals will preferentially fall on the high-power end of the ordered list. Somewhere in the list is the dividing line between pure noise and signal plus noise.

The spectral bin at the center of the rank-ordered list has the median power. If the spectrum of the signal is sufficiently narrow, the median power could be a reasonable estimate of the noise power level (whereas the average power would certainly be an overestimate).

308   Weather Radar

This illustrates the insensitivity of order statistics to values at the extremes. Noise estimates based on a weighted first quartile are insensitive to even broader signal spectra.

The weighted first-quartile method has been succeeded by superior methods that exploit the expected statistical behavior of noise. One simple method makes use of the fact that noise is a zero-mean Gaussian random process. Different spectral bins containing only noise represent different realizations of the same process. We expect the variance of the power in any subset of spectral bins to be approximately equal to the mean power of the subset divided by the number of statistically independent samples, i.e., the number of incoherent integrations applied to the spectra. This is actionable information.

The proposition above constitutes a test that can be applied to subsets of individual spectra. Beginning with a rank-ordered spectrum, high-power frequency bins are discarded, one by one, until the remaining bins satisfy the test. At this point, the remaining bins are interpreted as being pure noise, and their mean power is taken as an estimate of the noise power. The overall method is adaptive and robust. Other tests have been developed that offer slightly superior performance at the expense of greater complexity.

11.3.3 Wind Field Estimation

We consider the problem of deducing the wind field from line-of-sight Doppler shift measurements. If the winds were uniform in the vicinity of the radar, then three Doppler-shift measurements taken along three different bearings would be sufficient to completely specify the winds. The wind field is generally much more interesting than this, of course. Measurements along more bearings are necessary to specify a more complex wind field. Increasing the number of bearings comes at the cost of the cadence of the overall measurements and statistical confidence in the measurements made along any one bearing. Optimizing the tradeoffs is an ongoing problem in weather-radar research.

Figure 11.3 shows the geometry for the wind field estimation problem. The law of cosines gives the altitude of a scattering volume z in terms of the range r and elevation angle θ.

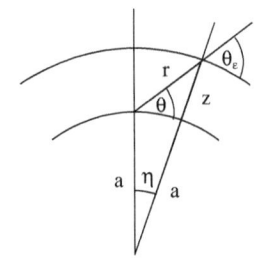

Figure 11.3 Geometry for Doppler-shift measurements. The scattering volume is at range r and has altitude z. The elevation angle of a ray to the scattering volume is θ, and the effective elevation angle is θ_e. The Earth's radius is a. The radar azimuth angle ϕ (not shown) is measured clockwise from northward, looking down on the radar.

11.3 Doppler Radar

$$(z+a)^2 = r^2 + a^2 + 2ra\sin(\theta).$$

A ray directed at the scattering volume has an elevation angle θ at the radar but θ_e at the scattering volume. The two angles are related by

$$\theta_e - \theta = \eta = \sin^{-1}\left(\frac{r\cos\theta}{a+z}\right).$$

If the region being monitored is small, such that the Earth's curvature can be neglected, one may simply take $\theta_e = \theta$ and $z = r\sin\theta$.

Suppose the wind field can be reasonably represented in the vicinity of a radar by the linear terms in a Taylor-series expansion,

$$(u,v,w) \qquad (11.25)$$
$$= (u_\circ, v_\circ, w_\circ) + (u_x, v_x, w_x)(x-x_\circ) + (u_y, v_y, w_y)(y-y_\circ) + (u_z, v_z, w_z)(z-z_\circ),$$

where u, v and w are the zonal, meridional and vertical components of the wind, respectively, x, y and z, are displacements in the zonal, meridional and vertical directions, respectively, $(x_\circ, y_\circ, z_\circ)$ is the point about which the expansion is being taken, and the subscripts denote partial derivatives. The coordinates probed by the radar at a given azimuth and zenith angle are

$$x = r\sin\phi\cos\theta_e \qquad (11.26)$$
$$y = r\cos\phi\cos\theta_e \qquad (11.27)$$
$$z = r\sin\theta_e. \qquad (11.28)$$

The expansion point can be considered to be adjustable, leaving twelve other parameters to be determined experimentally.

The Doppler shift along a given azimuth and elevation angle gives the line-of-sight projection of the winds in a scattering volume, viz.

$$v_r(r,\theta,\phi) = u\sin\phi\cos\theta_e + v\cos\phi\cos\theta_e + w\sin\theta_e. \qquad (11.29)$$

We can now substitute the expansion in (11.25) and the geometric factors in (11.26)–(11.28) into (11.29) and consider measurements taken along n different ranges and bearings denoted by the subscript $i = 1\ldots n$. Most commonly, the radar is swept in azimuth while the elevation is held constant. With some effort, the result can be formulated as a matrix equation $d = Gm$, where $d \in \mathbb{R}^n$ is a column vector of Doppler shift measurements and $G \in \mathbb{R}^{n\times 11}$ and $m \in \mathbb{R}^{11}$ are given below:

$$G_{i,1} = \cos\theta_{ei}\sin\phi_i$$
$$G_{i,2} = \cos\theta_{ei}\sin\phi_i(r_i\cos\theta_{ei}\sin\phi_i - x_\circ)$$
$$G_{i,3} = \cos\theta_{ei}\sin\phi_i(z_i - z_\circ)$$
$$G_{i,4} = \cos\theta_{ei}\cos\phi_i$$
$$G_{i,5} = \cos\theta_{ei}\cos\phi_i(r_i\cos\theta_{ei}\cos\phi_i - y_\circ)$$
$$G_{i,6} = \cos\theta_{ei}\cos\phi_i(z_i - z_\circ)$$
$$G_{i,7} = \cos\theta_{ei}[r_i\cos\theta_{ei}\sin\phi_i\cos\phi_i$$
$$\qquad -(1/2)(x_\circ\cos\phi_i + y_\circ\sin\phi_i)]$$
$$G_{i,8} = \sin\theta_{ei}$$
$$G_{i,9} = \sin\theta_{ei}(r_i\cos\theta_{ei}\sin\phi_i - x_\circ)$$
$$G_{i,10} = \sin\theta_{ei}(r_i\cos\theta_{ei}\cos\phi_i - y_\circ)$$
$$G_{i,11} = \sin\theta_{ei}(z_i - z_\circ)$$

$$m = \begin{bmatrix} u'_\circ \\ u_x \\ u_z \\ v'_\circ \\ v_y \\ v_z \\ u_y + v_x \\ w_\circ \\ w_x \\ w_y \\ w_z \end{bmatrix}.$$

By convention, the solution vector m includes the "modified wind components" $u'_\circ$ and $v'_\circ$ (defined by Easterbrook, 1974; Doviak and Zrnić, 1993)

$$u'_\circ \equiv u_\circ + \frac{y_\circ}{2}(v_x - u_y)$$
$$v'_\circ \equiv v_\circ - \frac{x_\circ}{2}(v_x - u_y)$$

as well as the sum $u_y + v_x$.

Note in particular that there are eleven independent parameters in the vector m to be found, rather than twelve. This can be traced to the fact that the quantities u_y and v_x cannot generally be independently determined from the information available, regardless of the number of measurements. Instead, the sum of the two parameters appears as a parameter. Furthermore, $u_\circ$ and $v_\circ$ have the same influence on the data as the difference $v_x - u_y$. If $u_\circ$ is known and $y_\circ$ is chosen to be nonzero, then u_y and v_x can be determined since their sum and difference appear as two parameters. (Likewise if $v_\circ$ is known and $x_\circ$ is chosen to be nonzero.) Otherwise, it is impossible to determine the vorticity from measurements from a single radar.

11.3.4 Least-Squares Estimation

So long as $n \geq 11$, the parameter vector m can be estimated from the data using linear least-squares estimation. This is the vector that minimizes the discrepancy between the data d and its prediction Gm in the sense of the L2 norm or the χ^2 parameter. The optimal estimator can be found readily using the method of normal equations. Let C_d be the data error covariance matrix for normally distributed sampling errors. This sets a metric for the error norm. Then the estimate of m is found from the linear optimization problem

$$m_{\text{est}} = \arg\min_{m} (Gm-d)^T C_d^{-1}(Gm-d)$$
$$= \arg\min_{m} \|C_d^{-1/2}(Gm-d)\|_2^2,$$

where we make use of the fact that C_d and its inverse are symmetric and can be factored in terms of the square-root information matrix as $C_d^{-1} = C_d^{-T/2} C_d^{-1/2}$.

That the norm of the vector $C_d^{-1/2}(Gm-d)$ cannot be forced to zero with the optimal choice of m is evidence that the column space of the matrix $C_d^{-1/2} G$ is incomplete and does not contain the residual vector. This means that $C_d^{-1/2} G$ is normal to the residual vector, which can be stated as

$$G^T C_d^{-T/2} \left[C_d^{-1/2}(Gm_{\text{est}}-d) \right] = 0$$
$$G^T C_d^{-1} G m_{\text{est}} = G^T C_d^{-1} d.$$

Finally, the parameter vector estimator is given by

$$m_{\text{est}} = (G^T C_d^{-1} G)^{-1} G^T C_d^{-1} d$$
$$= \tilde{G} d,$$

where $\tilde{G}$ is called the pseudoinverse operator. So long as $n \geq 11$, the pseudoinverse will exist and the matrix $(G^T C_d^{-1} G)$ will be nonsingular, although it may be near singular if the n measurements are not sufficiently independent. Care must be taken in experiment design to assure that the method is well conditioned. The error covariance matrix for the parameter estimates is given finally by ordinary error propagation,

$$C_m = \tilde{G} C_d \tilde{G}^T.$$

Note that the parameter error covariance matrix C_m will not in general be diagonal, even if the data error covariance matrix C_d is. Overall optimization of the winds measurement can be achieved by taking into consideration the L2 norm of C_m under different pointing and sampling scenarios.

11.4 Polarimetry

Contemporary weather radars rely increasingly on polarimetry to extract incisive information from meteorological echoes and resolve ambiguities that conventional reflectivity and Doppler spectral measurements cannot. (Although dual-polarization radars date back to the 1950s, the WSR-88D network was only upgraded for polarimetric measurements between 2011 and 2013.) Polarimetry exploits the fact that reflectivity for vertically and horizontally polarized waves generally differ. This is in part due to the fact that hydrometeors are not spherical as a rule. Small raindrops are very nearly spherical but become increasingly oblate with size, as drag from the air through which they fall compresses their leading edges. (Note especially well that raindrops do not have a teardrop shape as is sometimes represented in artwork!) Consequently, the reflectivity for horizontally polarized waves is greater than for vertically polarized waves. Other kinds of hydrometeors, especially icy particles, also have non-spherical shapes. Differences in the reflectivity of vertically and horizontally polarized radar waves, together with relative phase measurements, convey information about the size and shape, composition and dynamics of scatterers. Polarization diversity can also be exploited to combat certain kinds of radar clutter.

In polarimetry, co- and cross-polarization refer to signals received in the same and the opposite polarization as the transmitted signal, respectively. This language applies equally to forward and backscattered signals. Reflectivity is expressed as Z_{RT}, where "R" and "T" denote reception and transmission and can be either "H" or "V" depending on the polarization. For example, Z_{HV} refers to horizontally polarized signals received from vertically polarized transmissions. The ratio of horizontally to vertically polarized backscatter co-power is denoted Z_{DR} (for differential reflectivity), and the phase difference between the two polarizations is denoted ϕ_{DP} (for differential phase). The correlation between the two co-polarized signals is called ρ_{HV}, and the ratio of the cross-polar power to the co-polar power is the linear depolarization ratio, or L_{DR}. These are the important observables, which can also be expressed at the signal level in terms of complex matrices:

$$\begin{pmatrix} E_{sV} \\ E_{sH} \end{pmatrix} = \begin{pmatrix} S_{VV} & S_{VH} \\ S_{HV} & S_{HH} \end{pmatrix} \begin{pmatrix} E_{iV} \\ E_{iH} \end{pmatrix}.$$

Not only backscatter, but also absorption and depolarization, reflect information about the material in the radiation-illuminated volume.

11.4.1 Z_{DR}

The definition of the differential reflectivity of radar backscatter is

$$\begin{aligned} Z_{DR}\,(\mathrm{dB}) &= 10\log_{10}\frac{|S_{HH}|^2}{|S_{VV}|^2} \\ &= 10\log_{10}\frac{Z_{HH}}{Z_{VV}}. \end{aligned}$$

Differential reflectivity depends on the diameter and axial ratio of the hydrometeors. This can be shown by generalizing the results for Rayleigh–Mie scatter to oblate and prolate spherical targets. For raindrops, the axial ratio in turn depends on drop size. Overall, the relationship between Z_{DR} and mean drop size can be recovered empirically in field experiments. Typical values of Z_{DR} in rain range from about 0 to 5 dB and go approximately as a constant times the square of the mean drop size. The measurements are invertible and can be used to update drop-size distributions that feed into conventional reflectivity-based rainfall-rate estimates.

By comparison, snowfall exhibits relatively small differential reflectivity between about 0 and 1 dB that is not a strong function of size. Similar comments hold for hail and ice crystals, which can exhibit small positive and negative Z_{DR}.

Some non-meteorological reflectors can also be identified through differential reflectivity, including different kinds of biota as well as ground clutter. One of the important roles of polarimetry in weather radar is classification and removal of different kinds of clutter.

11.4.2 ρ_{HV}

This is the normalized correlation of the co-polar horizontally and vertically polarized signals, i.e.,

$$\rho_{HV} = \frac{\langle S_{HH} S_{VV} \rangle}{\sqrt{\langle |S_{HH}|^2 \rangle - N_H} \sqrt{\langle |S_{VV}|^2 \rangle - N_V}},$$

where $N_{H,V}$ refers to noise power in the given polarization. This parameter $|\rho_{HV}|$ turns out to be indicative of hydrometeor size and shape, type, orientation and regularity. It also turns out to be inversely proportional to Z_{DR}. This is in part due to the fact that hydrometeors with larger average diameters tend at once to be more oblate (increasing Z_{DR}) and less homogeneous and regular (decreasing $|\rho_{HV}|$).

Typical values for $|\rho_{HV}|$ are 0.98+ for light rain, 0.95 for driven rain, 0.9 for mixed rain and hail, 0.75 for mixed rain and snow, and 0.5− for debris fields such as occur during tornadoes. The parameter is therefore most useful for characterizing precipitation types. The analysis is complicated by other influences on $|\rho_{HV}|$, including the mixing ratio and scattering angle and geometry.

Feature with unambiguous signatures in $|\rho_{HV}|$ include bright bands where reflectivity maximizes at the layer where frozen particles begin to melt and to be coated in liquid water. Below the bright band where a layer of frozen and melted precipitation exists, $|\rho_{HV}|$ exhibits a deep minimum.

11.4.3 LDR

Formally, the linear depolarization ratio is

$$\text{LDR}_{VH} = 10 \log_{10} \frac{|S_{VH}|^2}{|S_{HH}|^2},$$

which measures the ratio of cross-polar to co-polar backscatter power. Depolarization is vanishingly small for spherical hydrometeors or oblate or prolate hydrometeors with their major or minor axes aligned with the radar wavevector. Only irregular or eccentric, skew-aligned targets give rise to significant depolarization. Depolarization therefore becomes detectable when the targets are falling at oblique angles with respect to the radar wave vector, tumbling, or irregular. For backscatter from precipitation, LDR is generally ranges from about −25 dB or less for light rain to −10 to −25 dB for hail/rain mixtures.

LDR maximizes at or a little below bright bands where tumbling, water-coated ice particles exist. Otherwise, specific relationships between measured LDR and hydrometeor characteristics are elusive. Moreover, LDR measurements are easily contaminated by noise, clutter and returns from antenna sidelobes.

11.4.4 ϕ_{DP}

Finally, ϕ_{DP} is the difference in phase angles between the co-polar horizontally and vertically polarized signals scattered from some range. This is the accumulated two-way phase delay between the radar and the target for the two propagation modes. The phase delays tend to be different if the intervening particles are anisotropic. Oblate hydrometeors produce larger phase delays in the horizontal polarization than in the vertical. The reverse is true of prolate particles.

Since ϕ_{DP} is an accumulated quantity, it is the derivative of ϕ_{DP} versus range (in degrees/km) that conveys information about the scattering medium at a given range. Values of $d\phi_{DP}/dr$ can be found to be as large as about $1/4°$/km for flat, plate-like ice crystals. For needle-like ice crystals, the sign of $d\phi_{DP}/dr$ is negative.

The measurement has the advantage of being independent of signal intensity variations due to range, attenuation and obstruction. It has the disadvantage of noisiness associated with differentiated data.

11.5 Clear-Air Echoes

This chapter opened with the remark that the strongest radar echoes from the troposphere are due to precipitation and the visible phenomena that accompany severe weather. Since the earliest days of weather radar, however, it was recognized that weaker echoes also arise when conditions are clear. These are alternatively called anomalous echoes, angel echoes and, more recently, clear-air echoes. The echoes are thought to arise from multiple sources including ground clutter, biota (e.g., insects and the birds that feed on them), atmospheric dust, cloud droplets and (predominantly) gradients in the index of refraction associated with irregularities in atmospheric temperature, pressure and composition (i.e., moisture content). These irregularities are mainly found at air-mass boundaries, sea-breeze fronts and where mixing gives rise to turbulence. The physical mechanism that allows them to be detected is a combination of Rayleigh–Mie and Bragg scatter. One of the most critical applications for clear-air echoes is the identification of sudden downdrafts in the vicinity of airports.

General-purpose meteorological radars like WSR-88D have specialized, high-resolution, high-sensitivity modes for monitoring clear air. Others are dedicated to the purpose. A class of high-power VHF radars exists for measuring winds and turbulence. These are called MST radars (mesosphere, stratosphere, troposphere) or simply "wind profiles." The index-of-refraction variations are larger at VHF frequencies than at UHF and

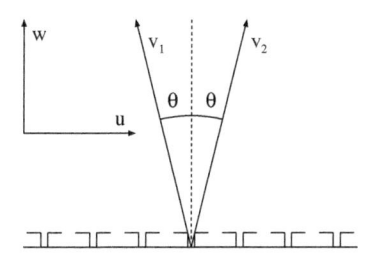

Figure 11.4 Vertical wind profiling mode for studying clear-air echoes. Line-of-sight Doppler velocities V_1 and V_2 are measured at equal but opposing zenith angles θ. The beams project into the horizontal u and vertical w wind speeds. Another pair of beam pointing positions would be used to investigate the third wind component v.

higher frequencies. Reflector antennas are impractical at VHF frequencies, so these radars typically employ large, planar, phased-array antennas looking upward. The radars observe Doppler-frequency shifts along high-elevation beams pointed in the cardinal directions to make inferences about atmospheric winds, momentum fluxes, and turbulence. Boundary-layer radars operate using similar principles, but at UHF and microwave frequencies because of the very high bandwidth necessary to resolve the thin layers.

11.5.1 Wind Profiling and Momentum Flux

The geometry for a radar wind profiling mode is illustrated in Figure 11.4. For the sake of simplicity, the diagram is confined to the plane that includes the horizontal and vertical winds u and w, respectively. In practice, two more beam pointing positions would be used to infer information about the remaining wind component v. Line-of-sight Doppler velocities V_1 and V_2 are measured on opposing beam pointing positions. We assume that u and w are the same in both beams. This assumption holds when the zenith angles θ are small, although this limits the observability of u.

The relationship between the measured Doppler velocities and the underlying winds is (neglecting noise and statistical fluctuations),

$$V_1 = w\sin\theta - u\sin\theta$$
$$V_2 = w\sin\theta + u\sin\theta.$$

It is obvious from the figure and the equations that the sum $V_1 + V_2$ is an estimator for w, while the difference is an estimator for u, given appropriate geometric scaling. Estimates like these can be made as functions of altitude, yielding wind profile measurements. The accuracy of the estimates depends on the accuracy of the Doppler velocity measurements and the homogeneity of the winds across the radar field-of-view.

Measurements of V_1 and V_2 will exhibit systemic variability as well as natural variability present in u and v. The natural variability is indicative of waves and instabilities riding on top of the background winds. If the measurement is stationary, then incoherent integration can reduce the systemic variability to tolerable levels without eliminating the natural variability. Still more averaging will eliminate the natural variability as well, leaving behind information about the long-term average winds only.

Suppose incoherent integration has been performed and systemic variability in the Doppler-velocity measurements is negligible. Then we may write

$$\overline{V}_1 + \delta V_1 = (\overline{w} + \delta w)\sin\theta - (\overline{u} + \delta u)\sin\theta \qquad (11.30)$$

$$\overline{V}_2 + \delta V_1 = (\overline{w} + \delta w)\sin\theta + (\overline{u} + \delta u)\sin\theta, \qquad (11.31)$$

where overbars imply long-term average winds and the δs indicate variations indicative of waves and instabilities. The variations are zero-mean random processes. Averaging (11.30) and (11.31) in time eliminates the variations, yielding equations and estimators for the average winds. Subtracting the averaged equations from the originals gives equations and estimators for the variations.

The variances of the estimated Doppler-speed variations can furthermore be found to obey the following relationships:

$$\overline{\delta V_1^2} = \overline{\delta w^2}\sin^2\theta + \overline{\delta u^2}\cos^2\theta - 2\overline{\delta w \delta u}\sin\theta\cos\theta$$

$$\overline{\delta V_2^2} = \overline{\delta w^2}\sin^2\theta + \overline{\delta u^2}\cos^2\theta + 2\overline{\delta w \delta u}\sin\theta\cos\theta.$$

Subtracting these equations yields a recipe for estimating the cross-term in the quadratic products:

$$\overline{\delta w \delta u} = \frac{\overline{\delta V_2^2} - \overline{\delta V_1^2}}{2\sin 2\theta}. \qquad (11.32)$$

This is the vertical flux of horizontal momentum, a key parameter for understanding the influence of propagating waves on the background average flow. Measuring momentum flux is one of the most important objectives of experimental clear-air atmospheric science.

If the radar beams are too far apart, then different winds u and w will be sampled by the two beams, and the interpretation of the measurements outlined above will break down. The sum and the difference of V_1 and V_2 will be affected by u and w, and the momentum-flux estimator prescribed by (11.32) will likewise become biased. However, if the beams are very close together, errors in the estimation of $\overline{\delta V_1^2}$ and $\overline{\delta V_2^2}$ due to systemic variations will grow, requiring more incoherent integration and rendering the measurement non-stationary. Experimental demands posed by momentum-flux measurements are, consequently, considerable.

11.5.2 Scattering from Irregular Media

This chapter concludes with a theoretical description of scattering from inhomogeneous media and the interpretation of the scattered signal. This will be an extension of diffraction

11.5 Clear-Air Echoes

theory for media with highly irregular and possibly stochastic variations in the index of refraction. The discussion will also serve as a bridge to the next chapter, which concerns radar imaging. The goal of imaging is not only to interpret but also to invert the measurements to reconstruct the intervening medium. To solve the inverse problem, one must first be able to solve the forward problem.

In Chapter 2, the inhomogeneous Helmholtz equation was developed for electromagnetic waves in inhomogeneous media:

$$\nabla^2 \mathbf{E} + k_\circ^2 n^2 \mathbf{E} = -\nabla \left[\nabla (\ln n^2) \cdot \mathbf{E} \right], \tag{11.33}$$

in which n is the index of refraction and $k_\circ = \omega/c$ is the free-space wavenumber. If the right side of the equation were zero and n were uniform, we would have the homogeneous Helmholtz equation for waves in a source-free region with well-known solutions: waves propagating along straight-line rays emanating from distant sources that enter the problem through the boundary conditions. Allowing gradual variations in n in just the left side of the equation would result in propagation along gently sloping rays and the other refractive effects described by geometric optics.

The term in the right side of (11.33) occupies the position of a source term. Abrupt variations in the index of refraction in a medium illuminated by radiation therefore behave like secondary sources of radiation. The secondary radiation excites tertiary radiation where it encounters other index-of-refraction variations, and so on. The problem considered here is one of solving (11.33) self-consistently for the net radiation when the characteristics of the medium are known. The more difficult problem left for Chapter 12 is one of inferring the characteristics of the medium from the signals measured by sensors outside of it.

The analysis can be simplified by considering a scalar Helmholtz equation such as applies to each component of the vector magnetic potential or the scalar potential in a source-free region:

$$(\nabla^2 + k^2)\psi = 0,$$

where the wavenumber $k = n\omega/c$ and where $n(\mathbf{x})$ varies in space. Let the average value of n be n_m and the corresponding wavenumber be $k_m = n_m \omega/c$. Then the Helmholtz equation can be expressed in inhomogeneous form,

$$\begin{aligned}(\nabla^2 + k_m^2)\psi &= (k_m^2 - k^2)\psi \\ &= -k_m^2 \left[(n/n_m)^2 - 1 \right] \psi \\ &= -f(\mathbf{x})\psi. \end{aligned} \tag{11.34}$$

As before, irregularity in the index of refraction plays the role of a source term in what has become the inhomogeneous Helmholtz equation.

Formally, the solution to linear equations such as (11.34) can be found using the Green's function, the solution to the inhomogeneous problem where the source term is a point source. The boundary conditions also need to be specified; for the sake of simplicity, we will merely require that the solution vanished at infinity. As has been shown previously, the corresponding time-independent Green's function is given by

$$(\nabla^2 + k_m^2)G(\mathbf{x}-\mathbf{x}') = -\delta(\mathbf{x}-\mathbf{x}')$$
$$G(\mathbf{x}-\mathbf{x}') = \frac{\exp(ik_m|\mathbf{x}-\mathbf{x}'|)}{4\pi|\mathbf{x}-\mathbf{x}'|}.$$

The solution is found by exploiting the translational property of Dirac's delta:

$$f(\mathbf{x})\psi(\mathbf{x}) = \int d^3\mathbf{x}' \, \delta(\mathbf{x}-\mathbf{x}')f(\mathbf{x}')\psi(\mathbf{x}')$$
$$= -\int d^3\mathbf{x}' \, (\nabla^2 + k_m^2)G(\mathbf{x}-\mathbf{x}')G(\mathbf{x}-\mathbf{x}')f(\mathbf{x}')\psi(\mathbf{x}')$$
$$= -(\nabla^2 + k_m^2)\int d^3\mathbf{x}' \, G(\mathbf{x}-\mathbf{x}')f(\mathbf{x}')\psi(\mathbf{x}').$$

Comparison with (11.34) indicates that the overall (incident plus scattered) potential obeys the integral equation

$$\psi(\mathbf{x}) = \int d^3\mathbf{x}' \, G(\mathbf{x}-\mathbf{x}')f(\mathbf{x}')\psi(\mathbf{x}'), \qquad (11.35)$$

which is exact but implicit. Note that we have made use of the fact that, since $G = G(|\mathbf{x}-\mathbf{x}'|)$, $\nabla^2 G = \nabla'^2 G$.

Born Approximation

Equation (11.35) cannot be solved exactly analytically. It can be solved approximately by considering solutions of the form $\psi(\mathbf{x}) = \psi_\circ(\mathbf{x}) + \psi_s(\mathbf{x})$, where $\psi_\circ$ is the incident field and ψ_s is the scattered field. In formulating the Born approximation, (11.35) is used to calculate the scattered field from the total field in view of information about the scattering medium, viz.

$$\psi_s(\mathbf{x}) = \int d^3\mathbf{x}' \, G(\mathbf{x}-\mathbf{x}')f(\mathbf{x}')\psi(\mathbf{x}')$$
$$\psi(\mathbf{x}) = \psi_\circ(\mathbf{x}) + \int d^3\mathbf{x}' \, G(\mathbf{x}-\mathbf{x}')f(\mathbf{x}')\psi(\mathbf{x}'), \qquad (11.36)$$

which is the Lippmann–Schwinger equation for the field ψ, given the perturbation f. Iterative solution of (11.36) yields a series of successive approximations – the Born series. The Born approximation actually corresponds to the first term in the series:

$$\psi(\mathbf{x}) = \psi_\circ(\mathbf{x}) + \int d^3\mathbf{x}' \, G(\mathbf{x}-\mathbf{x}')f(\mathbf{x}')\psi_\circ(\mathbf{x}').$$

The meaning of the Born approximation is that the contribution to the net field by the scattered field is small enough that it can be neglected in the computation of the scattered field. Thus, the self-consistency problem is avoided. Specifically, the Born approximation requires that the amplitude of the scattered field be small compared to that of the incident field and that the total accumulated phase angle induced by scattering be small compared to 2π.

Throughout the text, the Born approximation has been invoked wherever multiple-scattering effects have been neglected. Here we have the formal meaning of the approximation and the limits of its applicability.

Rytov Approximation

Another approximation also involves the decomposition of the field into incident and scattered components. Take the incident and total fields to have the form

$$\psi(\mathbf{x}) = \psi_\circ(\mathbf{x}) + \psi_s(\mathbf{x})$$
$$= \exp(\phi(\mathbf{x}))$$
$$\psi_\circ(\mathbf{x}) = \exp(\phi_\circ(\mathbf{x})),$$

where the complex exponentials imply both amplitude A and phase Φ:

$$\phi(\mathbf{x}) = i\Phi(\mathbf{x}) + \ln(A(\mathbf{x}))$$
$$\phi_\circ(\mathbf{x}) = i\Phi_\circ(\mathbf{x}) + \ln(A_\circ(\mathbf{x}))$$
$$\phi(\mathbf{x}) = \phi_\circ(\mathbf{x}) + \phi_s(\mathbf{x}),$$

Note now that both ψ and ϕ have been decomposed into incident and scattered components. The two decompositions compel the specific form of the scattered field:

$$\psi_s(\mathbf{x}) = \exp(\phi_\circ(\mathbf{x}))\left[\exp(\phi_s(\mathbf{x})) - 1\right].$$

Using the above expansion for $\psi(\mathbf{x})$, the inhomogeneous Helmholtz equation corresponding to it can be written

$$(\nabla^2 + k_m^2)\exp(\phi(\mathbf{x})) = -f(\mathbf{x})\exp(\phi(\mathbf{x}))$$
$$\left[\nabla^2\phi(\mathbf{x}) + (\nabla\phi(\mathbf{x}))^2 + k_m^2\right]\exp(\phi(\mathbf{x})) = -f(\mathbf{x})\exp(\phi(\mathbf{x}))$$
$$\left[\nabla^2\phi(\mathbf{x}) + (\nabla\phi(\mathbf{x}))^2 + k_m^2\right] = -f(\mathbf{x}),$$

which is a nonlinear partial differential equation. (That any kind of simplification is underway must be far from evident at this point!) Likewise, the equation corresponding to $\psi_\circ(\mathbf{x})$ is

$$\left[\nabla^2\phi_\circ(\mathbf{x}) + (\nabla\phi_\circ(\mathbf{x}))^2 + k_m^2\right] = 0, \tag{11.37}$$

which is a homogeneous equation in view of the fact that the incident field has no source term, by definition. Yet a third Helmholtz equation can be written by utilizing the original decomposition $\psi = \psi_\circ + \psi_s$,

$$\nabla^2(\psi_\circ(\mathbf{x}) + \psi_s(\mathbf{x})) + [\nabla(\psi_\circ(\mathbf{x}) + \psi_s(\mathbf{x}))]^2 + k_m^2 = -f(\mathbf{x}). \tag{11.38}$$

Subtracting (11.37) from (11.38) eliminates a number of terms, leaving behind

$$\nabla^2 \phi(\mathbf{x}) + 2\nabla\phi_s(\mathbf{x}) \cdot \nabla\phi_\circ(\mathbf{x}) + (\nabla\phi_s(\mathbf{x}))^2 = -f(\mathbf{x}).$$

Multiplying this equation by $\psi_\circ(\mathbf{x})$ and working backward through the chain rule yields the desired result:

$$(\nabla^2 + k_m^2)\psi_\circ(\mathbf{x})\phi_s(\mathbf{x}) = -\psi_\circ(\mathbf{x})\left[(\nabla\phi_s(\mathbf{x}))^2 + f(\mathbf{x})\right]. \tag{11.39}$$

Equation (11.39) is still exact, and is known at the Riccati equation. Under the Rytov approximation, the perturbation $f(\mathbf{x})$ is taken to dominate the phase-gradient term $(\nabla\phi_s(\mathbf{x}))^2$ beside it, which can then be neglected. The problem is therefore made linear again in the scattered field quantities. Given a known incident field $\psi_\circ$, (11.39) gives a prescription for calculating the scattered phase ϕ_s using the method of Green's function once more:

$$\phi_s(\mathbf{x}) = \frac{\int d^3x' G(\mathbf{x}-\mathbf{x}') f(\mathbf{x}') \psi_\circ(\mathbf{x}')}{\psi_\circ(\mathbf{x})}.$$

Recall that the complex scattered phase ϕ_s includes both amplitude and phase information for the scattered field and is therefore a complete solution.

The Born approximation applies to weak scattering, a condition that may be violated if multiple-scattering events can occur, for example. It is not suitable for modeling forward scatter since it does not update the incident field. The tendency will be for the scattered fields to add constructively in the forward direction, leading to growth and the violation of the conservation of energy over long-distance propagation. The Born approximation is consequently used almost exclusively to analyze backscatter.

The Rytov approximation meanwhile applies to smooth perturbations and correspondingly gradual changes in propagation direction. It is essentially a small scattering-angle approximation, well suited for studying long-distance forward propagation and scatter but ill-suited for modeling backscatter. In the case where the accumulated phase delay of the scattered field is small compared to 2π, it can be shown that the Rytov approximation becomes the Born approximation.

11.6 Notes and Further Reading

Weather radar is a mature discipline with numerous references covering everything from established doctrine to frontier research. Comprehensive texts on the subject include those by R. J. Doviak and D. S. Zrnić, V. N. Bringi and V. Chandrasekar, and F. Fabry. Scattering from clear air is taken up in the text by E. E. Gossard and R. G. Straugh. Voluminous data pertaining to radio propagation through the atmosphere is given in L. J. Battan's text.

Radar measurements of momentum fluxes were reviewed early on by R. A. Vincent and I. M. Reid and more recently by D. M. Riggin et al. Statistical issues related to momentum flux measurements were investigated in detail by E. Kudeki and S. J. Franke.

A recent review paper deriving the Clausius–Mossotti relation is cited below, as is a paper reviewing the Maxwell–Garnett theory and its extensions. (Original sources for the theories are provided by the respective references.) Additional background theory is given in the more general remote-sensing text by L. Tsang et al.

For a review of the first Born and first Rytov approximation see the paper by B. Cairns and E. Wolf.

D. Atlas and C. W. Ulbrich. Path- and area-integrated rainfall measurement by microwave attenuation in the 1–3 cm band. *J. Appl. Meteor.*, 16:1322–1331, 1977.

L. J. Battan. *Radar Observations of the Atmosphere*. The University of Chicago Press, Chicago, 1973.

V. N. Bringi and V. Chandrasekar. *Polarimetric Doppler Weather Radar*. Cambridge University Press, Cambridge, UK, 2001.

B. Cairns and E. Wolf. Comparison of the Born and Rytov approximations for scattering from quasi-homogeneous media. *Optics Comm.*, 74(5):284–289, 1990.

R. J. Doviak and D. S. Zrnić. *Doppler Radar and Weather Observations*, 2nd ed. Academic Press, San Diego, 1993.

C. C. Easterbrook. Estimating horizontal wind fields by two-dimensional curve fitting of single Doppler radar measurements. In *16th Conference on Radar Meteorology (preprints)*, pages 214–219. American Meteorological Society, Boston, 1974.

F. Fabry. *Radar Meteorology*. Cambridge University Press, Cambridge, UK, 2015.

E. E. Gossard and R. G. Strauch. *Radar Observations of Clear air and Clouds*. Elsevier, Amsterdam, 1983.

E. Kudeki and S. J. Franke. Statistics of momentum flux estimation. *J. Atmos. Sol. Terr. Phys.*, 60(16):1549–1553, 1998.

V. Markel. Introduction to the Maxwell Garnett approximation: tutorial. *J. Opt. Soc. Am.*, 33(7):1244–1256, 2016.

J. S. Marshall and W. Mck. Palmer. The distribution of raindrops with size. *J. Meteor*, 5:165–166, 1948.

D. M. Riggin, T. Tsuda and A. Shinbori. Evaluation of momentum flux with radar. *J. Atmos. Sol. Terr. Phys.*, 142:98–107, May 2016.

E. Talebian and M. Talebian. A general review of the derivation of the Clausius–Mossotti relation. *Optik*, 124:2324–2326, 2013.

L. Tsang, J. A. Kong and R. T. Shin. *Theory of Microwave Remote Sensing*. John Wiley & Sons, Wiley-Interscience, New York, 1985.

R. A. Vincent and I. M. Reid. HF Doppler measurements of mesospheric gravity wave momentum fluxes. *J. Atmos. Sci.*, 40(5):1321–1333, 1983.

Exercises

11.1 Derive the radar equation in (11.4) from scratch, being sure to account for the L_r and $2\ln 2$ factors. It may be helpful to review Sections 6.7 and 8.3.2.

11.2 Consider radio waves propagating through a cloud of water droplets with a water content of 1 g/m^3. The cloud is 3 km thick. The absorption coefficient at 3 cm (10 cm) wavelength is $-\Im(K) = 3.67 \times 10^{-5}$ (1.1×10^{-5}). Compare the one-way absorption at both frequencies. For these calculations, assume uniform drop sizes, choose a reasonable drop diameter, and estimate the corresponding drop density accordingly.

11.3 Calculate the electric field inside a Lorentz sphere, i.e., perform the integral given in (11.15). It is trivial to calculate the electric field at the center of the sphere but more difficult to show that the electric field is uniform within the sphere. Section 12 of the Appendix B offers a helpful hint.

11.4 Certain hydrometeors have a reflectivity of 50 dBZ. Calculate η at wavelengths of 3 cm and 10 cm. For the 10-cm case, take the transmitted power to be 100 kW, the pulse width to be 2 μs, the antenna gain to be 40 dB, and the beamwidth to be 1°. If the two-way losses due to attenuation are 3 dB, how much scattered power will be received from a range of 50 km?

11.5 Fill in the details leading up to the $Z - R$ relationship given in (11.24) starting from the gamma drop-size distribution. You will need to be able to evaluate the Gamma function to work this problem.

11.6 A Doppler spectrum is computed using an 8-point discrete Fourier transform (DFT) as shown in the adjoining figure. Estimate the Doppler shift and spectral width from the measurement. A noise measurement (i.e., with the transmitter off) made separately yields an integrated noise power estimate of 12 units. Using this information, recalculate the Doppler shift and spectral width and discuss the bias in both quantities introduced by the noise (see Fig. 11.5).

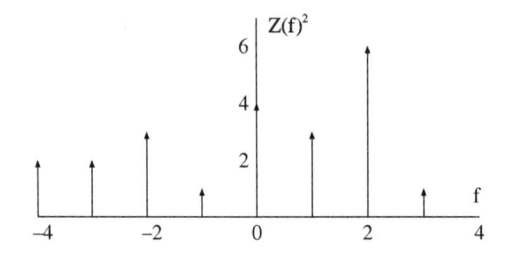

Figure 11.5 Discrete power spectrum for Exercise 11.6.

Note that the spectral power assigned to the −4 frequency bin could equally well be assigned to the +4 bin. What is the best way to handle the power in this bin?

12
Radar Imaging

In this final chapter, we review more recent advances in radar remote sensing and radar imaging. Additional information about our environment can be extracted from radar signals using spatial diversity, transmitters of opportunity, synthetic apertures, multiple antenna pointing positions, and even the near field. Growth in the radar art in the near term is expected to take place mainly in these areas. As interferometry lies at the root of the imaging problem, the chapter opens with an introduction to interferometry.

12.1 Radar Interferometry

Radar interferometry utilizes multiple spaced receivers to derive information about the spatial organization of the radar targets in the direction(s) transverse to the radar line of sight. We assume that the radar target is sufficiently far away to be in the far field of the entire interferometry array, i.e., so that the rays from all the antennas to the target are parallel. There are two classes of interferometry: additive and multiplicative. Additive interferometry, where signals from distant antennas are added prior to detection, is really just another name for beam forming, something that was addressed in Chapter 3. The discussion here concerns multiplicative interferometry, where the outputs from spaced antennas undergo multiplication.

Figure 12.1 (left) shows the geometry for an interferometer constructed from a single pair of antennas. The antennas are separated by a baseline distance d. The only coordinate of importance is the angle ψ, the angle between the interferometry baseline and the rays to the target. The path length difference from the target to the two antennas is just $d\cos\psi$. We can determine this angle for backscatter from a single target by beating or multiplying the signals from the two antennas,

$$v_1 = v_\circ e^{j(\mathbf{k}\cdot\mathbf{x}_1 - \omega t)}$$
$$v_2 = v_\circ e^{j(\mathbf{k}\cdot\mathbf{x}_2 - \omega t)},$$

where $\mathbf{x}_{1,2}$ are the displacements of the two antennas from a fixed reference, and

$$v_1^* v_2 = |v_\circ|^2 e^{j\mathbf{k}\cdot(\mathbf{x}_2 - \mathbf{x}_1)}$$
$$= |v_\circ|^2 e^{j\mathbf{k}\cdot\mathbf{d}_{21}}$$

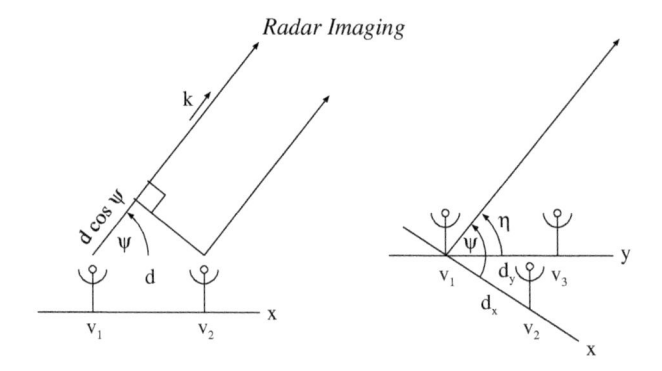

Figure 12.1 Geometry for radar interferometry in two (left) and three (right) dimensions.

$$= |v_\circ|^2 e^{jkd\cos\psi}$$
$$= |v_\circ|^2 e^{j\phi}, \quad \phi = \frac{2\pi}{\lambda} d\cos\psi.$$

The cosine of the elevation angle ψ is the sine of the zenith angle θ, and for targets nearly overhead, $\sin\theta \sim \theta$. In any event, the phase of the interferometric cross-correlation gives the target bearing.

Determining the bearing of a target in three dimensions is accomplished using interferometry with multiple baselines, as illustrated in Figure 12.1 (right),

$$v_1^* v_2 = |v_\circ|^2 e^{jkd_x\cos\psi}$$
$$v_1^* v_3 = |v_\circ|^2 e^{jkd_y\cos\eta}.$$

Thus, the interferometer provides complete range and bearing (and Doppler frequency) information for a single target. Note that information from the third baseline formed by antennas 2 and 3 is redundant but available as a consistency check, for example.

12.1.1 Ambiguity

As with range and Doppler processing, ambiguity is inherent in interferometry. The root cause is the periodicity of phase angles, i.e.,

$$\phi = \frac{2\pi}{\lambda} d\cos\psi + 2\pi n,$$

where n is any integer. Solving for the direction cosine gives

$$\cos\psi = \frac{\lambda}{d}\left(\frac{\phi}{2\pi} - n\right).$$

12.1 Radar Interferometry

If $d \leq \lambda/2$, it is impossible to solve this equation with a real valued ψ with any value of n other than zero, regardless of the measured phase angle ϕ. Ambiguity arises when baseline lengths are longer than $\lambda/2$. In such cases, however, ambiguity can be resolved by using multiple baselines, each one satisfying

$$\cos\psi = \frac{\lambda}{d_j}\left(\frac{\phi_j}{2\pi} - n_j\right)$$

for the given index j. One need only determine the n_j that point to a consistent angle ψ. Ambiguity can also be avoided or mitigated by transmitting with a narrow-beam antenna, thereby limiting the possible values of ψ.

12.1.2 Volume Scatter

Next, suppose interferometry is applied when multiple, discrete targets exist in the scattering volume. The received voltages and their product in that case are

$$v_1 = \sum_{i=1}^{m} v_i e^{j(\mathbf{k}_i \cdot \mathbf{x}_1 - \omega t)}$$

$$v_2 = \sum_{j=1}^{m} v_j e^{j(\mathbf{k}_j \cdot \mathbf{x}_2 - \omega t)}$$

$$v_1^* v_2 = \sum_{i=1}^{m} \sum_{j=1}^{m} v_i^* v_j e^{j(\mathbf{k}_j \cdot \mathbf{x}_2 - \mathbf{k}_i \cdot \mathbf{x}_1)}.$$

The voltages and their product have the properties of random variables when the number of targets in the volume is large. In that case, it is the expectation of the product that is of interest. In many applications, we can usually take the components of the received voltages from different targets to be statistically uncorrelated. Then

$$\langle v_1^* v_2 \rangle = \langle \sum_{i=1}^{m} |v_i|^2 e^{j\mathbf{k}_i \cdot \mathbf{d}_{21}} \rangle$$
$$= m\langle |v_i|^2 e^{jkd\cos\psi_i} \rangle$$
$$= m\langle |v_i|^2 \rangle \langle e^{jkd\theta_i} \rangle.$$

As before, the angle brackets imply the expected value. The last step was performed with the assumption that the scatterer amplitudes and positions are also uncorrelated. The direction cosine was replaced with the zenith angle, θ_i, for simplicity.

The interferometric cross-correlation evidently depends on the number of targets in the volume and on the power scattered by each. It also depends on the expectation of the phasor

326   Radar Imaging

$\exp(jkd\theta)$, which would be unity in the case of collocated antennas. Let us decompose the zenith angle θ into a mean angle $\langle\theta\rangle$ and a deviation from the mean $\delta\theta = \theta - \langle\theta\rangle$. Then

$$\langle e^{jkd\theta}\rangle = e^{jkd\langle\theta\rangle}\langle e^{jkd\delta\theta}\rangle$$
$$\approx e^{jkd\langle\theta\rangle}\left(1 - \frac{k^2d^2}{2}\langle\delta\theta^2\rangle\right),$$

where we have expanded the exponential in a Taylor series and made use of the fact that $\langle\delta\theta\rangle \equiv 0$. The expansion is legitimate so long as the scattering volume is sufficiently narrow, either because the antenna beams are narrow or the targets are confined spatially for some reason. In practice, this approximation is usually, but not always, valid, since kd may be large.

Properly normalized, the interferometry cross product has the form

$$\frac{\langle v_1^* v_2\rangle}{\sqrt{\langle|v_1|^2\rangle - N_1}\sqrt{\langle|v_2|^2\rangle - N_2}} = e^{jkd\langle\theta\rangle}\left(1 - \frac{k^2d^2}{2}\langle\delta\theta^2\rangle\right), \tag{12.1}$$

where N_1 and N_2 are noise power estimators for the two channels and where the noise signals from the two channels are uncorrelated. (This assumption could be violated if some of the noise in question is actually interference masquerading as noise.) Equation (12.1) shows that the phase angle of the normalized cross product is associated with the mean arrival angle of the backscatter, the scattering center of gravity. The magnitude or *coherence*, meanwhile, is associated with the mean-square deviation or spread of arrival angles. Similar remarks hold in three dimensions. Interferometry with two or more non-collinear baselines therefore completely determines the first three moments of the spatial distribution of scatterers (the total power, bearing and spread).

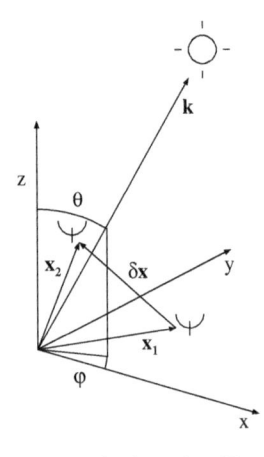

Figure 12.2 Geometry for an aperture-synthesis radar. Two antennas are shown, but multiple antennas with multiple pairings would be used for radar imagery.

More insight into the problem comes from considering the formal definition of the normalized cross product

$$\langle e^{jkd\theta} \rangle \equiv \frac{\int F(\theta) e^{jkd\theta} d\theta}{\int F(\theta) d\theta},$$

where F is the backscatter or *brightness* distribution introduced earlier in the discussion of adaptive arrays. Interferometry with a large number of non-redundant baselines can specify the normalized cross product or *visibility* in detail, albeit incompletely. This function is in turn related to the brightness by Fourier transform. Various strategies exist for inverting the transformation and recovering the brightness on the basis of available data. This is the field of aperture synthesis radar imaging.

12.2 Aperture Synthesis Imaging

Aperture synthesis imaging involves image reconstruction based on interferometry with multiple baselines. This is the foundation of very long baseline interferometry (VLBI) but applies equally well in radar applications where the image is illuminated by a radar. While the method relies on interferometric concepts introduced in the preceding section of this chapter, it is instructive to redevelop it in a more general way from first principles.

Consider the voltage received at location $\mathbf{x}_1$ as being due to a collection of plane waves. Each plane wave has a wavevector $\mathbf{k}$ with a wavenumber $k = 2\pi/\lambda = |\mathbf{k}|$. The radio receiver is tuned to a particular wavelength λ and is therefore sensitive to a particular wavenumber. Suppose that the amplitude of each plane wave is given by $E(\mathbf{k})$, and the phase by the product $\mathbf{k} \cdot \mathbf{x}$. Consequently, the received voltage can be expressed as

$$v(\mathbf{x}) = \int d\Omega E(\mathbf{k}) e^{i\mathbf{k} \cdot \mathbf{x}},$$

where the integral is over all differential solid angles in the sky, $d\Omega$.

Suppose next that the signals from two spaced receivers are correlated according to

$$\langle v(\mathbf{x}_1) v^*(\mathbf{x}_2) \rangle = \left\langle \int d\Omega E(\mathbf{k}) e^{i\mathbf{k} \cdot \mathbf{x}_1} \int d\Omega' E^*(\mathbf{k}') e^{-i\mathbf{k}' \cdot \mathbf{x}_2} \right\rangle,$$

where the angle brackets denote an ensemble average in principle and a time average in practice. Here, we can regard the amplitudes of the various radio sources as random variables. Furthermore, we can regard the amplitudes of signals arriving from different bearings as being statistically uncorrelated. Consequently, one of the integrals in the above equation can be performed trivially, leaving

$$\langle v(\mathbf{x}_1) v^*(\mathbf{x}_2) \rangle = \int d\Omega \langle |E(\mathbf{k})|^2 \rangle e^{i\mathbf{k} \cdot (\mathbf{x}_1 - \mathbf{x}_2)}.$$

Finally, define the displacement of the two receiving stations as $\delta\mathbf{x} \equiv \mathbf{x}_1 - \mathbf{x}_2$. Further define the correlation of the two signals as the visibility $V(\delta\mathbf{x}$, a function of the displacement, and the signal power arriving from a given bearing as the brightness distribution $B(\mathbf{k})$. Then,

$$V(\delta\mathbf{x};k) = \int d\Omega B(\mathbf{k})e^{i\mathbf{k}\cdot\delta\mathbf{x}}, \qquad (12.2)$$

and the functional relationship between the brightness (the image) and the visibility (the measurement) can be seen to be a linear integral transform. By measuring the visibility using multiple pairs of receivers, information necessary to infer the brightness distribution of the sky is gathered. The brightness distribution is an image of the radar target or radio sources at the given wavelength.

At first glance, the integral transform in (12.2) might appear to be a Fourier transform. This would be the case if the integral were over all wavevectors, i.e., $d\mathbf{k}$. In fact, the integral is only over wavevector directions, with the wavenumber held constant. To clarify matters, it is helpful to expand (12.2) in Cartesian and spherical coordinates:

$$V(\delta\mathbf{x};k) = \int \sin\theta d\theta d\phi B(\theta,\phi)\exp(ik(\sin\theta\cos\phi\delta x + \sin\theta\sin\phi\delta y + \cos\theta\delta z))$$
$$= \int d\eta d\xi \frac{B(\eta,\xi)}{\sqrt{1-\eta^2-\xi^2}} \exp\left(ik\left(\eta\delta x + \xi\delta y + \sqrt{1-\eta^2-\xi^2}\delta z\right)\right), \qquad (12.3)$$

where we have defined $\eta \equiv \sin\theta\cos\phi$ and $\xi \equiv \sin\theta\sin\phi$ to be the direction cosines of the wavevector with respect to the cardinal x and y directions and used the Jacobian of the transformation ($J = \cos\theta\sin\theta$) in writing (12.3) after noting that $\cos\theta = \sqrt{1-\eta^2-\xi^2}$.

Note that the preceding calculation neglected the effect of the shape of the individual antenna radiation pattern(s) on the visibilities. To remedy the problem, the brightness distribution $B(\eta,\xi)$ above should be reinterpreted as the product of the true brightness distribution and the receiving antenna element pattern (and also the transmitting antenna element pattern in the case of radar), i.e., $B(\eta,\xi) = B_\circ(\eta,\xi)P_r(\eta,\xi)P_t(\eta,\xi)$. The appropriate patterns can be divided from brightness estimates in the final analysis, if need be. In many cases, the individual element pattern or patterns are very wide compared to the targets being imaged, in which case the correction is not critical. The correction is not considered in the treatment that follows.

In the case of all the receivers being in a plane such that all the $d\delta z = 0$ and that the brightness is confined to regions where η and ξ are small, we have

$$V(\delta x, \delta y; k) \approx \int\int d\eta d\xi B(\eta,\xi)\exp(ik(\eta\delta x + \xi\delta y)), \qquad (12.4)$$

which is in the form of a 2-D Fourier transform. This is the Cittert–Zernike theorem. The formula for an inverse Fourier transform is well known. However, how should the inverse

12.2 Aperture Synthesis Imaging

problem of estimating the brightness be carried out in view of the fact that only a discrete set of visibilities are known, presumably on a sparse, irregular grid of points?

Equation (12.4) has the form of a linear inverse problem. A number of strategies exist for solving such problems. Most of these have been designed for the discrete domain. Posed in terms of discrete quantities, the imaging problem looks like

$$d = Gm + e, \tag{12.5}$$

where d is a column vector containing visibility measurements, m is a column vector containing brightness estimates (in different pixels), G is the point spread function in matrix form, and e contains experimental errors/observation noise. One of the methods was already discussed in Chapter 3 in the context of adaptive beamforming. The LCMV method (Capon's method) was an optimization method in which the gain of the synthetic antenna array was minimized in all directions except the direction of interest, where it was held fixed. The image was formed by making every possible direction the direction of interest. The algorithm is fast, stable, and simple to implement.

Two other algorithms are popular in radar and radio astronomy. They have different philosophies and are known to produce somewhat different results. These methods are reviewed below.

12.2.1 CLEAN

CLEAN is a local, iterative deconvolution procedure that is relatively fast and easy to implement. It can be applied to any deconvolution problem, not just problems in radio imaging. Although its mathematical properties are not thoroughly understood, the efficacy of the procedure has been demonstrated in a number of contexts, particularly in radio astronomy. CLEAN is in widespread use and continues to undergo development.

Some preliminary definitions are in order. We have seen (Equation (12.4)) that the brightness function is related to the visibility function by a 2-D Fourier transform. However, we cannot simply use another Fourier transform to estimate the former, the latter being sampled at a few discrete points only. Let us regard the measured visibility function as a product of the complete visibility function and a mask or sampling function $S(\delta x, \delta y; k)$ which is only nonzero at the coordinates where visibility samples exist. According to the convolution theorem, the brightness function (called the "dirty map") resulting from the Fourier transform of this product is the true brightness function (called the "clean map") convolved with the Fourier transform of the sampling function (called the "dirty beam"). The dirty beam is the array pattern of the entire imaging array which could well have multiple, irregular sidelobes. The CLEAN algorithm attempts to account for these sidelobes and to recover the clean map or something close to it from the dirty map. The imaging problem is essentially a deconvolution problem.

To complete the set of definitions, the "clean beam" is an idealized beam shape. It is often taken to have an elliptical shape that fits inside the main lobe of the dirty beam.

The original CLEAN algorithm was introduced by Högbom in 1974. It was intended for images with a small number of point targets but can also be applied to images with extended targets. It starts with the specification of an initial guess for the clean map. The residual, the difference between the dirty map and the clean map convolved with the dirty beam, is computed at each iteration. The point where the residual is largest is identified. This point signifies a potential target. A copy of the dirty beam, shifted to this point and multiplied by the size of the residual and by a constant, the loop gain, is subtracted from the dirty beam. (Typical loop gains are between about 0.1 and 0.3.) A delta function, centered on the same point and scaled by the same factor, is also stored for later use. Iteration continues until the a norm of the residual falls below a preset critical level. The remaining contents of the residual are taken to be noise at this point. The final image is then formed from the convolution of the clean map, to which all the stored, shifted, weighted delta functions have been added, with the clean beam. This last step limits the resolution of the image but helps to suppress artifacts.

It can be shown that CLEAN will converge so long as the beam is symmetric, the Fourier transform of the dirty beam is non-negative, and there are no Fourier components in the dirty image that are not also present in the dirty beam. If the number of targets does not exceed the number of visibility measurements, CLEAN converges to the least-squares fit of sinusoidal functions to the measured visibilities, and CLEAN may be seen as being equivalent to such a fitting procedure. The only way to perform error analysis with CLEAN is using Monte Carlo methods, which increase computational cost greatly.

A number of modifications have been made to the basic procedure to improve speed and stability and suppress artifacts. CLEAN can be confined to small boxes or windows. The Clark algorithm improves speed by working in small windows (in so-called minor cycles) and then shifting the results (in major cycles) back to perform global dirty-beam corrections with the aid of FFTs. The Cotton–Schwab method uses Clark cycles but performs major-cycle corrections directly on the visibility data rather than on the dirty beam. Cornwell introduced a stabilized variant of CLEAN meant to help suppress spurious stripes associated with undersampled visibilities and extended targets. The CLEAN 2 algorithm achieves better performance with extended targets by using a dirty beam that is somewhat narrower than the Fourier transform of the sampling function.

In addition to a tendency to artifacts and difficulty with error analysis, CLEAN suffers from a high degree of sensitivity to its control parameters: the loop gain, number of iterations, box locations, beam shapes etc. Noticeable differences in outcomes often result from varying these parameters. More consistent performance can often be obtained, at the cost of increased computation time, from more complicated deconvolution algorithms such as the one described below.

12.2.2 Maximum Entropy Method

Another popular approach to aperture synthesis imaging is a global optimization method rooted in statistical inverse theory. This is a computationally expensive method but one with well-understood mathematical properties. The goal of this method is to minimize an objective function which has two components. One is the chi-square parameter, the square of the L2 norm of the discrepancy between the predicted and the measured visibilities. This term can be formulated such that the penalty reflects the anticipated properties of the observation error, i.e., the theoretical error covariance matrix C_d:

$$\chi^2 = e^T C_d^{-1} e,$$

where e is a column vector containing the discrepancy vector or residual, $Gm - d$.

The other component of the objective function is Shannon's entropy, applied here to the brightness of the pixels in the image,

$$S = -\sum_{j=1}^{n} m_j \ln \frac{m_j}{M}, \quad (12.6)$$

where n is the number of pixels, m_j is the brightness of the jth pixel, and M is the total image brightness, summed over all pixels. The entropy S is related to the number of ways that a given image could be realized from the random deposition of brightness quanta (photons) indistinguishably into a finite number of image pixels or bins. The greater the entropy, the more "generic" the image. The vast majority of realizable images have entropies clustered around the maximum.

High entropy reflects low information content, and candidate images with entropy lower than the maximum imply the possession of information. Such candidates can only be adopted to the extent they have support in the data, i.e., that the chi-square parameter demands it. In practice, one typically seeks the brightness distribution that minimizes the negative of the entropy (negentropy) with the constraint that the error norm remains equal or close to its expected value. Since the entropy metric represents prior information, the maximum entropy method is a Bayesian method. It can also be thought of as a regularization method, with entropy serving as the regularization parameter. Since the entropy metric is insensitive to the arrangement of the pixels in the image, the method has no preference for smoothness and is edge preserving. (The method does have a preference for uniformity.) The application of the Shannon–Hartley theorem can be used to show that the method is not bound by the diffraction limit and yields the greatest spatial resolution of any nonparametric method. The spatial resolution is related to the logarithm of the signal-to-noise ratio, and so the performance is best in the high-SNR limit.

Although the relationship between brightness and visibility is linear, the introduction of entropy makes the optimization problem nonlinear, requiring an iterative solution method and the specification of an initial guess. This being a Bayesian method, additional prior information can easily be incorporated.

The maximum entropy solution is the extremum of the objective function

$$f_1 = -S + \lambda^T (d - Gm) + \Lambda (M - 1^T m), \quad (12.7)$$

where λ is a vector of Lagrange multipliers which enforce model/data congruity, Λ is another Lagrange multiplier enforcing the normalization condition, $M = \sum_j m_j$, and 1 is a column vector made up of 1s. Differentiating with respect to m_j gives

$$-\ln\frac{m_j}{M} - 1 - \lambda^T G^{[,j]} - L = 0, \qquad (12.8)$$

in which $G^{[,j]}$ stands for the jth column of G. This can be solved for m_j, making use of the normalization condition,

$$m_j = M\frac{e^{-\lambda^T G^{[,j]}}}{Z} \qquad (12.9)$$

$$Z = \sum_j e^{-\lambda^T G^{[,j]}}, \qquad (12.10)$$

with Z playing the role of a partition function. Indeed, the approach to optimization being pursued here is the same one that leads to the Gibbs partition function that lies at the foundation of statistical mechanics.

It is expedient to substitute this expression for m_j into the original objective function and to reformulate it, adopting a bootstrapping strategy. There is, furthermore, a fundamental flaw in (12.7), which insists on an exact match between the data and the model prediction. In fact, we expect discrepancies arising from sampling errors. The refined objective function is therefore

$$f_2 = s + \lambda^T(d + e - Gm) + \Lambda(e^T C_d^{-1} e - \Sigma) \qquad (12.11)$$
$$= \lambda^T(d + e) - M\ln Z + \Lambda(e^T C_d^{-1} e - \Sigma), \qquad (12.12)$$

where (12.9) has been utilized. Here, Σ is the expected value of the chi-square parameter, and the Lagrange multiplier Λ is now enforcing the fixed error constraint.

Finally, the problem is solved by minimizing f_2 with respect to the Lagrange multipliers and the undetermined error terms,

$$\frac{\partial f_2}{\partial \lambda_j} = d + e - Gm = 0 \qquad (12.13)$$

$$\frac{\partial f_2}{\partial e_j} = C_d\lambda + 2\Lambda e = 0 \qquad (12.14)$$

$$\frac{\partial f_2}{\partial \Lambda} = e^T C_d^{-1} e - \Sigma = 0, \qquad (12.15)$$

which is a coupled system of nonlinear equations. Equation (12.13) restates the forward problem. Equation (12.14) relates the error terms to the Lagrange multipliers in λ. Equation (12.15) restates the error constraint. Substituting (12.14) into (12.15) gives a relationship between Λ and the other Lagrange multipliers:

12.2 Aperture Synthesis Imaging

$$\Lambda^2 = \frac{\lambda^T C_d \lambda}{4\Sigma}. \tag{12.16}$$

The model m above is expressible in terms of the Lagrange multipliers λ through (12.9), which are the only parameters left to be determined.

Eliminating m, e and Λ as described above leaves two real equations for every complex visibility measurement that must be solved as a system iteratively. The system can be solved using Newton's method or a variant, given a data vector, a setting for Σ and an initial guess for the λ vector. The best initial guess is one that is consistent with a uniform model and a small Λ. Note that not only the equations but also their derivatives can be expressed analytically for incorporation into the iteration scheme. Iteration can proceed until errors are limited by machine precision. Once found, the Lagrange multipliers λ imply the image pixel brightnesses through (12.9).

The maximum entropy method has only one adjustable parameter: the Σ parameter. Making this term smaller than (larger than) the expectation of the chi-squared parameter produces images that are sharper (less sharp). The changes are gradual and quantitative rather than qualitative. Various rationales for setting the Σ parameter high or low exist in the literature. For an appropriate choice of Σ, the maximum entropy method has a unique solution on which the algorithm will converge, so long as the data are physically realizable. The solution thus found is the image with the greatest entropy consistent with the data.

A demonstration of aperture synthesis imaging is presented in Figure 12.3. The figure depicts in grayscale format (on a logarithmic scale) the signal-to-noise ratio of echoes received from plasma turbulence in the Earth's ionosphere. The scattering mechanism at work is essentially Bragg scatter, and the backscattered power arrives in the plane perpendicular to the geomagnetic field. The backscatter is therefore quasi 2-D.

The top panel of the figure presents backscatter vs. altitude and time in conventional range-time-intensity (RTI) format. This panel depicts the passage of a number of "plumes" of intense backscatter. It is not a true image in the sense that the horizontal axis represents time rather than space, and the morphology of the plumes is not represented faithfully in general in such a presentation. For this measurement, the radar antenna beam was "spoiled" or intentionally broadened in the east-west direction, the direction of motion of the plumes. This had the effect of extending the duration of time across which each plume could be observed. The purpose was to illuminate the broadest volume possible for imaging purposes.

The four panels below the RTI plot are aperture-synthesis images computed at four different local times when backscatter plumes were directly over the radar. Each of the images was computed on the basis of just a few seconds of data using the methods described in this section of the text. These are true images that reveal the actual morphology of the turbulence and its evolution over time. The effective angular resolution of the method is about 0.1°, which is about one-tenth the width of the narrowest radar beam the radar antenna

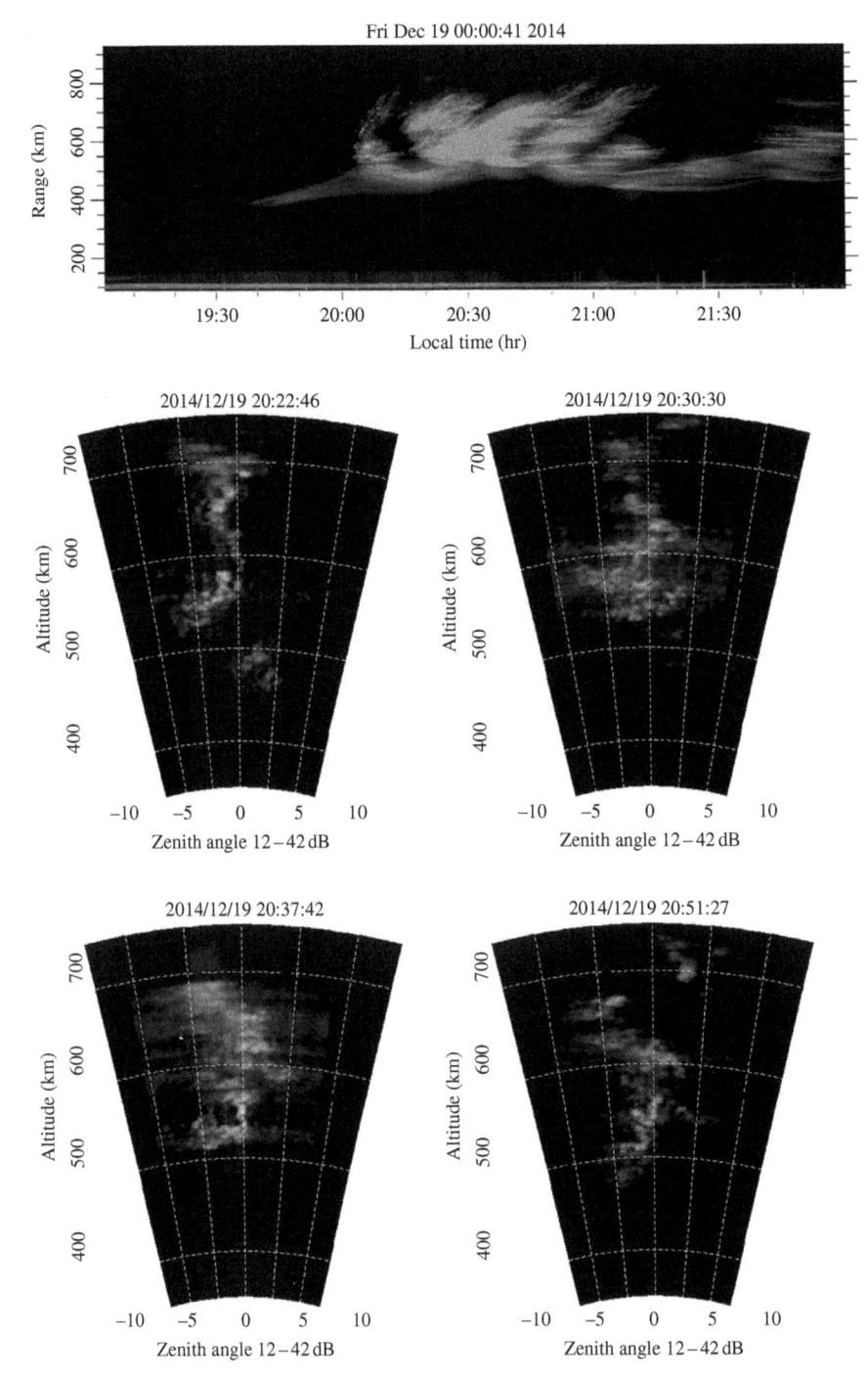

Figure 12.3 Radar imagery of turbulence in the Earth's ionosphere observed overhead by the Jicamarca Radio Observatory near Lima, Peru, close to the magnetic equator. The top panel is a conventional range-time-intensity plot. The four panels beneath are aperture synthesis images computed at four different times.

is capable of producing. Some artificial broadening of the imagery due to radio scintillation that occurs as the radar pulse propagates through the irregular medium is evident, particularly at the highest altitudes.

12.3 Synthetic Aperture Radar (SAR)

The basic principles behind synthetic aperture radar were already discussed in the context of virtual antenna arrays in Chapter 3. Here, we delve into the details of SAR data processing and image formation. As in aperture synthesis imaging, the details are complicated, and the data processing burden can be enormous. New methods for processing SAR data more optimally and expediently continue to be developed. The impacts on planetary exploration, geophysical research and intelligence gathering to name but a few applications justify the effort.

SAR is similar to planetary radar in that the 2-D locations of objects in the radar field of view are determined through a combination of range and Doppler analysis. In SAR, surfaces of constant range on the ground are concentric circles centered on the spot below the vehicle carrying the radar. Surfaces of constant Doppler shift are hyperbolas, symmetric about the path of travel of the vehicle. In planetary radar, the Doppler shift is due to the motion of the target, whereas in synthetic aperture radar, it is due partly or entirely to the motion of the vehicle itself. Objects in the field of view must be stationary in some SAR applications and moving in others. Range and Doppler ambiguities can be avoided through the careful selection of radar beam shapes, frequencies, and interpulse periods, in any case. Objects to the left and right of a SAR will exhibit the same Doppler shifts, just like objects in the northern and southern hemispheres do in planetary radar, but ambiguity can be avoided in the former case by using side-looking radars.

The frequency for space-borne SAR in terrestrial applications is usually between the L and X bands. Below L-band frequencies, ionospheric irregularities severely degrade the imagery. Above X-band, meteorological effects are detrimental. Different constraints apply for planetary radar, radars on aircraft, etc. SAR signal processing can also be applied to acoustic signals including sonar.

The radar range resolution in SAR can be enhanced through pulse compression which most often takes the form of LFM chirp sounding in practice. We recall that pulse compression utilized the fact that echoes from targets at different ranges will exhibit different phase histories that can precisely identify the range. Likewise, echoes from targets at different azimuths in the SAR context will also exhibit different phase histories (frequency chirps, in fact) that can be used to identify the azimuth. This phenomenon is known as "azimuth compression," which is completely analogous to pulse compression. Image resolution in the direction of vehicle motion is improved by incorporating very long time series in the determination of the echo Doppler shifts, since long time series imply fine spectral resolution. In this way, broad-beam antennas can be advantageous, since they permit targets on the ground to be observed for long periods of time as the vehicle moves.

SAR data analysts refer to "beamwidth compression," which is also analogous to pulse compression in its attempt to make a small antenna perform like a large one. The price to be paid for using small antennas is reduced sensitivity, and, as always, an optimal balance between resolution and sensitivity must be struck.

Signal processing invariably involves coherent processing of long time series, but can take several forms, depending on the objective and sophistication of the imaging algorithm. Most often, it begins with matched filter decoding of the chirped transmitter waveform. This produces the standard data matrix encountered back in Chapter 7, with each row representing data from a different pulse and each column data from a different range. Pulse compression is usually accomplished using Fourier transforms rather than direct convolution for computational expediency. The range resolution is determined by the pulse width and compression ratio in the usual way. An important complication in SAR arises from the fact that the range to an individual target varies or "migrates" according to a hyperbolic curve during the time of data acquisition. This is both the key to and a complicating factor in the processing or "focusing" of SAR data.

SAR observations are made mainly in one of three modes: spotlight, stripmap or scanning. In the first, the radar beam is constantly redirected so that the same scene is always illuminated. In the second, the beam is directed at a constant angle (squint angle) with respect to a straight-line vehicle trajectory. Stripmap mode can be used to observe much larger scenes at the cost of lower spatial resolution. Whereas the far-field approximation can generally be made in spotlight-mode data analysis, it generally cannot with stripmap-mode data. Finally, systematic scanning can be incorporated in SAR data analysis, further increasing the coverage area at additional cost in resolution. The discussion that follows mainly pertains to stripmap-mode SAR.

In the analysis that follows, details regarding determining and controlling the trajectory of the vehicle carrying the SAR, the characteristics of the radar beam and the surface topology will be neglected for simplicity. The attenuation of the signal amplitude with range will also be neglected, and certain multiplicative constants will be omitted. Figure 12.4 illustrates the geometry of an idealized SAR experiment. Reflectivity on the ground at $u(x,r)$ is specified in terms of cylindrical coordinates x (downrange distance) and r (distance of closest approach to the vehicle track). Data $d(x,t)$ are acquired at position x along the flight track and time delay t since the last pulse. The radar is regarded as being essentially stationary during the time it takes to send a pulse and receive its echoes.

The angle $\theta_\circ$ is the squint angle, the angle between the vehicle track and the main beam of the radar. For simplicity, we take $\theta_\circ = 90°$ in what follows. The results below can easily be generalized for other squint angles, although complications arise when $\theta_\circ$ is very small.

Fundamental to SAR signal processing is the impulse response or point spread function which maps a point target in the reflectivity or image space into the corresponding signal in the data space. Let the emitted radar pulse be $p(t)\exp(j\omega_\circ t)$, where $p(t)$ is the modulation envelope and $\omega_\circ$ is the carrier frequency. The echo received from a point reflector at distance R after removal of the carrier will then be $p(t-2R/c)\exp(-j\omega_\circ 2R/c)$. The range

12.3 Synthetic Aperture Radar (SAR)

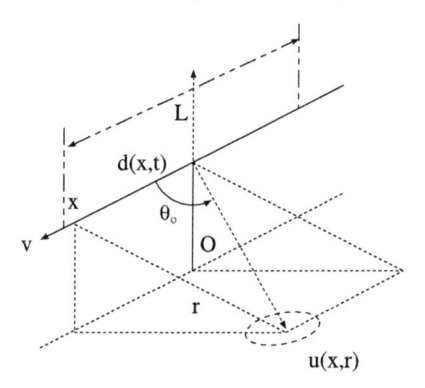

Figure 12.4 Geometry of a stripmap SAR experiment. The vehicle carrying the radar moves with constant velocity **v**. The image space contains scatterers with reflectivities given by $u(x,r)$. The data space contains signals $d(x,t)$ acquired at position x and delay time t (after the last pulse).

R depends on the location of the vehicle and the target. If the vehicle is at position x' and the point target is at the coordinates (x,r), then the distance R follows the rule

$$R(x';x,r) = \sqrt{(x'-x)^2 + r^2} \approx r + \frac{(x'-x)^2}{2r},$$

where the approximation holds for very distant targets and will be used throughout what follows. The range has a hyperbolic (approximately parabolic when $x' \sim x$) dependence on x' and so migrates throughout the observation at a rate that depends on r. It will be expedient to divide the distance into two parts, viz.

$$R(x',x;r) = R(x'-x;r) = r + \Delta R(x'-x;r),$$

where we note that the correction ΔR is a (quadratic) function of the difference between x' and x. This term represents range migration, which both complicates and facilitates precision radar imaging.

After matched filtering in time and with the application of pulse compression, the pulse envelope (the autocorrelation of $p(t)$) should become narrow and can be considered to be a Dirac delta function for present purposes. The point spread function can subsequently be expressed as

$$h(x',x,t;r) \equiv \delta\left(t - \frac{2}{c}R\right)\exp\left(-j\omega_\circ \frac{2}{c}R\right)$$
$$= \tilde{h}(x'-x, t - 2r/c; r)\exp\left(-j\omega_\circ \frac{2}{c}r\right)$$
$$\tilde{h}(x'-x,t;r) \equiv \delta\left(t - \frac{2}{c}\Delta R\right)\exp\left(-j\omega_\circ \frac{2}{c}\Delta R\right),$$

where a normalized version of the point spread function is defined by the last equation. This version will help make certain relationships clear in what follows.

The point spread function is the response in the data space at (x',t) to a point source in the image space at (x,r). Formally, the reflectivity image $u(x,r)$ can be constructed from the data $d(x',t)$ through correlation with the point spread function.

The prescription for reconstructing the SAR image is

$$u(x,r) = \int\int_{-\infty}^{\infty} d(x',t) h^*(x',x,t;r) dx' dt \qquad (12.17)$$

$$= \exp(-2j\omega_\circ r/c) \int\int_{-\infty}^{\infty} d(x',t) \tilde{h}^*(x'-x, t-2r/c; r) dx' dt \qquad (12.18)$$

$$= \exp(-2j\omega_\circ r/c) \int_{-\infty}^{\infty} d(x,t) \odot_x \tilde{h}(x, t-2r/c; r) dt. \qquad (12.19)$$

In (12.17), to infer the reflectivity at the coordinates (x,r), we correlate the data we expect to see for a reflector at those coordinates with the data that were actually acquired. The operation is analogous to matched filtering as discussed in the context of signal-to-noise ratio optimization, except that the point spread function in this case is an inseparable function of the coordinates x and r. It is now evident how range migration has complicated the focusing problem. Not only is the point spread function two-dimensional, it also inhomogeneous. Although (12.19) has the form of a correlation in x, the explicit dependence of $\tilde{h}$ on r means that it does not have the form of a correlation in r, meaning that direct solutions in terms of 2-D Fourier transforms are not obviously unavailable. As written, (12.17) and (12.18) would seem to be computationally intractable for all but very simple circumstances in which range migration can be neglected. While there are applications where this is permissible, they are limited.

Note that the $\exp(-2j\omega_\circ r/c)$ phase term in (12.18) and (12.19) is immaterial to the defocusing problem and can be absorbed into the definition of $u(x,r)$ for present intents and purposes. The phase term is critical for InSAR, which will be discussed later.

There are two main approached for forming images, which are referred to as range-Doppler (RD) and wavenumber-domain (WD) processing, respectively. The former came first and is more intuitive. It involves 1-D Fourier analysis of data columns (i.e., in the "azimuth" direction). Range migration is corrected using interpolative resampling in time, and azimuth compression is accomplished with the application of a 1-D linear filter. The approach is effective but approximate and somewhat *ad hoc*. The second is a more contemporary development and renders focusing performance closer to ideal. The key to WD (sometimes called "$\omega - k$") processing is Fourier analysis of the data in two dimensions (i.e., in both the "azimuth" and "range" direction). A key change of variables permits the evaluation of (12.17) using Fourier transforms alone. WD processing has its roots in wave mechanics but can be formulated in a straightforward way using the formalism of signal processing. Below, the WD strategy is briefly outlined.

12.3.1 Wavenumber-Domain (WD) Processing

We seek a reformulation of (12.19) amenable to solution using the machinery of Fourier transforms (discrete Fourier transforms in practice). Transforming the equation from the spatial (x) domain to the wavenumber (k_x) domain is straightforward and converts the correlation to a product,

$$U(k_x, r) = \int_{-\infty}^{\infty} D(k_x, t) H^*(k_x, t - 2r/c; r) \, dt,$$

where $U(k_x, r)$, $D(k_x, t)$ and $H(k_x, t; r)$ are the Fourier transforms of $u(x, r)$, $d(x, t)$ and $\tilde{h}(x, t; r)$ with respect to x. In order to transform the equation from the time to the frequency domain, Parseval's theorem may be invoked,

$$U(k_x, r) = \frac{1}{2\pi} \int_{-\infty}^{\infty} D(k_x, \omega) H^*(k_x, \omega; r) e^{j(2r/c)\omega} \, d\omega, \tag{12.20}$$

where U, D and H now denote quantities transformed to (k_x, ω) space.

At this point, it is useful to carry out the calculation of $H(k_x, \omega; r)$ from $\tilde{h}(x, t; r)$. The transformation from the time to the frequency domain is straightforward and gives

$$H(x, \omega; r) = \exp\left\{ -j\omega' \left(\frac{2\Delta R(x, r)}{c} \right) \right\}, \tag{12.21}$$

where $\omega' = \omega_\circ + \omega$. The next step is to transform the equation from the spatial (x) to the wavenumber (k_x) domain. This can be accomplished using the method of stationary phase. The result can be stated simply (ignoring a slowly varying multiplicative factor) as

$$H^*(k_x, \omega; r) \propto \exp\left\{ -jr \left(2\omega'/c - \sqrt{(2\omega'/c)^2 - k_x^2} \right) \right\} \tag{12.22}$$

$$H^*(k_x, \omega; r) e^{j(2r/c)\omega} \propto \exp\left\{ -jr \left(2\omega_\circ/c - \sqrt{(2\omega'/c)^2 - k_x^2} \right) \right\}$$

$$\equiv E(k_z, \omega; r),$$

where the function $E(k_z, \omega; r)$ is the kernel of the integral we are trying to evaluate.

Finally, it is useful to perform one more Fourier transform – this time from the r to the k_r domain:

$$E(k_z, \omega; k_r) \propto \delta\left[k_r - \left\{ \sqrt{(2\omega'/c)^2 - k_x^2} - 2\omega_\circ/c \right\} \right].$$

The Dirac delta function here involving k_r suggests and indeed imposes a crucial change of variables known as the "Stolt mapping," viz.

$$k_r = \sqrt{(2\omega'/c)^2 - k_x^2} - 2\omega_\circ/c,$$

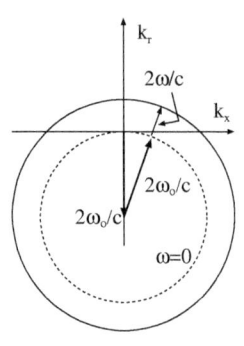

Figure 12.5 Graphical representation of the Stolt mapping from (k_r, k_z) to ω. Contours of constant frequency in k-space are circles centered at $k_r = -2\omega_\circ/c$, $k_x = 0$. The dashed curve is the zero-frequency circle. Points outside (inside) this line will have positive (negative) Doppler frequencies. The Doppler frequency ω for a point in k-space is given by the length of the short radial vector.

or, equivalently,

$$\omega = \frac{c}{2}\sqrt{(k_r + 2\omega_\circ/c)^2 + k_x^2} - \omega_\circ,$$

which leads to an intuitive form of the kernel function:

$$E(k_z, \omega; k_r) \propto \delta\left(\omega - \frac{c}{2}\sqrt{(k_r + 2\omega_\circ/c)^2 + k_x^2} + \omega_\circ\right).$$

The Stolt mapping is illustrated graphically in Figure 12.6. The mapping is analogous to an Ewald sphere in x-ray crystallography, except in two dimensions. The mapping corrects the Doppler shift of the echoes as the underlying targets migrate, essentially accounting for the fact that the wavefronts are spherical rather than cylindrical. The drawing and attendant calculations apply for squint angles of $\theta_\circ = \pi/2$. Different squint angles can be accommodated by rotating the axis of symmetry of the concentric circles. The change of variables permits the accommodation of the auxiliary variable r into the transformation between data and image space and the focusing solution to be expressed entirely in terms of Fourier analysis.

Finally, (12.20) (or rather its Fourier transform in r–k_r space) can be evaluated, giving an exact relationship between the data and the image, each in different, 2-D Fourier spaces:

$$U(k_x, k_r) \propto D\left(k_x, \frac{c}{2}\sqrt{(k_r + 2\omega_\circ/c)^2 + k_x^2} - \omega_\circ\right).$$

The image itself is recovered by inverse transforming $U(k_x, k_r)$ back into physical space.

In summary, stripmap data analysis for an idealized set of SAR observations using basic WD formalism involves three steps:

12.3 Synthetic Aperture Radar (SAR)

Kilauea (wrapped)

(unwrapped)

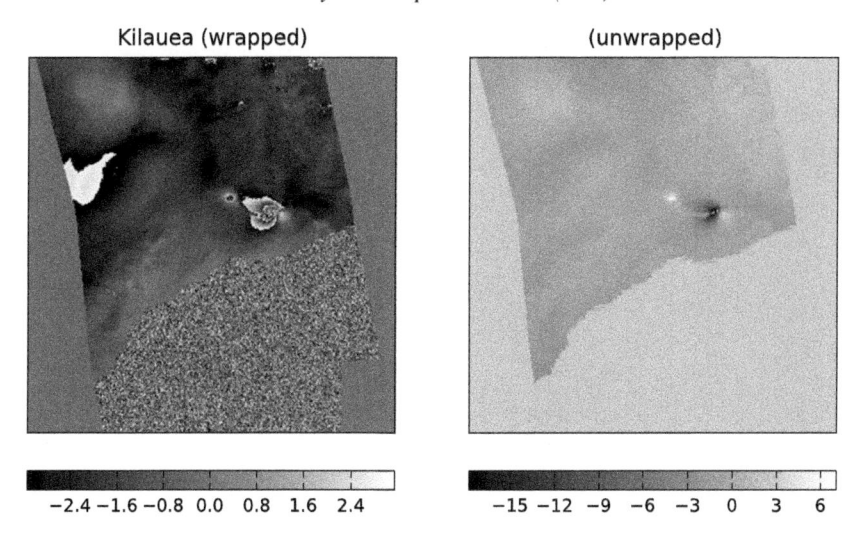

−2.4 −1.6 −0.8 0.0 0.8 1.6 2.4   −15 −12 −9 −6 −3 0 3 6

Figure 12.6 Interferogram for ALOS satellite track 291, May 5, 2007 to June 20, 2007, doi:10.7283/S21596, http://dx.doi.org/10.7283/S21596. The image shows the southern part of the Island of Hawaii and features the Kilauea volcano. Phases in the ocean south of the island are random. The images at left and right show wrapped and unwrapped phase angles, respectively.

1. Transform the data from $d(x,t) \to D(k_x, \omega)$ using a 2-D Fourier transform.
2. Perform a change of variables according to the Stolt mapping:
   $U(k_x, k_r) \propto D(k_x, \frac{c}{2}\sqrt{(k_r + 2\omega_\circ/c)^2 + k_x^2} - \omega_\circ)$.
3. Transform back to image space using an inverse 2-D Fourier transform: $U(k_x, k_r) \to u(x, r)$.

The image power is the square modulus of the reflectivity. Although the derivation involves wavenumbers in the image space, it has been carried out within the formalism of signal processing rather than wave mechanics.

12.3.2 Practical Implementation Issues

Regardless of the focusing method, a number of considerations apply to most SAR applications. The most important of these are discussed below.

Image Resolution The resolution of the recovered imagery in the cross-track direction is governed by the experimental geometry together with the radar pulse length and the compression ratio in the usual way. For a given range resolution, the spatial resolution of the image deteriorates as the incidence of the radar rays approach the vertical. Nearly horizontal observing geometries are therefore favored.

The resolution in the along-track direction is limited by the granularity of the data in the Doppler frequency domain. Given a total observing time $T = L/v$, where v is the vehicle speed and L is the length of the vehicle track, the width of the Doppler bins will be $\delta f =$

$1/T = v/L$. The longer the observing time, the longer the synthetic antenna array, and the finer the frequency resolution. Near the time of closest approach to a specific target when the range is r_o, we have (combining the formula for the Doppler shift with the expression for the range to the target),

$$\delta f \approx -\frac{2v^2}{\lambda r_o}\delta t$$

$$\delta x = -v\delta t$$

$$= \frac{\lambda r_o}{2L},$$

which is the same resolution estimate obtained earlier on the basis of synthetic aperture size arguments. Note that the present treatment has been made in the limit $r_o \gg L$, although a more general analysis can be performed.

Sample Rates and Aliasing Since fine spatial resolution depends on long observation times and therefore large L, the beamwidth of the illuminating radar should be correspondingly large, implying that the dimensions of the antenna, measured in wavelengths, should be small. Broad angular coverage also implies that the target will be broadband, requiring a high pulse rate for the Nyquist sampling criterion to be met. Roughly speaking, the spread of Doppler shifts will be

$$\Delta f \approx 2f_o \frac{v}{c}\sin\theta = \frac{2c}{\lambda}\frac{v}{c}\frac{\lambda}{l} = \frac{2v}{l}, \qquad (12.23)$$

where θ is the span of angles illuminated and l is the size of the radar aperture. To avoid frequency aliasing, pulses should therefore be emitted at intervals not longer than $l/2$ along the vehicle track.

Broad-beam illumination also invites backscatter from a wide span of ranges, however, increasing the potential for range aliasing. The observing geometry and radar frequency need to be considered in tandem to assure that the target remains underspread. The problem becomes more challenging as interest in observations at ever higher frequencies grows.

Range Delay Due to Pulse Chirp We have seen that pulse compression using linear FM chirps introduces a systematic time delay due to a skewing of the ambiguity function. The magnitude of the delay is given by $|\tau| = |f_D t_p / \Delta f|$, where f_D is the Doppler shift, t_p is the pulse length and Δf is the frequency excursion of the chirp. This can also be written as $|\tau| = |f_D t_p t_{\text{eff}}|$, where t_{eff} is the effective pulse length after compression. Since it is this last expression that determines the cross-track resolution of the images, it is important that the artificial range delay $|\tau|$ be small by comparison to t_{eff}.

The Doppler shift of a target on the ground will be $f_D = (2v/\lambda)\cos(\theta_o)$, where v is the vehicle speed. Combining these factors, the range delay introduced by chirping can be neglected if the following ratio is small compared to unity:

$$\frac{|\tau|}{\tau_{\text{eff}}} = \frac{2v}{\lambda} t_p \cos\theta.$$

As an example, consider an L-band radar operating at 1.07 GHz with a 4 μs pulse length on a spacecraft traveling at 7.1 km/s observing a target at a squint angle of 60°. The ratio in this case would be precisely 0.1 and small enough for the artificial delay to be neglected.

Noise and Clutter Random errors in SAR imagery are mainly the consequence of self clutter rather than noise. Self clutter gives rise to "speckle noise" in individual images. This can be reduced by incoherently processing imagery from several radar passes of the same scene. Interference as well as signal degradation due to propagation effects also limit image quality.

12.3.3 InSAR

InSAR combines interferometry with synthetic aperture radar processing to yield two-dimensional imagery with relief (see Figure 12.6 for a vivid example). Instead of forming a single virtual 1-D antenna array from data from a moving vehicle, InSAR forms multiple virtual arrays or, more generally, a virtual 2-D array. This can be done either by deploying two or more radars on a vehicle or by combining data from two or more passes of the same radar and vehicle past a given scene. The former case requires a very large platform but was demonstrated on early flights of the space shuttle. In the latter case, the target should remain invariant between the passes. If not, coherence will be lost, and relief information with it, although the loss of coherence would by itself be itself useful, indicative of something interesting like an earthquake or landslide. InSAR most often involves one radar on one vehicle, which could be a planetary or terrestrial satellite, airplane, surface ship or ground vehicle.

The basic principles behind interferometry were discussed earlier. In the InSAR case, the samples are the reflectivity estimates $u(x,r)$ discussed in the preceding section. These are complex quantities, each element with modulus and phase. We regard each as the sum of the reflectivities of all the reflectors in the spatial interval dx, dr. Cross products of two comparable estimates, properly normalized, will also have a modulus (the coherence) and a phase (the interferometer phase angle). The interferometry baseline for a two-pass InSAR observation is the distance between the passes, which can be regarded as a constant if the passes are nearly parallel. The interferometer phase angle is indicative of the direction cosine to the reflectors in dx, dr with respect to the baseline. Variations in the direction cosine can be mapped to variations in relief. Problems involving ambiguity and phase wrapping apply.

As mentioned above, the loss of coherence accompanied by fluctuations in phase sound signify evolution of the scene in the time between successive samples. However, interferometric coherence and phase estimates are the realization of random processes and exhibit fluctuations even when there is no physical cause for decorrelation. Assessing the significance of phase fluctuations in particular requires careful error analysis. Below, we analyze the statistical performance of interferometric measurements. For simplicity, we consider only the high signal-to-noise ratio case.

The estimator for the normalized interferometric correlation function ρ can be expressed as

$$\hat{\rho} = \frac{\frac{1}{n}\sum_{i=1}^{n} u_{1i}u_{2i}^*}{\sqrt{\frac{1}{n}\sum_{i=1}^{n}|u_{1i}|^2 \frac{1}{n}\sum_{j=1}^{n}|u_{2j}|^2}} \qquad (12.24)$$
$$= \frac{A}{\sqrt{B}},$$

where u_{1i} and u_{2i} are complex signals corresponding, respectively, to the first and second passes over a given pixel of a scene. Each pixel receives contributions from a number of randomly situated scatterers, and the sums in (12.24) represent sums over those scatterers. Echoes from different scatterers may be regarded as being statistically uncorrelated. The effective number of scatterers n can be augmented by averaging the contributions from adjacent pixels, albeit at the cost of spatial resolution. The total number of azimuth and range bins considered is call the number of looks.

The A and B factors in (12.24) will each exhibit statistical fluctuations and may be expressed in terms of their expectations and the deviations from them:

$$A = \langle A \rangle (1 + \varepsilon_A)$$
$$B = \langle B \rangle (1 + \varepsilon_B).$$

It can readily be verified that $\langle A \rangle = S\rho$ where S is the signal power and ρ is the interferometric correlation. At the same time, one can easily show that $\langle B \rangle = S^2(1+|\rho|^2/n)$. Whereas A is an unbiased estimator, the normalization factor B is biased and tends toward overestimation. The bias can be traced to the $i = j$ terms in the denominator of (12.24) and implies that $\hat{\rho}$ is also biased and tends toward underestimation. The root cause of the bias is related to the fact that $|\hat{\rho}|$ mathematically cannot exceed unity. When the coherence is high, fluctuations will necessarily favor reductions in coherence. The bias is small if the fluctuations are small which occurs when n is large. In what follows, we restrict the error analysis to cases where the bias in B may be neglected.

Combining terms, the normalized correlation estimator can be written as

$$\hat{\rho} = \frac{\langle A \rangle}{\sqrt{\langle B \rangle}} \left(1 + \varepsilon_A + \frac{1}{2}\varepsilon_B - \cdots\right),$$

12.3 Synthetic Aperture Radar (SAR)

where only the dominant 1st-order corrections have been retained. The associated mean-square error or MSE is therefore

$$\delta^2 = \langle |\hat{\rho} - \langle \hat{\rho} \rangle|^2 \rangle$$
$$\approx |\rho|^2 \left\langle \left| \varepsilon_A - \frac{1}{2}\varepsilon_B \right|^2 \right\rangle, \qquad (12.25)$$

where we do not differentiate between ρ and $\langle A \rangle / \sqrt{\langle B \rangle}$ here or in what follows. Although the expectations of the deviations ε_A and ε_B are zero, the expectations of their products are not.

What remains is to calculate the expectations of those products. Consider the following analysis:

$$\langle |A|^2 \rangle = |\langle A \rangle|^2 (1 + \langle |\varepsilon_A|^2 \rangle) \qquad (12.26)$$
$$= \frac{1}{n^2} \sum_{i,j=1}^{n} \langle v_{1i} v_{2i}^* v_{1j}^* v_{2j} \rangle$$
$$= \frac{1}{n} \langle v_{1i} v_{2i}^* v_{1i}^* v_{2i} \rangle + \frac{n(n-1)}{n^2} \langle v_{1i} v_{2i}^* v_{1j}^* v_{2j} \rangle_{i \neq j}$$
$$= \frac{1}{n} S^2 (1 + |\rho|^2) + \left(1 - \frac{1}{n}\right) S^2 |\rho|^2$$
$$= S^2 (|\rho|^2 + 1/n),$$

where the even-moment theorem for Gaussian random variables has been used. Combining this result with (12.26) leads, after some rearranging, to the result:

$$\langle |\varepsilon_A|^2 \rangle = \frac{1}{n|\rho|^2}.$$

Similar reasoning leads to the following additional results:

$$\langle |\varepsilon_B|^2 \rangle = \frac{2}{n}(1 + |\rho|^2)$$
$$\langle \varepsilon_A \varepsilon_B^* \rangle = \frac{2}{n}.$$

Substituting all of these results into (12.25) then yields an expression for the mean-square error of the normalized correlation function estimate:

$$\delta^2 = \frac{1}{n}\left(1 - \frac{3}{2}|\rho|^2 + \frac{1}{2}|\rho|^4\right). \tag{12.27}$$

The result is that the estimator becomes a perfect one as the coherence approaches unity even with a finite number of statistically independent samples. For low-coherence signals, the mean-square error is simply $1/n$.

The error analysis is still incomplete; also desirable is a means of quantifying estimation errors in the interferometric phase angle. For this, it is necessary to perform separate error analyses for the real and imaginary parts of the normalized correlation function estimators. A helpful auxiliary variable here is

$$\delta'^2 = \langle(\hat{\rho} - \langle\hat{\rho}\rangle)^2\rangle$$
$$= \frac{1}{2n}(-\rho^2 + \rho^2|\rho|^2),$$

which is like δ^2 only without the modulus operator, and which can be evaluated using the same procedure as the one followed for δ^2 above. If the normalized correlation estimation error is divided into real and imaginary parts, i.e., $\hat{\rho} - \langle\hat{\rho}\rangle = e_r + ie_i$, then it is clear that $\delta^2 = \langle e_r^2\rangle + \langle e_i^2\rangle$, $\delta'^2 = \langle e_r^2\rangle - \langle e_i^2\rangle + 2i\langle e_r e_i\rangle$, and therefore

$$\langle e_r^2\rangle = \frac{1}{2}\Re(\delta^2 + \delta'^2)$$
$$= \frac{1}{2n}(1 - (\Re\rho)^2)(1 - |\rho|^2) \tag{12.28}$$

$$\langle e_i^2\rangle = \frac{1}{2}\Re(\delta^2 - \delta'^2)$$
$$= \frac{1}{2n}(1 - (\Im\rho)^2)(1 - |\rho|^2) \tag{12.29}$$

$$\langle e_r e_i\rangle = \frac{1}{2}\Im(\delta'^2)$$
$$= \frac{1}{2n}\Re\rho\Im\rho(|\rho|^2 - 1). \tag{12.30}$$

It is noteworthy that the real and imaginary parts of the errors are correlated in general except when the phase angle is aligned with the real or imaginary axis. We can surmise that the errors in the direction of and normal to the phase angle are uncorrelated. Finally, on the basis of (12.28)–(12.30), the variances in the coherence $\gamma = \sqrt{(\Re\rho)^2 + (\Im\rho)^2}$ and phase angle $\phi = \mathrm{atan}(\Im\rho/\Re\rho)$ estimates are now calculable:

$$\langle\delta\gamma^2\rangle = \frac{1}{2n}(1 - |\rho|^2)^2 \tag{12.31}$$

$$\langle\delta\phi^2\rangle = \frac{1}{2n}(1 - |\rho|^2)/|\rho|^2. \tag{12.32}$$

Equations (12.31) and (12.32) are Cramér–Rao Lower Bound (CRLB) for the experimental variance and phase angle, respectively. The formulas highlight the importance of high correlation and a large number of "looks" n on estimator accuracies. Note that fluctuations in coherence and phase are correlated. The formulas apply only in the high signal-to-noise ratio case although they can easily be generalized for arbitrary SNRs.

12.3.4 ISAR

Inverse synthetic aperture radar, or "ISAR," is a method for imaging moving targets using a stationary radar. Is is more like planetary radar than stripmap SAR, and the analysis is simpler than what was presented above. The methodology is similar to spotlight SAR, however, except that the target is stationary and the radar is in motion in the latter case. Since it is the relative motion that matters, the analysis below also applies to the spotlight-SAR problem, given the appropriate shift in the frame of reference.

Consider a target moving with respect to a fixed radar. From the point of view of the radar, which shall be trained on the target, the target will generally appear to undergo rotation about a central axis as it moves. The Doppler shift of echoes from different parts of the target will depend linearly on the distance from the axis of rotation, projected on the radar field of view. This is precisely the situation for planetary radar. Images in two dimensions can consequently be computed using range-Doppler processing, although the mechanics are different due to the small sizes of the targets of interest in the ISAR case (small compared to practical pulse lengths).

The range resolution of the imagery will depend on the bandwidth of the emitted waveform as usual. Enormous bandwidth is required to generate useful images of compact targets like aircraft and surface ships, and ISAR work is consequently confined to microwave frequencies and higher. While LFM-chirps might be utilized, a more common practice in ISAR is to emit periodic bursts of pulses that span a wide range of discrete frequencies. This is referred to as a swept-frequency waveform. The span of frequencies is the effective bandwidth of the waveform and may be a significant fraction of the center frequency in practice.

Each pulse in a burst is long enough for a single receiver sample following it to contain echoes from all parts of the compact target. Every such sample will moreover be indicative of the target reflectivity at the given pulse frequency. Since the radar wavenumber $k = \omega/c$ is proportional to frequency, samples from different pulses in a burst will signify reflectivity as a function of wavenumber. If n pulses with different frequencies are sent in a burst, the resulting n samples will contain information about the target reflectivity at n different wavenumbers.

Furthermore, since the target is rotating in the radar field of view, as if on a turntable, its reflectivity will also be interrogated as a function of rotation angle θ. If the bursts are short enough, the samples from each burst can be regarded as corresponding to a single angle. Over time, echoes from m sequential bursts will then give information about the reflectivity at m angles.

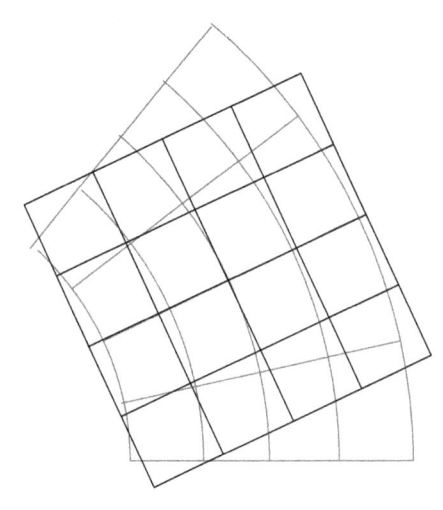

Figure 12.7 Graphical representation of ISAR target sampling in wavevector space. Data are acquired on cylindrical coordinates (blue lines) but may be interpolated onto a rectangular grid (black lines) for subsequent Fourier analysis.

Since the angle in question is the angle of the scattered radar waves with respect to a coordinate system fixed on the target, what is being interrogated is the reflectivity as a function of the scattering wave vector **k** at n different wavenumbers and m different angles.

The specification of reflectivity in physical space is related to the specification in wavevector space by a two-dimensional Fourier transform. This suggests that images can be formed by computing the two-dimensional discrete inverse Fourier transform of the $n \times m$ data matrix. This is nearly the case, but neglects the coordinate system in which the data have been accumulated. Figure 12.7 illustrates that coordinate system. Here, blue circles and lines are surfaces of constant wavenumber and angle, respectively. Samples in the $n \times m$ data matrix lie at the vertices of these lines.

In order to transform the data from wavevector to physical space using a two-dimensional discrete Fourier transform, the samples need to lie at the vertices of a Cartesian coordinate system like the one illustrated in Figure 12.7 with black lines. Two strategies for proceeding exist. The first is to extrapolate the samples from the polar grid to the Cartesian grid. This strategy is referred to as a polar format algorithm or PFA. The second strategy is to correct the data by multiplying each sample by its corresponding wavenumber prior to application of the inverse DFT. This is a filter that emphasizes high-frequency components in the data. The application of this filter converts the $dk\,d\theta$ differential term implicit in the DFT to $k\,dk\,d\theta$, the correct differential for transformation in polar coordinates. This strategy is known as filtered backprojection and is related to the Radon transform and its inverse, a transformation that appears in the closely related problem of absorption tomography.

12.4 Notes and Further Reading

An exhaustive description of interferometry in the context of radio astronomy is given in the text by A. R. Thompson et al., and another in the context of synthetic aperture radar is given in R. F. Hanssen's text. Two of the earliest treatments of ground-based radar interferometry were provided by R. F. Woodman and D. T. Farley et al.

General information about SAR image processing can be found in the text by G. Franceschetti and R. Lanari. The treatment of SAR image focusing outlined in this chapter follows those presented by R. Bamler and M. P. Nguyen.

Key references in the development of the CLEAN algorithm are listed below. Likewise, some of the most important contributions in the area of the maximum entropy algorithm are also listed. Error analysis in radar interferometry is analyzed in some detail in the paper by D. L. Hysell and J. L. Chau.

J. G. Ables. Maximum entropy spectral analysis. *Astron. Astrophys. Suppl. Ser.*, 15:383, 1974.

R. Bamler. A systematic comparison of SAR focusing algorithms. In *Proc. Int. Geoscience and Remote Sensing Symp., IGARSS'91*, volume 2, pages 1005–1009. Expoo, Finland, June 1991.

A comparison of range-Doppler and wavenumber domain SAR focusing algorithms. *IEEE Trans. Geosci. Remote Sensing*, 30(4):706–713, 1992.

J. Capon. High-resolution frequency-wavenumber spectrum analysis. *Proc. IEEE*, 57:1408, 1969.

B. G. Clark. An efficient implementation of the algorithm "CLEAN." *Astron, Astrophys.*, 89:377–378, 1980.

T. J. Cornwell. A method of stabilizing the CLEAN algorithm. *Astron. Astrophys.*, 121:281–285, 1983.

T. J. Cornwell and K. F. Evans. A simple maximum entropy deconvolution algorithm. *Astron. Astrophys.*, 143:77, 1985.

G. J. Daniell. Of maps and monkeys. In B. Buck and V. A. Macaulay, editors, *Maximum Entropy in Action*, chapter 1, pages 1–18. Clarendon, Oxford, 1991.

D. T. Farley, H. M. Ierkic and B. G. Fejer. Radar interferometry: A new technique for studying plasma turbulence in the ionosphere. *J. Geophys. Res.*, 86:1467–1472, 1981.

G. Franceschetti and R. Lanari. *Synthetic Aperture Radar Processing*. CRC Press, Boca Raton, FL, 1999.

R. F. Hanssen. *Radar Interferometry*. Klewer Academic Publishers, Dordrecht, 2001.

J. A. Högbom. Aperture synthesis with a non-regular distribution of interferometer baselines. *Astron. Astrophys. Supp.*, 15:417–426, 1974.

D. L. Hysell and J. L. Chau. Optimal aperture synthesis radar imaging. *Radio Sci.*, 41:10.1029/2005RS003383, RS2003, 2006.

E. T. Jaynes. Where do we stand on maximum entropy? In R. D. Levine and M. Tribus, editors, *The Maximum Entropy Formalism*, pages 15–118. Cambridge, MA: MIT Press, 1979.

On the rationale of maximum-entropy methods. *Proc. IEEE*, 70:939, 1982.

N. P. Nguyen. Range cell migration correction for phase error compensation of highly squinted SAR. Proc. 10th Eur. Conf. Synthetic Aperture Radar, 2014.

F. R. Schwab. Relaxing the isoplanatism assumption in self-calibration; applications to low-frequency radio interferometry. *Astron. J.*, 1984.

A. Segalovitz and B. R. Frieden. A "CLEAN"-type deconvolution algorithm. *Astron. Astrophys.*, 70:335–343, 1978.

J. Skilling and R. K. Bryan. Maximum entropy image reconstruction: General algorithm. *Mon. Not. R. Astron. Soc.*, 211:111, 1984.

O. J. Souvers, J. L. Fanselow and C. S. Jacobs. Astrometry and geodesy with radio interferometry: Experiments, models, results. *Rev. Mod. Phys.*, 70:1393–1454, 1998.

R. H. Stolt. Migration by Fourier transform. *Geophysics*, 43:23–48, 1978.

A. R. Thompson, J. M. Moran and G. W. Swenson. *Interferometry and Synthesis in Radio Astronomy*. John Wiley & Sons, New York, 1986.

R. Wilczek and S. Drapatz. A high accuracy algorithm for maximum entropy image restoration in the case of small data sets. *Astron. Astrophys.*, 142:9, 1985.

R. F. Woodman. Inclination of the geomagnetic field measured by an incoherent scatter technique. *J. Geophys. Res.*, 76:178, 1971.

Exercises

12.1 Obtain the expression for Shannon's entropy given in the text. Consider that a total of M quanta distributed among n different but equivalent bins. Let the number of quanta falling into the jth of n bins is m_j, and let the quanta in each bin be indistinguishable.

Calculate the number of possible ways that a given arrangement of quanta could occur. To get the result into the form of (12.6), make use of Stirling's approximation:

$$\ln n! = n\ln n - n + O(\ln n).$$

Argue that n be regarded as a large number in this problem.

12.2 Perform the Fourier transformation from the spatial (x) to the wavenumber (k_x) domain, i.e., go from (12.21) to (12.22) in the text by making use of the method of stationary phase. This method is useful for evaluating integrals of the form

$$I = \int_{-\infty}^{\infty} F(x)e^{-j\phi(x)}\,dx,$$

where the phase function $\phi(x)$ varies rapidly over most of the domain of integration and where $F(x)$ varies slowly by comparison. This implies that important

Exercises 351

contributions to the integrand will come only from regions where the phase function varies slowly, from stationary phase points defined by $\phi'(x_\circ) = 0$. Near such points, the phase function can be approximated by the first two nonzero terms in the Taylor-series expansion, giving

$$I \approx F(x_\circ)e^{-j\phi(x_\circ)} \int_{-\infty}^{\infty} e^{-j\phi''(x_\circ)(x-x_\circ)^2/2} dx$$

$$\approx \sqrt{\frac{2\pi}{j\phi''(x_\circ)}} F(x_\circ)e^{-j\phi(x_\circ)}.$$

For the problem in question, the only part of the solution of interest is the trailing $\exp(-j\phi(x_\circ))$ term.

12.3 Following the same procedure leading up to the expression for $\langle|\varepsilon_A|^2\rangle$ in the text, compute the corresponding expressions for $\langle|\varepsilon_B|^2\rangle$ and $\langle\varepsilon_A\varepsilon_B\rangle$. Next, show how to obtain (12.31) and (12.32) from the expressions preceding them.

Appendix A
Radio Frequency Designations

Table A.1 *Common definitions of radio and radar frequency bands.*

Band	frequency	wavelength
ELF	< 3 Hz	> 1000 km
VLF	3 – 30 Hz	10 – 100 km
LF	30 – 300 kHz	1 – 10 km
MF	300 – 3000 kHz	100 – 1000 m
HF	3 – 30 MHz	10 – 100 m
VHF	30 – 300 MHz	1 – 10 m
UHF	300 – 1000 MHz	30 – 100 cm
L	1 – 2 GHz	15 – 30 cm
S	2 – 4 GHz	7.5 – 15 cm
C	4 – 8 GHz	3.75 – 7.5 cm
X	8 – 12 GHz	2.5 – 3.75 cm
Ku	12 – 18 GHz	1.67 – 2.5 cm
K	18 – 27 GHz	1.11 – 1.67 cm
Ka	27 – 40 GHz	0.75 – 1.11 cm
millimeter	40 – 100 GHz	< 0.75 cm

Appendix B
Review of Electromagnetics

We develop Maxwell's equations from their experimental foundations in differential and integral form for both static and dynamic applications. We consider first the microscopic and then the macroscopic forms of the equations. Maxwell's equations are numbered in what follows.

Coulomb's Law

Stationary point charges exert a mutual central force on one another. The force exerted on charge q_2 under the influence of q_1 in Figure B.1 is governed by the following empirical law:

$$\mathbf{F}_{21} = kq_1q_2 \frac{\mathbf{x}_2 - \mathbf{x}_1}{|\mathbf{x}_2 - \mathbf{x}_1|^3}$$
$$\equiv q_2 \mathbf{E}_{21},$$

and is attributed to the electric field established by q_1:

$$\mathbf{E}(\mathbf{x}) = kq_1 \frac{\mathbf{x} - \mathbf{x}_1}{|\mathbf{x} - \mathbf{x}_1|^3}.$$

In MKS units, $k = \frac{1}{4\pi\varepsilon_\circ}$ and $\varepsilon_\circ = 8.854 \times 10^{-12}$ F/m. The electric (Coulomb) force is a central force with a $1/r^2$ dependence on radial distance. The contributions from n point charges add linearly and exert the total force on the ith particle:

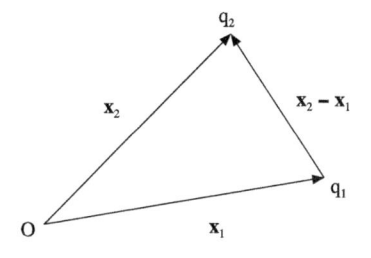

Figure B.1 Forces exerted by static charges.

$$\mathbf{F}_i = \frac{q_i}{4\pi\varepsilon_\circ} \sum_{\substack{j=1 \\ j\neq i}}^{n} q_j \frac{\mathbf{x}_i - \mathbf{x}_j}{|\mathbf{x}_i - \mathbf{x}_j|^3}$$

$$\mathbf{E}(\mathbf{x}_i) = \frac{1}{4\pi\varepsilon_\circ} \sum_{\substack{j=1 \\ j\neq i}}^{n} q_j \frac{\mathbf{x}_i - \mathbf{x}_j}{|\mathbf{x} - \mathbf{x}_j|^3}$$

$$\equiv \lim_{q_i \to 0} \frac{\mathbf{F}_i}{q_i},$$

where we note that a point charge exerts no self force. For a continuous charge distribution, the electric field becomes

$$\mathbf{E}(\mathbf{x}) = \frac{1}{4\pi\varepsilon_\circ} \int \rho(\mathbf{x}') \frac{\mathbf{x} - \mathbf{x}'}{|\mathbf{x} - \mathbf{x}'|^3} d^3 x',$$

where the primed and unprimed coordinates denote the source and observation points, respectively, and where $\rho(\mathbf{x})$ is the volume charge density, with MKS units of Coulombs per cubic meter. Substituting $\rho(\mathbf{x}) = \sum_{j=1}^{n} q_j \delta(\mathbf{x} - \mathbf{x}_j)$ recovers the electric field due to n discrete, static point charges.

Gauss' Law

Evaluating Coulomb's law for any but the simplest charge distributions can be cumbersome and requires specification of the charge distribution *a priori*. An alternative and often more useful relationship between electric fields and charge distributions is provided by Gauss' law. Gauss' law will become part of a system of differential equations that permit the evaluation of the electric field, even before the charge distribution is known under some circumstances.

Consider the electric flux through the closed surface S surrounding the volume V containing the point charge q, shown in Figure B.2. According to Coulomb's law, the flux through the differential surface area element da is

$$\mathbf{E} \cdot \hat{\mathbf{n}} da = \frac{q}{4\pi\varepsilon_\circ} \frac{\cos\alpha}{r^2} da,$$

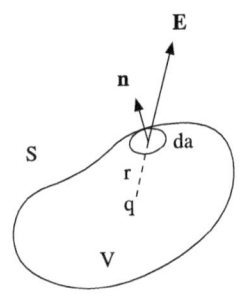

Figure B.2 Gaussian surface S enclosing a volume V with a point charge q within.

where r is the distance from the point charge to the surface area element, $\hat{\mathbf{n}}$ is the unit normal of da and α is the angle between $\hat{\mathbf{n}}$ and $\mathbf{E}$. Note next that $\cos\alpha\, da$ is the projection of the area element da on the surface of a sphere of radius r, concentric with the point charge. This projection could be expressed as $ds = r^2 \sin\theta d\theta d\phi$ in spherical coordinates or, more generally, as $ds = r^2 d\Omega$, where $d\Omega$ is the differential solid angle spanned by da as viewed from the point charge. Consequently,

$$\mathbf{E}\cdot\hat{\mathbf{n}}\, da = \frac{q}{4\pi\varepsilon_\circ} d\Omega.$$

Integrating over the surface of S then yields (noting that the total solid angle subtended by any exterior closed surface is 4π Sr)

$$\oint_S \mathbf{E}\cdot\hat{\mathbf{n}}\, da = \frac{q}{\varepsilon_\circ}.$$

This result was obtained for a single point charge residing within the closed surface. The electric flux due to multiple enclosed charges would be just the sum of the fluxes due to the individual charges. The net flux due to a charge outside the closed surface, meanwhile, would be zero, since all the flux entering the volume must also exit. In integral form, we may then write Gauss' law as

$$\oint_S \mathbf{E}\cdot\hat{\mathbf{n}}\, da = \frac{Q}{\varepsilon_\circ}, \tag{B.1}$$

where Q is the total charge enclosed by the surface through which the flux passes.

Writing the right side of this expression as the volume integral of the charge density, invoking the divergence theorem, and combining terms, then yields

$$\int_V \left(\nabla\cdot\mathbf{E} - \frac{1}{\varepsilon_\circ}\rho\right) d^3x = 0,$$

where the integral is over the volume V. Since this volume is an arbitrary one, the integrand must vanish at all points in space. This leads to the differential form of Gauss' law:

$$\nabla\cdot\mathbf{E} = \rho/\varepsilon_\circ. \tag{B.2}$$

While it is often stated that (B.1) and (B.2) are equivalent, the former is actually more general, since the differential operator in (B.2) is not defined in regions of space where the electric field is discontinuous. The same statement holds for all of Maxwell's equations.

Faraday's Law (Statics)

From the continuous form of Coulomb's law above, we have

$$\mathbf{E}(\mathbf{x}) = \frac{1}{4\pi\varepsilon_\circ} \int \rho(\mathbf{x}') \frac{\mathbf{x}-\mathbf{x}'}{|\mathbf{x}-\mathbf{x}'|^3} d^3x',$$

but note that
$$\nabla_x \frac{1}{|\mathbf{x} - \mathbf{x}'|} = -\frac{\mathbf{x} - \mathbf{x}'}{|\mathbf{x} - \mathbf{x}'|^3},$$
which can readily be verified by expressing the vectors in Cartesian coordinates so that, for instance, $|\mathbf{x} - \mathbf{x}'|$ becomes the radical $\sqrt{(x-x')^2 + (y-y')^2 + (z-z')^2}$, etc. In this notation, ∇_x means the gradient operating on the $\mathbf{x}$ coordinate, treating the $\mathbf{x}'$ coordinate as a constant. In view of this identity, the electric field can be expressed as

$$\begin{aligned}\mathbf{E}(\mathbf{x}) &= -\frac{1}{4\pi\varepsilon_\circ}\nabla_x \int \frac{\rho(\mathbf{x}')}{|\mathbf{x} - \mathbf{x}'|} d^3x' \\ &= -\nabla\phi\end{aligned}$$

$$\phi(\mathbf{x}) = \frac{1}{4\pi\varepsilon_\circ}\int \frac{\rho(\mathbf{x}')}{|\mathbf{x} - \mathbf{x}'|} d^3x',$$

where it is noted that the gradient does not operate on the charge density, a function of the primed (source) coordinates here.

This transformation shows that the electrostatic field can be written as (minus) the gradient of a scalar potential, the electrostatic potential $\phi(\mathbf{x})$. This potential is related to the potential energy of a charge embedded in the potential field, as illustrated in Figure B.3. The work done in moving the charge q from A to B against the Coulomb force is

$$\begin{aligned}W &= -\int_A^B \mathbf{F} \cdot \mathbf{dl} \\ &= -q\int_A^B \mathbf{E} \cdot \mathbf{dl} \\ &= q\int_A^B \nabla\phi \cdot \mathbf{dl} \\ &= q\int_A^B d\phi \\ &= q(\phi_B - \phi_A),\end{aligned}$$

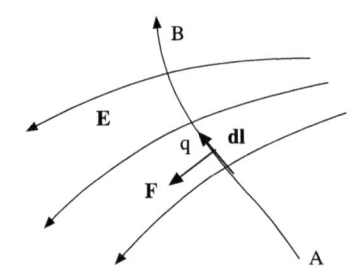

Figure B.3 Diagram illustrating the work done in moving a point charge q quasi-statically from A to B in a background electric field.

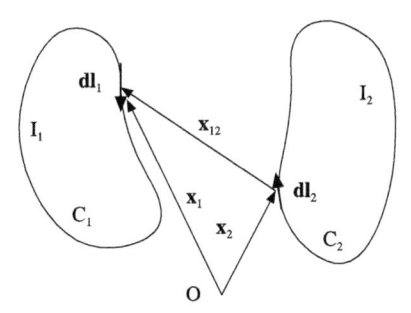

Figure B.4 Mutual forces exerted by current elements.

and so the potential energy is given by the product of the charge of the particle and the potential at its location. Note that the work done is path independent, and the force is therefore conservative. The line integral of the electric field around any closed path C is consequently identically zero,

$$\oint_C \mathbf{E} \cdot \mathbf{dl} = 0, \tag{B.3}$$

which is a statement of the integral form of Faraday's law for statics. To derive the differential form, apply Stoke's theorem to (B.3), expressing the line integral as the flux of the curl of the electric field through any surface S bounded by the contour C:

$$\int_S (\nabla \times \mathbf{E}) \cdot \hat{\mathbf{n}} da = 0.$$

Since the contour C and the surface S are arbitrary, the integrand in the expression above must be identically zero, i.e.,

$$\nabla \times \mathbf{E} = 0. \tag{B.4}$$

This is the differential form of Faraday's law in the static limit. Note that it follows immediately from the realization above that the electrostatic field is expressible as the gradient of a scalar potential.

Biot Savart Law

As Coulomb's law is the experimental foundation for electrostatics, the Biot Savart law is the foundation for magnetostatics. This law expresses the mutual forces exerted by steady current elements on one another. Note that net current and net charge can exist independently; in the domain of static charges and currents, these two sources and the fields that result function independently.

The force on current element $\mathbf{dl}_1$ due to its proximity to current element $\mathbf{dl}_2$ is found experimentally to be

$$\mathbf{F}_{12} = kI_1I_2 \oint_{C_1} \oint_{C_2} \frac{\mathbf{dl}_1 \times (\mathbf{dl}_2 \times \mathbf{x}_{12})}{|\mathbf{x}_{12}|^3},$$

where $\mathbf{x}_{12} = \mathbf{x}_1 - \mathbf{x}_2$. Like the Coulomb force, this force has a $1/r^2$ dependence on distance. The force is not a central force, however. In the MKS system, $4\pi k = \mu_\circ = 4\pi \times 10^{-7}$ H/m. Attributing the force on $\mathbf{dl}_1$ to the magnetic induction established by $\mathbf{dl}_2$ allows us to write

$$\mathbf{F}_{12} = I_1 \oint_{C_1} \mathbf{dl}_1 \times \mathbf{B}(\mathbf{x}_1),$$

where

$$\mathbf{B}(\mathbf{x}_1) = \frac{\mu_\circ}{4\pi} I_2 \oint_{C_2} \frac{\mathbf{dl}_2 \times \mathbf{x}_{12}}{|\mathbf{x}_{12}|^3}$$

is the magnetic induction at the location where the force is exerted. Note that both results can be generalized for volumetric current density distributions, i.e.,

$$\mathbf{F}_{12} = \int_V \mathbf{J}(\mathbf{x}) \times \mathbf{B}(\mathbf{x})$$

and

$$\mathbf{B}(\mathbf{x}) = \frac{\mu_\circ}{4\pi} \int_V \mathbf{J}(\mathbf{x}') \times \frac{\mathbf{x} - \mathbf{x}'}{|\mathbf{x} - \mathbf{x}'|^3} d^3x'.$$

Previously, a vector identity was applied to Coulomb's law to show how the electrostatic field can be written as the gradient of a scalar potential. Here, we perform a similar operation to show how the magnetic induction can be written as the curl of a vector potential. It is once again easily shown (by utilizing Cartesian coordinates for example) that the preceding equation can be expressed as

$$\mathbf{B}(\mathbf{x}) = \frac{\mu_\circ}{4\pi} \nabla_x \times \int_V \frac{\mathbf{J}(\mathbf{x}')}{|\mathbf{x} - \mathbf{x}'|} d^3x'.$$

Another Equation

We will delay interpreting the integrand in this equation. For now, it is sufficient to note that the divergence of a vector field expressible as the curl of another field is identically equal to zero, i.e.,

$$\nabla \cdot \mathbf{B} = 0. \tag{B.5}$$

Integrating this expression over an arbitrary volume V enclosed by a surface S with an outward unit normal $\hat{\mathbf{n}}$ then yields, with the application of Stoke's theorem,

$$\int_V \nabla \cdot \mathbf{B} d^3x = \oint_S \mathbf{B} \cdot \hat{\mathbf{n}} da \;\; = \;\; 0. \tag{B.6}$$

Appendix B: Review of Electromagnetics

In view of Gauss' law, which governs the electric field in the vicinity of charges (electric monopoles), these equations can be interpreted as a statement of the absence of magnetic monopoles. The actual presence or absence of magnetic monopoles in nature has some profound consequences and is an area of research and debate.

Ampère's Law (Statics)

Finally, just as Coulomb's law led to the differential form of Gauss' law through the application of a few identities of vector calculus, so the law of Biot and Savart leads to Ampère's law of magnetostatics. The derivation follows from the calculation of the curl of **B**:

$$\nabla \times \mathbf{B} = \frac{\mu_\circ}{4\pi} \nabla_x \times \nabla_x \times \int \frac{\mathbf{J}(\mathbf{x}')}{|\mathbf{x}-\mathbf{x}'|} d^3 x'.$$

Applying the triple vector product identity, $\nabla \times (\nabla \times \mathbf{A}) = \nabla(\nabla \cdot \mathbf{A}) - \nabla^2 \mathbf{A}$, then, gives

$$\nabla \times \mathbf{B} = \frac{\mu_\circ}{4\pi} \nabla_x \int \mathbf{J}(\mathbf{x}') \cdot \nabla_x \left(\frac{1}{|\mathbf{x}-\mathbf{x}'|}\right) d^3 x' - \frac{\mu_\circ}{4\pi} \int \mathbf{J}(\mathbf{x}') \nabla_x^2 \left(\frac{1}{|\mathbf{x}-\mathbf{x}'|}\right) d^3 x'.$$

Next, we note that

$$\nabla_x \left(\frac{1}{|\mathbf{x}-\mathbf{x}'|}\right) = -\nabla_{x'} \left(\frac{1}{|\mathbf{x}-\mathbf{x}'|}\right).$$

Furthermore, it is straightforward to show that

$$\nabla^2 \left(\frac{1}{|\mathbf{x}-\mathbf{x}'|}\right) = -4\pi \delta(\mathbf{x}-\mathbf{x}'),$$

meaning we can write

$$\nabla \times \mathbf{B} = -\frac{\mu_\circ}{4\pi} \nabla_x \int \mathbf{J}(\mathbf{x}') \cdot \nabla_{x'} \left(\frac{1}{|\mathbf{x}-\mathbf{x}'|}\right) d^3 x' + \mu_\circ \mathbf{J}.$$

The remaining integral can be evaluated through integration by parts. This yields a volume integral over the divergence of the current density, which is identically zero in the magnetostatic (steady state) limit. Consequently, the curl of the magnetic induction reduces to

$$\nabla \times \mathbf{B} = \mu_\circ \mathbf{J}, \tag{B.7}$$

which is the differential form of Ampère's law for static (steady) currents. The integral form is readily found by computing the flux of both sides of this equation through an arbitrary surface bounded by a closed contour C and applying Stoke's theorem,

$$\oint_C \mathbf{B} \cdot d\mathbf{l} = \mu_\circ I, \tag{B.8}$$

where I in this case refers to the total current linking the contour C.

Faraday's Law of Induction

At this point, forces and fields due to static charges and steady currents alone have been examined. In this domain, the electric and magnetic fields, charge and current densities are essentially uncoupled, and wave phenomena are not evident. The electromagnetic field emerges as a coupled quantity, able to transport intelligence and power, when Faraday's law of induction is introduced. This is an experimental law that relates changes in the magnetic flux linking a closed loop with the electromotive force (EMF or ε). Like the Coulomb force, the electromotive force can drive a current in a conductor in accordance with Ohm's law ($I = \varepsilon/R$). Unlike the Coulomb force, the EMF is not conservative.

The electromotive force is the line integral of the electric field along a closed contour. If a conducting wire is coincident with the contour, a current can be driven by the EMF. Faraday's law of induction states that the EMF is proportional to the total time rate of change of the magnetic induction linking the contour, i.e.,

$$\oint_C \mathbf{E}' \cdot \mathbf{dl} = -k\frac{d}{dt}\int_S \mathbf{B} \cdot \hat{\mathbf{n}} da. \tag{B.9}$$

In MKS units, the constant k is simply unity. Two subtle points merit attention. The first is that the $\mathbf{E}'$ notation refers to the electric field in the frame or reference of the contour or the loop. The electromagnetic field must undergo transformation when evaluated in different frames of reference. The second point is that the total time derivative above may reflect variations in the magnetic flux linking the contour due to (1) local variations in $\mathbf{B}$, (2) motion of the contour and (3) deformation of the contour. Any of these phenomena can induce an EMF and drive a current in the loop.

If the special case where the contour C is fixed and stationary, the total time derivative above can be replaced with a partial time derivative, and the frame of reference of the contour becomes the laboratory frame. In this case, we have (with the application of Stoke's theorem)

$$\int_S \left(\nabla \times \mathbf{E} + \frac{\partial \mathbf{B}}{\partial t}\right) \cdot \hat{\mathbf{n}} da = 0.$$

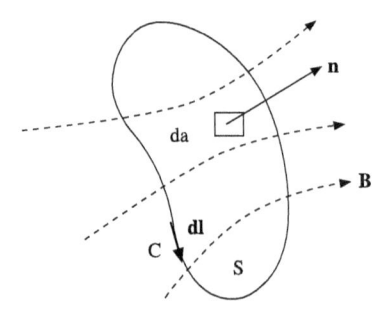

Figure B.5 Faraday's law of induction.

Since the above holds for arbitrary contours and their enclosed surfaces, it must be true that

$$\nabla \times \mathbf{E} + \frac{\partial \mathbf{B}}{\partial t} = 0. \tag{B.10}$$

Ampére's Law (Revisited)

Finally, in order to complete Maxwell's equations, we must reconsider Ampère's law, which so far has only been considered in the static limit. Taking the divergence of (B.7) results in the statement that the current density is divergence-free (solenoidal). This is tantamount to writing that current runs in closed loops and is the basis for Kirchhoff's current law. The requirement for divergence-free current density is not general, however, and neglects the fact that unclosed currents can converge on and diverge from charge reservoirs acting like capacitors. In fact, the more general principle is the conservation of charge, which can be expressed with the following continuity equation:

$$\nabla \cdot \mathbf{J} + \frac{\partial \rho}{\partial t} = 0.$$

Incorporating Gauss' law into the above, we may write

$$\nabla \cdot \left(\mathbf{J} + \varepsilon_\circ \frac{\partial \mathbf{E}}{\partial t} \right) = 0,$$

which implies that it's the divergence of the quantity in brackets that should be taken to be solenoidal rather than the current density. Replacing the current density with this quantity, which is composed of the conduction current density and a new term, the displacement current density, we arrive at Ampère's law for electrodynamics in differential form,

$$\nabla \times \mathbf{B} = \mu_\circ \mathbf{J} + \mu_\circ \varepsilon_\circ \frac{\partial \mathbf{E}}{\partial t}. \tag{B.11}$$

The integral form of this equation is written by computing the flux of (B.11) through a surface bounded by a closed contour.

Potentials and the Wave Equation

While essentially complete, Maxwell's equations are not especially compact or amenable to analysis. They are redundant; any fields that satisfy the rotation equations automatically satisfy the divergence equations. The close relationship between components of the electric and magnetic field suggest that it should be possible to formulate electrodynamics with fewer free parameters and fewer, higher-order equations. A number of formulations exist utilizing different kinds of potential functions. The most common one involves the scalar potential ϕ and vector potential $\mathbf{A}$.

Since $\nabla \cdot \mathbf{B}$ is identically equal to zero, we can define a vector scalar potential such that $\mathbf{B} = \nabla \times \mathbf{A}$. This does not uniquely define the field $\mathbf{A}$, however, since the $\nabla \cdot \mathbf{A}$ remains unspecified. Inserting the vector potential into Faraday's law shows that $\nabla \times (\mathbf{E} + \partial \mathbf{A}/\partial t) = 0$. Consequently, a scalar potential can also be defined such that $-\nabla \phi = \mathbf{E} + \partial \mathbf{A}/\partial t$. Now, both the electric and magnetic field are defined in terms of the vector and scalar potentials.

The remaining two equations, Ampère's and Gauss' laws, then acquire the forms

$$\nabla \nabla \cdot \mathbf{A} - \nabla^2 \mathbf{A} = \mu_\circ \mathbf{J} + \mu_\circ \varepsilon_\circ \frac{\partial}{\partial t}\left(-\nabla \phi - \frac{\partial \mathbf{A}}{\partial t}\right)$$

$$\nabla \cdot \left(-\nabla \phi - \frac{\partial \mathbf{A}}{\partial t}\right) = \frac{\rho}{\varepsilon_\circ},$$

where the triple vector product has been invoked. Being still at liberty to define the divergence of the vector potential, we can do so in such a way as to decouple the two equations. The Lorenz gauge is the definition that does so:

$$\nabla \cdot \mathbf{A} = -\frac{1}{c^2}\frac{\partial \phi}{\partial t}.$$

Substituting above produces the following coupled, second-order, inhomogeneous partial differential equations:

$$\nabla^2 \mathbf{A} - \frac{1}{c^2}\frac{\partial^2 \mathbf{A}}{\partial t^2} = -\mu_\circ \mathbf{J}$$

$$\nabla^2 \phi - \frac{1}{c^2}\frac{\partial^2 \phi}{\partial t^2} = -\frac{\rho}{\varepsilon_\circ}.$$

The equations have the form of the wave equation and, coupled with appropriate boundary conditions, represent well-posed problems with numerous avenues of solution. The equations are coupled by their sources, the current and charge density, which are themselves related by the continuity equation. It is frequently necessary to solve only a subset of the four scalar equations represented here in order to completely solve a problem in electrodynamics.

Macroscopic Equations

The equations developed above are in so-called "microscopic" form; except for quantum mechanical and relativistic phenomena, they provide a complete and exact description of electromagnetic phenomena so far as we know. However, their strict application in regions with materials is impractical. For one thing, one would have to specify the location of every source (point charge, current element) in order to apply the equations. There are too many sources to count in a macroscopic material sample, and we are ill equipped mathematically to deal with the singular behavior of the fields in the neighborhood of each of the sources. For another, materials respond to applied fields in such a way as to change the total field. Reformulating Maxwell's equations in so-called "macroscopic" form provides a means of accounting for the response self-consistently while also avoiding the problem of singularities near sources.

The first step in the reformulation of Maxwell's equations is a reinterpretation of the field quantities, which we now regard as spatial averages of the microscopic fields over a finite-sized domain. The domain should be large enough to average out the singular behavior of the fields near sources but small enough to retain phenomena of interest (short-wavelength waves). For most intents and purposes, we can consider the averaging domain to be the size of probes we might use to measure the electric and magnetic fields.

The second step is to account for material responses to applied electric and magnetic fields. There are three important possibilities. The first is that free current might flow, as happens in conductors. This possibility can be taken into account by invoking Ohm's law. The second is that electric dipoles might arise, as happens in dielectric materials. The third is that magnetic dipoles might arise, as happens in magnetic materials. The electric and magnetic properties of materials will be addressed in detail later. A formal treatment is given here.

In the case of dielectrics, electric dipoles can either be induced by an applied electric field, or if the material is already composed of permanent but randomly aligned dipoles, those can be aligned by the applied field to produce a net dipole moment.

Consider the situation in Figure B.6. A known charge distribution $\rho(\mathbf{x}')$ generates an electric field that acts on a dielectric material sample. Dipole moments are induced in the sample, and the net potential $\phi(\mathbf{x})$ everywhere represents contributions from both the original charge distribution and the induced dipole moments. To know the material response, one must know the net field in the sample. To know the net field, one must know the material response. However, it turns out that the problem can be closed. Note that the free charge $\rho(\mathbf{x}')$, the dielectric, and the observer can be co-located.

A dipole moment $\mathbf{p}$ is made of equal and opposite charges $\pm q$ displaced a small distance $\mathbf{dx}$ such that $\mathbf{p} = q\mathbf{dx}$. It can readily be shown that the potential at $\mathbf{x}$ due to a dipole moment at $\mathbf{x}'$ is

$$\phi(\mathbf{x}) = \frac{1}{4\pi\varepsilon_\circ} \frac{\mathbf{p} \cdot (\mathbf{x} - \mathbf{x}')}{|\mathbf{x} - \mathbf{x}'|^3}.$$

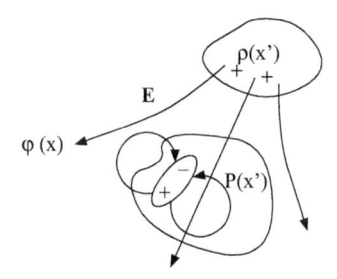

Figure B.6 A free charge distribution $\rho(\mathbf{x}')$ induces polarization $P(\mathbf{x}')$ in a dielectric sample. An observer perceives the potential $\phi(\mathbf{x})$ due to both the free charge and the induced dipole moments.

For dealing with continuous materials, we define the dipole moment per unit volume or the polarization $P(\mathbf{x})$ such that $d\mathbf{p} = \mathbf{P}d^3x$. Let us write an expression for the potential that includes contributions from the charge distribution $\rho(\mathbf{x}')$ as well as the dipole moments induced in the material:

$$d\phi(\mathbf{x}) = \frac{1}{4\pi\varepsilon_\circ} \left(\frac{\rho(\mathbf{x}')}{|\mathbf{x}-\mathbf{x}'|} + \frac{\mathbf{P}\cdot(\mathbf{x}-\mathbf{x}')}{|\mathbf{x}-\mathbf{x}'|^3} \right) d^3x'.$$

The total potential then comes from the volume integral of this expression over all regions containing sources. Note again that

$$\nabla_{x'} \frac{1}{|\mathbf{x}-\mathbf{x}'|} = \frac{\mathbf{x}-\mathbf{x}'}{|\mathbf{x}-\mathbf{x}'|^3},$$

so that

$$\phi(\mathbf{x}) = \frac{1}{4\pi\varepsilon_\circ} \int \left(\frac{\rho(\mathbf{x}')}{|\mathbf{x}-\mathbf{x}'|} + \mathbf{P}\cdot\nabla_{x'}\frac{1}{|\mathbf{x}-\mathbf{x}'|} \right) d^3x'.$$

Integrating the second term in the brackets by parts then yields

$$\phi(\mathbf{x}) = \frac{1}{4\pi\varepsilon_\circ} \int \frac{1}{|\mathbf{x}-\mathbf{x}'|} \left(\rho(\mathbf{x}') - \nabla_{x'}\cdot\mathbf{P}(\mathbf{x}') \right) d^3x'.$$

This reveals that the dipole moments induced in the material contribute to the potential only to the extent that the polarization is divergent. Where divergence is present, an effective charge density equal to $\rho_{\text{eff}} = -\nabla\cdot\mathbf{P}$ is also present and should be included, for instance, in evaluating Gauss' law, i.e.,

$$\nabla\cdot\mathbf{E} = (\rho_f - \nabla\cdot\mathbf{P})/\varepsilon_\circ$$

or

$$\nabla\cdot(\varepsilon_\circ\mathbf{E}+\mathbf{P}) = \rho_f.$$

It is easy to understand why effective charge density is associated with divergent or convergent polarization. Imagine a collection of electric dipoles arranged with their moments converging on (diverging from) a point. In the vicinity of that point, there will be an overabundance of positive (negative) charge. In the absence of convergence or divergence, the charges of the dipoles will tend to be compensated for by the charges of neighboring dipoles such that the total effect is negligible.

We next combine the terms in brackets above, defining the electric displacement as $\mathbf{D} = \varepsilon_\circ\mathbf{E}+\mathbf{P}$. Making the substitution above produces the macroscopic form of Gauss' law:

$$\nabla\cdot\mathbf{D} = \rho_f. \tag{B.12}$$

Appendix B: Review of Electromagnetics

The notation ρ_f refers to free charge, which exists independent of the material response and should be differentiated from the effective or bound charge associated with polarization. (This notation is nonstandard and will not be used consistently.) We see from (B.12) that the electrical displacement behaves like the part of the electric field (within a constant) arising purely from free charge.

We are not done, as it is generally the electric field and not the displacement that is of interest. For many simple materials, however, the polarization and the electric field are related by a constant, the electric susceptibility, such that $\mathbf{P} = \varepsilon_\circ \chi_e \mathbf{E}$. In such cases, the displacement and the electric fields are also related by a constant, viz $\mathbf{D} = \varepsilon_\circ (1 + \chi_e) \mathbf{E} = \varepsilon \mathbf{E}$. This constitutive relationship makes closure possible. In many (but not all) dielectrics, the overall effect of polarization is to reduce or shield the electric field within the material.

In the case of inhomogeneous dielectrics, the susceptibility is a function of position, $\chi(\mathbf{x})$. In anisotropic dielectrics, the susceptibility becomes a tensor. In dispersive dielectrics, the susceptibility is a function of frequency. These conditions can occur together. Each condition implies unique characteristics in radio waves propagating through the media.

Faraday's law contains no source terms and so need not be modified except for interpreting the field quantities as spatial averages.

The derivation of the macroscopic equations for magnetic materials parallels the one for dielectrics. In magnetic materials, the primary response to the application of a magnetic field is the emergence of magnetic dipole moments, $\mathbf{m} = IA\hat{\mathbf{n}}$, which represent small current loops enclosing an area A with unit normal $\hat{\mathbf{n}}$. Magnetic dipoles arise mainly from the orbits and spins of electrons. In most materials, thermal agitation randomizes the dipole moment vectors, but an applied magnetic field organizes them, yielding a net dipole moment.

A free current distribution somewhere in space, within or nearby a magnetic material, gives rise to a magnetic field and causes the material to become magnetized. The net magnetic induction observed is the sum of the contributions from the free current and the magnetized medium. Knowledge of the material response requires knowledge of the net magnetic field, and vice versa. Even so, the problem can be closed.

We saw earlier while discussing the Biot Savart law that the magnetic induction can be expressed as the curl of another vector field, which is called the magnetic vector potential, $\mathbf{A}$. One can readily show that the vector potential due to a magnetic dipole moment is

$$\mathbf{A}(\mathbf{x}) = \frac{\mu_\circ}{4\pi} \frac{\mathbf{m} \times (\mathbf{x} - \mathbf{x}')}{|\mathbf{x} - \mathbf{x}'|^3}.$$

For continuous media, we consider the dipole moment per unit volume or magnetization $\mathbf{M}$, which is defined by $d\mathbf{m} = \mathbf{M} d^3 x$. The total differential magnetic vector potential arising from a free current distribution and a magnetized material is

$$d\mathbf{A}(\mathbf{x}) = \frac{\mu_\circ}{4\pi} \left(\frac{\mathbf{J}(\mathbf{x}')}{|\mathbf{x} - \mathbf{x}'|^3} + \frac{\mathbf{M} \times (\mathbf{x} - \mathbf{x}')}{|\mathbf{x} - \mathbf{x}'|^3} \right) d^3 x'.$$

The total vector potential is the volume integral of this quantity over all space. Making use of the vector identities applied earlier in this review and integrating the magnetization term by parts yields the following expression for the vector potential:

$$\mathbf{A}(\mathbf{x}) = \frac{\mu_\circ}{4\pi} \int \frac{\mathbf{J}(\mathbf{x}') + \nabla_{x'} \times \mathbf{M}(\mathbf{x}')}{|\mathbf{x} - \mathbf{x}'|} d^3 x'.$$

The point of this exercise is to show that magnetic vector potential and, by extension, magnetic induction arise both from free current density and from an effective current density that flows in a material when the magnetization is rotational, i.e., $\mathbf{J}_{\text{eff}} = \nabla \times \mathbf{M}$. It is easy to see why this should be the case. Imagine a chain of magnetic dipoles connected together to form a small loop. All the currents associated with the dipoles add constructively at the center of the loop. In the absence of a rotational magnetization, all these currents would tend to add destructively, having no net effect.

The effective current density should be included in Ampère's law alongside the free conduction current density. In a statics context,

$$\nabla \times \mathbf{B} = \mu_\circ \mathbf{J}_f + \mu_\circ \nabla \times \mathbf{M},$$

where the $\mathbf{J}_f$ notation has once again been used to emphasize the distinction between free current density, which flows independently of any material effects, and effective current density due to magnetization. Let us now define the magnetic field $\mathbf{H} = \mathbf{B}/\mu_\circ - \mathbf{M}$. Substituting gives

$$\nabla \times \mathbf{H} = \mathbf{J}_f.$$

This shows that the magnetic field acts (to within a constant) like that part of the magnetic induction that arises from free current density.

To obtain the macroscopic form of Ampère's law for electromagnetics, one must consider also the displacement current term. The following expression is evidently the desired equation; taking the divergence and invoking Gauss' law yields the continuity equation,

$$\nabla \times \mathbf{H} = \mathbf{J}_f + \frac{\partial \mathbf{D}}{\partial t}. \tag{B.13}$$

A constitutive relation for magnetic material comes from the fact that, in many materials, the magnetization and magnetic field are related by a proportionality constant, the magnetic susceptibility, such that $\mathbf{M} = \chi_m \mathbf{H}$. This means that the magnetic field and magnetic induction are also linearly related, i.e., $\mathbf{B} = \mu_\circ(1 + \chi_m)\mathbf{H} = \mu\mathbf{H}$. Note, however, that no such relationship holds in ferromagnetic media, an important class of materials. Note also that inhomogeneous, anisotropic, and dispersive magnetic materials also exist.

Boundary Conditions

Straightforward application of Stoke's theorem and the divergence theorem to the appropriate set of Maxwell's equations produces well-known boundary conditions that govern changes in the electric and magnetic field across material interfaces. These are summarized below.

$$(\mathbf{D}_2 - \mathbf{D}_1) \cdot \hat{\mathbf{n}} = \sigma_f$$
$$(\mathbf{B}_2 - \mathbf{B}_1) \cdot \hat{\mathbf{n}} = 0$$
$$\hat{\mathbf{n}} \times (\mathbf{H}_2 - \mathbf{H}_1) = \mathbf{k}_f$$
$$\hat{\mathbf{n}} \times (\mathbf{E}_2 - \mathbf{E}_1) = 0,$$

where the subscripts refer to materials 1 and 2, which form a planar interface, and $\hat{\mathbf{n}}$ is the unit normal of the interface directed into medium 2. Also, σ_f and $\mathbf{k}_f$ are the free surface charge density and free surface current density at the interface, respectively. These equations state that the normal component of magnetic induction and tangential components of electric field are continuous across the interface. The normal component of electric displacement and tangential components of magnetic field are discontinuous by an amount equal to the surface charge density and surface current density, respectively.

In a perfect conductor, we expect the electric field essentially to vanish to prevent the emergence of an infinite current density, which we will not tolerate. By Faraday's law, we therefore expect the magnetic induction to be a constant, which we take to be also zero for the purposes of this text. If one of the materials in question above is a perfect conductor, we therefore expect there to be no electric field tangential to and no magnetic induction normal to the interface in the other medium. Finite charge and current density at the interface can however support the presence of finite electric displacement normal to the interface and finite magnetic field tangential to it, respectively.

Three more boundary conditions can be derived from the preceding discussions. For dielectrics, we have $\nabla \cdot \mathbf{P} = -\rho_{\text{eff}}$. For magnetic materials, we have $\nabla \times \mathbf{M} = \mathbf{J}_{\text{eff}}$. For conductors in the electrostatic limit, we have $\nabla \cdot \mathbf{J} = 0$. From these, we can conclude that

$$(\mathbf{P}_2 - \mathbf{P}_1) \cdot \hat{\mathbf{n}} = -\sigma_{\text{eff}}$$
$$\hat{\mathbf{n}} \times (\mathbf{M}_2 - \mathbf{M}_1) = \mathbf{k}_{\text{eff}}$$
$$(\mathbf{J}_2 - \mathbf{J}_1) \cdot \hat{\mathbf{n}} = 0.$$

Even in the absence of free charge and current density at a material interface, electric and magnetic fields tend to deflect, the cause being the effective or bound charge and current density induced at the interface. Changes in conductivity from one material to the next must also be considered.

Boundary Value Problems

The utility of the formulations outlined above can be demonstrated by considering boundary value problems in electrostatics. The problem is one of solving for the electrostatic potential in 2-D and 3-D source-free spaces enclosed by a boundary on which the potential is specified. One need not specify (and generally cannot know) how charge is distributed *a priori*. The charge distribution can be determined *a posteriori*, however, with the application of boundary conditions.

An illustrative boundary value problem is that of a uniform dielectric sphere of radius r_o and dielectric constant ε immersed in a vacuum with a previously uniform background electric field $\mathbf{E} = E_o \hat{\mathbf{z}}$ as in Figure B.7. We can think of the background field as being established by a distant charge distribution that will remain unaffected by the sphere. The sphere itself contains no free charge. However, electric dipoles will be induced inside the sphere by the background field, and the resulting polarization P will be discontinuous across the $r = r_o$ boundary. Consequently, bound surface charge will arise there, contributing to the field inside and outside the sphere, particularly in its immediate vicinity.

Inside and outside the sphere, the dielectric constant is uniform, and the macroscopic form of Gauss' law says that the potentials must individually obey Laplace's equation, $\nabla^2 \phi = 0$. In cases such as this that are azimuthally symmetric, the method of separation of variables and series solution can be used to show that the potentials can be expressed in spherical coordinates as

$$\phi_{in}(r,\theta) = \sum_{l=0}^{\infty} (a_l r^l + b_l r^{-(l+1)}) P_l(\cos\theta)$$

$$\phi_{out}(r,\theta) = \sum_{l=0}^{\infty} (A_l r^l + B_l r^{-(l+1)}) P_l(\cos\theta),$$

where the coefficients are determined by the boundary conditions and where P_l are the Legendre polynomials. The definition of the Legendre polynomials, their identities, recursion relations, asymptotic behavior and orthogonality properties are explained in any number of mathematics and electrodynamics texts. As we shall see, we need not dwell extensively on the properties of the Legendre polynomials to be able to solve the current problem.

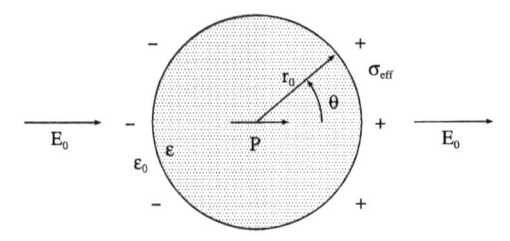

Figure B.7 A uniform dielectric sphere of radius a immersed in a previously uniform background electric field aligned in the $\hat{\mathbf{z}}$ direction.

Appendix B: Review of Electromagnetics

Since there are two distinct domains where Laplace's equation applies, distinct potentials must be allowed for inside and outside the sphere. The particular solution for this problem is constructed from this general solution through the determination of all the coefficients. The problem may appear challenging, but a little intuition resolves it quickly.

The potential at the origin would be singular if any of the b_l coefficients were nonzero. This counter-intuitive situation is therefore avoided by setting all the b_l to zero. Similar reasoning would appear to apply to the A_l coefficients, which would seem to lead to singular behavior as r goes to infinity. However, note that $P_1(x) = x$ and that the A_1 term is therefore $A_1 r \cos\theta = A_1 z$. Setting $A_1 = -E_\circ$ and the remaining A_l to zero therefore satisfies the boundary condition at $r = \infty$, namely that the electric field $-\nabla\phi|_{r=\infty}$ returns to $E_\circ \hat{z}$.

The problem is now

$$\phi_{in}(r,\theta) = \sum_{l=0}^{\infty} a_l r^l P_l(\cos\theta)$$

$$\phi_{out}(r,\theta) = \sum_{l=0}^{\infty} B_l r^{-(l+1)} P_l(\cos\theta) - E_\circ r P_1(\cos\theta).$$

Two conditions must be satisfied at the material boundary: the continuity of the normal electric displacement and the continuity of the tangential electric field:

$$-\varepsilon \frac{\partial \phi_{in}}{\partial r}\bigg|_{r=r_\circ} = -\varepsilon_\circ \frac{\partial \phi_{out}}{\partial r}\bigg|_{r=r_\circ}$$

$$-\frac{1}{r}\frac{\partial \phi_{in}}{\partial \theta}\bigg|_{r=r_\circ} = -\frac{1}{r}\frac{\partial \phi_{out}}{\partial \theta}\bigg|_{r=r_\circ}.$$

Since the conditions must be satisfied at all angles θ, not only must the series on the left and right sides of the equal signs equate, the coefficients on the Legendre polynomials of each order l must equate separately. For all of the $l \neq 1$ terms, the equalities can be satisfied by the trivial solution with $a_l = B_l = 0$. This leaves us with only a_1 and B_1 to determine. These cannot both be zero because of the $E_\circ$ term lingering above. Rewriting the boundary conditions for the $l = 1$ terms gives

$$\varepsilon a_1 = -\varepsilon_\circ (2B_1/r_\circ^3 + E_\circ)$$
$$a_1 = B_1/r_\circ^3 - E_\circ,$$

with solutions for a_1 and B_1 given by

$$a_1 = -\left(\frac{3}{2+\varepsilon_r}\right) E_\circ$$

$$B_1 = \left(\frac{\varepsilon_r - 1}{\varepsilon_r + 2}\right) E_\circ r_\circ^3,$$

giving the potentials the form

$$\phi_{in}(r,\theta) = -\left(\frac{3}{2+\varepsilon_r}\right) E_\circ r \cos\theta$$

$$\phi_{out}(r,\theta) = \left(\frac{\varepsilon_r-1}{\varepsilon_r+2}\right) E_\circ \frac{r_\circ^3}{r^2} \cos\theta - E_\circ r \cos\theta,$$

where we have defined $\varepsilon_r = \varepsilon/\varepsilon_\circ$.

The corresponding electric fields are calculated by taking the negative of the gradient of the potentials. Inside the sphere, the electric field is uniform, parallel to $\hat{z}$ and has the magnitude $3E_\circ/(\varepsilon+2)$. For most common materials, $\varepsilon > 1$, and the effect of polarization is to shield the inner material from the environmental field. Outside the sphere, the potential is that of the background electric field plus another component that is that of an electric dipole with a dipole moment given by

$$\mathbf{p} = 4\pi\varepsilon_\circ \left(\frac{\varepsilon_r-1}{\varepsilon_r+2}\right) r_\circ^3 E_\circ \hat{z}.$$

We can calculate the polarization inside the sphere by regarding it as the volume of the sphere multiplied by the dipole moment. Using the appropriate boundary conditions from the last section, it is then possible to calculate the effective surface charge density induced on the sphere:

$$\sigma_{eff} = \frac{3}{\varepsilon_\circ}\left(\frac{\varepsilon_r-1}{\varepsilon_r+2}\right) E_\circ \cos\theta.$$

This demonstrates how the conventional formulation of electrodynamics makes it possible to solve for the fields without an *a priori* specification of the sources.

Green's Functions and the Field Equivalence Principle

The preceding analysis was suitable for finding the potential inside homogeneous, source-free regions on the basis of the demands of the boundary conditions. We can extend the analysis to include sources with the method of Green's functions, which relies upon Green's theorem. We will consider time-harmonic fields with an explicit $exp(j\omega t)$ time dependence and define the free-space wavenumber $k = \omega/c$. Then, the wave equation for the potential in vacuum can be written

$$\nabla^2 \phi(\mathbf{x}) + k^2 \phi(\mathbf{x}) = -\frac{\rho(\mathbf{x})}{\varepsilon_\circ}. \tag{B.14}$$

A similar form holds for each of the three vector components of the vector potential, and the formulas can be generalized easily for any homogeneous medium. If we are seeking the solution for the potential in a bounded region, then it is also necessary to specify either φ or its normal derivative $\partial\phi/\partial n$ on the surface of that region according to the theory of differential equations.

To find $\phi(\mathbf{x})$, consider finding the solution of a simpler problem, where the source is concentrated at a point in space $\mathbf{x}'$. That problem is described by the equation

$$\nabla_x^2 G(\mathbf{x},\mathbf{x}') + k^2 G(\mathbf{x},\mathbf{x}') = -\delta(\mathbf{x}-\mathbf{x}'), \tag{B.15}$$

where δ is the Dirac delta function and where the function that solves the equation, G, is known as the Green's function. Further, let us insist that either G or its normal derivative vanish on the surface bounding the region. In fact, the Green's function that accomplishes this can be found using the method of separation of variables. Green's functions for bounded regions with common shapes (spheres, cylinders, etc.) have been tabulated and are readily available for our use. We assume that G is known in what follows. We also assert without proof that $G(\mathbf{x},\mathbf{x}') = G(\mathbf{x}',\mathbf{x})$, which is a statement of reciprocity.

The independent variables $\mathbf{x}$ and $\mathbf{x}'$ can be regarded as source and observing coordinates. Let us exchange the two variables in (B.14) and (B.15):

$$\nabla_{x'}^2 \phi(\mathbf{x}') + k^2 \phi(\mathbf{x}') = -\frac{\rho(\mathbf{x}')}{\varepsilon_\circ} \tag{B.16}$$

$$\nabla_{x'}^2 G(\mathbf{x}',\mathbf{x}) + k^2 G(\mathbf{x}',\mathbf{x}) = -\delta(\mathbf{x}'-\mathbf{x}). \tag{B.17}$$

Next, multiply (B.16) by $G(\mathbf{x}',\mathbf{x})$ and (B.17) bg $\phi(\mathbf{x}')$ and subtract

$$G(\mathbf{x}',\mathbf{x})\nabla_{x'}^2\phi(\mathbf{x}') - \phi(\mathbf{x}')\nabla_{x'}^2 G(\mathbf{x}',\mathbf{x}) = -G(\mathbf{x}',\mathbf{x})\frac{\rho(\mathbf{x}')}{\varepsilon_\circ} + \phi(\mathbf{x}')\delta(\mathbf{x}'-\mathbf{x}),$$

where the terms involving k^2 are seen to have canceled. Integrating the equation over the volume $V' = d^3 x'$ of interest bounded by the surface S' yields

$$\int_{V'} \left(G\nabla_{x'}^2\phi - \phi\nabla_{x'}^2 G\right) dV' = \int_{V'}\left(-G\frac{\rho}{\varepsilon_\circ} + \phi\delta(\mathbf{x}'-\mathbf{x})\right) dV'$$
$$= -\int_{V'} G\frac{\rho}{\varepsilon_\circ} dV' + \underbrace{\phi(\mathbf{x})}_{\mathbf{x}'\varepsilon V},$$

where the last term on the right side is nonzero only if $\mathbf{x}'$ is within the volume V'. The left side of the equation can be transformed using Green's theorem, which is essentially just the application of integration by parts,

$$\int_{V'} \left(G\nabla_{x'}^2\phi - \phi\nabla_{x'}^2 G\right) dV' = \int_{S'} \left(G\nabla_{x'}\phi - \phi\nabla_{x'} G\right)\cdot d\mathbf{S}',$$

which implies a surface integral of the normal component of the given vector field. Finally, solving for the potential somewhere within the volume of interest gives

$$\phi(\mathbf{x}) = \int_{V'} G \frac{\rho(\mathbf{x}')}{\varepsilon_\circ} dV' + \int_{S'} \left(G \frac{\partial \phi(\mathbf{x}')}{\partial n'} - \phi \frac{\partial G(\mathbf{x}',\mathbf{x})}{\partial n'} \right) \cdot dS'. \tag{B.18}$$

As has been the case throughout the derivation, primed and unprimed coordinates designate source and observation points, and the partial derivatives here are derivatives with respect to the coordinate normal to the surface bounding the volume of interest.

The result is profound as it gives a complete prescription for determining the potential in a bounded, homogeneous region on the basis of any sources within the volume along with information from the boundary. We have already demanded that either G or its normal derivative be zero on the boundary, making it sufficient to specify either ϕ or its normal derivative there to complete the problem, and not both.

If the homogeneous region of interest is all space, then the boundary term in (B.18) can be removed to infinity and neglected. In that case, it can easily be shown (see Chapter 2) that $G(\mathbf{x},\mathbf{x}') = \exp(-jk|\mathbf{x}-\mathbf{x}'|)/4\pi|\mathbf{x}-\mathbf{x}'|$, and the potential is given by the usual expression for a distribution of charges. This is the Green's function for radiation from a point source. In the limit $\omega = k = 0$, this reduces further to the electrostatic potential for a static charge distribution. Much more complicated problems in electrostatics and electrodynamics, including the physics of radiation, can of course be addressed with this formalism.

The field equivalence principle can now be inferred from the Green's function solution in (B.18). Field equivalence is a consequence of our freedom to select the volume of interest for the sake of expediency. In the case of radiation from an aperture in an infinite planar barrier, for example, the volume of interest can be chosen as the half space on the other side of the aperture from where the sources reside. The integral in (B.18) involving the sources need not in that case be evaluated, since there are no sources in the region of interest. This is of practical significance, since the source distribution in an aperture radiation problem usually cannot be specified easily.

Only the boundary surface integral need be integrated in the aperture radiation problem, and then only over the aperture, since the fields are presumed to vanish elsewhere on the planar boundary and also out at infinity, where the surface closes. One need only find the appropriate Green's function that solves (B.15) and that itself vanishes on the barrier so as to leave only one term inside the surface integral to evaluate. The Green's function for radiation from a point source was already given in terms of $|\mathbf{x}-\mathbf{x}'| \equiv R$. The desired Green's function here can be constructed using the method of images. Define $\mathbf{x}''$ as the mirror image of $\mathbf{x}'$ across the barrier and the scalar $R' \equiv |\mathbf{x}-\mathbf{x}''|$ as the distance from the observer to the image point. Then the Green's function

$$G(\mathbf{x},\mathbf{x}') = \frac{e^{-ikR}}{4\pi R} - \frac{e^{-ikR'}}{4\pi R'}$$

gives the potential due to a point source near the barrier. Whereas the image term negates the source term when evaluating G on the barrier, it doubles it when evaluating the normal derivative of G. What becomes of (B.18) is then essentially equivalent to the expression for aperture radiation given in Chapter 5.

Reciprocity

The translational invariance of Green's functions for electromagnetic fields in linear, isotropic, unbounded media can be inferred from the Lorentz reciprocity theorem, which will be derived here. This theorem also has important overtones for antenna performance in the context of reception. Consider Maxwell's rotation equations written in time-harmonic form:

$$\nabla \times \mathbf{E}_{a,b} = -j\omega \mathbf{B}_{a,b} \tag{B.19}$$

$$\nabla \times \mathbf{H}_{a,b} = j\omega \mathbf{D}_{a,b} + \mathbf{J}_{a,b}. \tag{B.20}$$

In these equations, the subscripts a and b refer to two arbitrary, distinct configurations where a current density $\mathbf{J}_a$ or $\mathbf{J}_b$ is imposed by itself, giving rise to the fields with the corresponding a or b subscript in the same linear (but not necessarily homogeneous or isotropic) medium. In view of the definition of the triple scalar product rule

$$\nabla \cdot (\mathbf{f} \times \mathbf{g}) = \mathbf{g} \cdot \nabla \times \mathbf{f} - \mathbf{f} \cdot \nabla \times \mathbf{g},$$

we may write

$$\begin{aligned}\nabla \cdot (\mathbf{E}_a \times \mathbf{H}_b - \mathbf{E}_b \times \mathbf{H}_a) &= \mathbf{E}_a \nabla \times \mathbf{H}_b - \mathbf{H}_b \nabla \times \mathbf{E}_a - \mathbf{E}_b \nabla \times \mathbf{H}_a + \mathbf{H}_a \nabla \times \mathbf{E}_b \\ &= \mathbf{J}_b \cdot \mathbf{E}_a - \mathbf{J}_a \cdot \mathbf{E}_b + j\omega(\mathbf{E}_a \cdot \mathbf{D}_b - \mathbf{E}_b \cdot \mathbf{D}_a) \\ &\quad + j\omega(\mathbf{H}_b \cdot \mathbf{B}_a - \mathbf{H}_a \cdot \mathbf{B}_b),\end{aligned}$$

where the last line was achieved with the substitution of (B.19) and (B.20) for configurations a and b. Now, if the medium in question is isotropic such that $\mathbf{B} = \mu \mathbf{H}$ and $\mathbf{D} = \varepsilon \mathbf{E}$, then the last two terms on the right side of this equation obviously vanish. More generally, the terms will vanish in any *reciprocal* media described by symmetric permeability and permittivity tensors ($\varepsilon = \varepsilon^t$, $\mu = \mu^t$), since $\mathbf{E}_b \varepsilon \mathbf{E}_a = \mathbf{E}_a \varepsilon^t \mathbf{E}_b$ and $\mathbf{H}_a \mu \mathbf{B}_b = \mathbf{H}_b \mu^t \mathbf{H}_a$. Examples of nonreciprocal media for which this cancellation does not take place are magnetized plasmas such as occur in the Earth's ionosphere, where the dielectric tensor is Hermitian rather than symmetric. Other gyrotropic materials like ferrites are nonreciprocal, as are nonlinear media in general.

Fore reciprocal media, we have

$$\nabla \cdot (\mathbf{E}_a \times \mathbf{H}_b - \mathbf{E}_b \times \mathbf{H}_a) = \mathbf{J}_b \cdot \mathbf{E}_a - \mathbf{J}_a \cdot \mathbf{E}_b \tag{B.21}$$

After integrating over some portion of the medium and applying the divergence theorem, this becomes

$$\oint_S (\mathbf{E}_a \times \mathbf{H}_b - \mathbf{E}_b \times \mathbf{H}_a) \cdot d\mathbf{s} = \int_V (\mathbf{J}_b \cdot \mathbf{E}_a - \mathbf{J}_a \cdot \mathbf{E}_b) dv, \tag{B.22}$$

where s is the surface bounding the volume v which may contain any fraction of the source currents. Equations (B.21) and (B.22) are the differential and integral forms of the Lorentz reciprocity theorem, respectively. At this point, the volume and bounding surface are completely general.

Consider next the special case where the current sources in question are localized and contained at the center of a Gaussian surface of infinite extent. Only the far-field radiation field components will be present at the surface, for which $\mathbf{H}_{a,b} = \hat{\mathbf{r}} \times \mathbf{E}_{a,b}/\eta$, where η is the impedance and $\hat{\mathbf{r}}$ is the radial unit vector. Under these circumstances, the integrand vanishes identically at each location on the surface, leaving behind

$$\int_v (\mathbf{J}_b \cdot \mathbf{E}_a - \mathbf{J}_a \cdot \mathbf{E}_b) dv = 0, \tag{B.23}$$

where the volume integral is over all space. This expression of reciprocity illustrates the interchangeability of source and observation points in unbounded, linear, isotropic media and the consequent reciprocity of the Green's function, viz. $G(\mathbf{x}, \mathbf{x}') = G(\mathbf{x}', \mathbf{x})$ in such media.

The identity (B.23) will clearly be unchanged if the domain of integration is restricted to portions of the volume where currents actually flow in configuration a or b. Let us restrict the domain of integration accordingly, substitute (B.21), and reapply the divergence theorem,

$$\oint_{s'} (\mathbf{E}_a \times \mathbf{H}_b - \mathbf{E}_b \times \mathbf{H}_a) \cdot d\mathbf{s}' = 0, \tag{B.24}$$

where s' is now a surface enclosing all source currents in either configuration considered.

Suppose those source currents flow on one of two antennas, labeled "1" and "2," which are surrounded by the surface as shown in Figure B.8. Each antenna is fed by a coaxial cable with a free end at infinity. Let the surface be a sphere with a radius that measures a large number of wavelengths. The sphere is deformed so that the surface runs tangent to

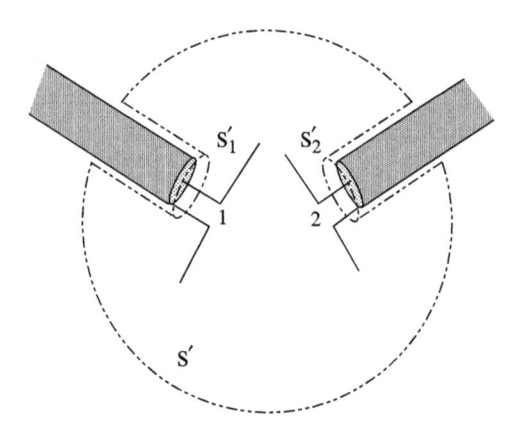

Figure B.8 Geometry illustrating reciprocity theorem.

Appendix B: Review of Electromagnetics   375

the outside of the coaxial feedlines all the way to the antenna feedpoints. At the feedpoints, the surface is tangent to the exposed ends of the feedlines.

In configuration a, antenna 1 is used for transmission and antenna 2 is used for reception. In configuration b, the roles are reversed: antenna 1 is used for reception and antenna 2 is used for transmission. During reception, we consider an antenna to be open circuited. Be certain not to confuse the roles of the letters and numbers in this thought experiment.

Because of the dimensions of the spherical part of the surface, only far-field radiation is present there in either configuration, and consequently there will be no contribution to the integral in (B.24) from the spherical part by the argument made above. A similar argument holds for the TEM wave fields present on the parts of the surface tangent to the coaxial feedlines. The only parts of the integrand of (B.24) that can possibly contribute are, therefore, the parts associated with the feedpoints.

The part of (B.24) associated with the feedpoint of antenna 1 is

$$\oint_{s_1'} (\mathbf{E}_a \times \mathbf{H}_b - \mathbf{E}_b \times \mathbf{H}_a) \cdot d\mathbf{s}_1'. \quad (B.25)$$

(A similar expression can of course be written for the feedpoint of antenna 2.) Let us express the area element in polar coordinates as $d\mathbf{s}_1' = d\mathbf{r}_1 \times d\mathbf{l}_1$, where $d\mathbf{r}_1$ is integrated from the inner to the outer conductor of the transmission line and $d\mathbf{l}_1$ is an arc length element circling the inner conductor. Using the quadruple scalar product rule, $(\mathbf{e} \times \mathbf{f}) \cdot (\mathbf{g} \times \mathbf{h}) = (\mathbf{e} \cdot \mathbf{g})(\mathbf{f} \cdot \mathbf{h}) - (\mathbf{e} \cdot \mathbf{h})(\mathbf{f} \cdot \mathbf{g})$, (B.25) becomes

$$\int \mathbf{E}_a \cdot d\mathbf{r}_1 \oint \mathbf{H}_b \cdot d\mathbf{l}_1 - \int \mathbf{E}_b \cdot d\mathbf{r}_1 \oint \mathbf{H}_a \cdot d\mathbf{l}_1, \quad (B.26)$$

where use has been made of the fact that the electric field (magnetic field) in the transmission line has only a radial (azimuthal) component. Now, the open-circuit voltage that appears across the terminals of antenna 1 due to the electric field $\mathbf{E}$ there is $V = -\int \mathbf{E} \cdot d\mathbf{r}_1$. Meanwhile, the driving current responsible for the magnetic field in between the conductors of transmission line 1 is $I_1 = \oint \mathbf{H} \cdot d\mathbf{l}_1$. Consequently, (B.26) can be written as $-V_{1a}I_{1b} + V_{1b}I_{1a}$. Combining this with the equivalent terms coming from the part of (B.24) associated with antenna 2 gives

$$V_{1a}I_{1b} - V_{1b}I_{1a} + V_{2a}I_{2b} - V_{2b}I_{2a} = 0.$$

Recall that the numerical subscripts refer to the antenna and the character subscripts refer to the configuration. Given that antennas used for reception are open circuited, we have $I_{2a} = I_{1b} = 0$, and

$$Z_{12} = \left.\frac{V_{1b}}{I_{2b}}\right|_{I_{1b}=0} = \left.\frac{V_{2a}}{I_{1a}}\right|_{I_{2a}=0} = Z_{21}, \quad (B.27)$$

which is the reciprocity theorem stated in circuit form. It says that the ratio of the open circuit voltage induced across the terminals of the receiving antenna to the current exciting the transmitting antenna is the same for either configuration. In terms of transconductances, it says that $Z_{12} = Z_{21}$ for coupled antennas in a reciprocal medium.

The derivation above is limited to the case of reciprocal media. In the case of nonreciprocal media with Hermitian permeability and permittivity tensors, a generalized form of reciprocity still holds. In this case, in addition to reversing the roles of the transmitting and receiving antennas, one must also make the substitution $\varepsilon \to \varepsilon^H$ and $\mu \to \mu^H$. In the case of magnetized ionospheric plasmas, this substitution is equivalent to reversing the direction of the background magnetic field.

References

J. G. Ables. Maximum entropy spectral analysis. *Astron. Astrophys. Suppl. Ser.*, 15:383, 1974.

M. Abramowitz and I. A. Stegun. *Handbook of Mathematical Functions with Formulas, Graphs, and Mathematical Tables*. US Government Printing Office, Washington, DC, 20402, 1972.

B. D. O. Anderson and J. B. Moore. *Optimal Filtering*. Springer-Verlag, New York, 1979.

E. V. Appleton. Wireless studies of the ionosphere. *J. Inst. Elec. Engrs.*, 71:642, 1932.

E. V. Appleton and M. A. F. Barnett. Local reflection of wireless waves from the upper atmosphere. *Nature*, 115:333–334, 1925.

D. Atlas and C. W. Ulbrich. Path- and area-integrated rainfall measurement by microwave attenuation in the 1–3 cm band. *J. Appl. Meteor.*, 16:1322–1331, 1977.

C. A. Balanis. *Antenna Theory, 3rd ed.* Wiley-Interscience, New Jersey, 2005.

R. Bamler. A systematic comparison of SAR focusing algorithms. In *Proc. Int. Geoscience and Remote Sensing Symp., IGARSS'91*, volume 2, pages 1005–1009. Espoo, Finland, June 1991.

R. Bamler. A comparison of range-Doppler and wavenumber domain SAR focusing algorithms. *IEEE Trans. Geosci. Remote Sensing*, 30(4):706–713, 1992.

R. H. Barker. Group synchronization of binary digital systems. In W. Jackson, editor, *Communication Theory*. Academic, New York, 1953.

L. J. Battan. *Radar Observations of the Atmosphere*. The University of Chicago Press, Chicago, 1973.

J. A. Bennet. The ray theory of Doppler frequency shifts. *Aust. J. Phys.*, 21:259–272, 1968.

I. B. Bernstein. Geometric optics in space- and time-varying plasmas. *Phys. Fluids*, 18(3):320–324, 1974.

R. B. Blackman and J. W. Tukey. *The measurement of Power Spectra*. Dover Press, New York, 1958.

R. E. Blahut. *Algebraic Codes for Data Transmission, 2nd ed.* Cambridge University Press, Cambridge, UK, 2003.

C. F. Bohren and D. R. Huffman. *Absorption and Scattering of Light by Small Particles*. Wiley, New York, 1983.

H. G. Booker. Oblique propagation of electromagnetic waves in a slowly varying non-isotropic medium. *Proc. Roy. Soc. A*, 237:411, 1936.

M. Born and E. Wolf. *Principles of Optics*. Pergamon Press, New York, 1970.

K. L. Bowles, G. R. Ochs, and J. L. Green. On the absolute intensity of incoherent scatter echoes from the ionosphere. *J. Res. NBS – D. Rad. Prop.*, 66D:395, 1962.

W. H. Bragg and W. L. Bragg. The reflexion of x-rays by crystals. *Proc R. Soc. Lond. A.*, 88(605):428–438, 1913.

G. Breit and M. A. Tuve. A radio method for estimating the height of the conducting layer. *Nature*, 116:357, 1925.

A test of the existence of the conducting layer. *Phys. Rev.*, 28:554, 1926.

V. N. Bringi and V. Chandrasekar. *Polarimetric Doppler Weather Radar*. Cambridge University Press, Cambridge, UK, 2001.

K. G. Budden. *The Propagation of Radio Waves*. Cambridge University Press, New York, 1985.

R. Buderi. *The Invention That Changed the World*. Simon and Schuster, 1996.

B. Cairns and E. Wolf. Comparison of the Born and Rytov approximations for scattering from quasi-homogeneous media. *Optics Comm.*, 74(5):284–289, 1990.

D. B. Campbell, J. K. Harmon, and M. C. Nolan. Solar system studies with the Arecibo planetary radar system. In *General Assembly and Scientific Symposium, 2011 XXXth URSI, August 13–20, 2011*. doi: 10.1109/URSIGASS.2011.6051217.

J. Capon. High-resolution frequency-wavenumber spectrum analysis. *Proc. IEEE*, 57:1408, 1969.

S. Chandrasekhar. *Radiative Transfer*. Oxford University Press, Oxford, UK, 1950.

D. K. Cheng. *Field and Wave Electromagnetics*, 2nd ed. Addison-Wesley, Reading, MA, 1989.

B. G. Clark. An efficient implementation of the algorithm "CLEAN." *Astron, Astrophys.*, 89:377–378, 1980.

C. Coleman. *An Introduction to Radio Frequency Engineering*. Cambridge University Press, Cambridge, UK, 2004.

R. E. Collin. *Antennas and Radiowave Propagation*. McGraw-Hill, New York, 1985.

J. W. Cooley and J. W. Tukey. An algorithm for the machine computation of complex Fourier series. *Math. Comput.*, 19:297–301, 1965.

T. J. Cornwell. A method of stabilizing the CLEAN algorithm. *Astron. Astrophys.*, 121:281–285, 1983.

T. J. Cornwell and K. F. Evans. A simple maximum entropy deconvolution algorithm. *Astron. Astrophys.*, 143:77, 1985.

J. P. Costas. A study of a class of detection waveforms having nearly ideal range-Doppler ambiguity properties. *Proc. IEEE*, 72(8):996–1009, 1984.

T. M. Cover and J. A. Thomas. *Elements of Information Theory*. John Wiley & Sons, New York, 2006.

G. J. Daniell. Of maps and monkeys. In B. Buck and V. A. Macaulay, editors, *Maximum Entropy in Action*, chapter 1, pages 1–18. Clarendon, Oxford, 1991.

K. Davies. *Ionospheric Radio*. Peter Peregrinus Ltd., London, 1990.

L. de Forest. Absorption (?) of undamped waves. *Electrician (letter to the editor)*, 69:369–370, 1912.

Recent developments in the work of the Federal Telegraph Company. *Proc. IRE*, 1:37–57, 1913.

R. J. Doviak and D. S. Zrnić. *Doppler Radar and Weather Observations*, 2nd ed. Academic Press, San Diego, 1993.

S. Drabowitch, A. Papiernik, H. D. Griffiths, J. Encinas, and B. L. Smith. *Modern Antennas*, 2nd ed. Springer, Dordrecht, The Netherlands, 2005.

C. C. Easterbrook. Estimating horizontal wind fields by two-dimensional curve fitting of single Doppler radar measurements. In *16th Conference on Radar Meteorology (preprints)*, pages 214–219. American Meteorological Society, Boston, 1974.

F. Fabry. *Radar Meteorology*. Cambridge University Press, Cambridge, UK, 2015.

D. T. Farley. Incoherent scatter correlation function measurements. *Radio Sci.*, 4:935–953, 1969.

Multiple-pulse incoherent-scatter correlation function measurements. *Radio Sci.*, 7:661, 1972.

On-line data processing for MST radars. *Radio Sci.*, 20:1177–1184, 1985.

D. T. Farley, H. M. Ierkic, and B. G. Fejer. Radar interferometry: A new technique for studying plasma turbulence in the ionosphere. *J. Geophys. Res.*, 86:1467–1472, 1981.

G. Franceschetti and R. Lanari. *Synthetic Aperture Radar Processing*. CRC Press, Boca Raton, FL, 1999.

R. L. Frank. Polyphase codes with good nonperiodic correlation properties. *IEEE Trans. Inform. Theory*, IT-9:43–45, 1963.

A. Fresnel. Memoir on the diffraction of light. In *The Wave Theory of Light: Memoirs by Huygens, Young and Fresnel*. American Book Company, Woodstock, GA, 1818.

H. Friedman. From ionosonde to rocket sonde. *J. Geophys. Res.*, 99(A10):19,143–19,153, 1994.

M. J. E. Golay. Notes on digital coding. *Proc. IRE*, 37:657, 1949.

Complementary series. *IRE Trans. Inform. Theory*, 7(2):82–87, 1961.

R. Gold. Optimal binary sequences for spread spectrum multiplexing (corresp.). *IEEE Trans. Inform. Theory*, 13(4):619–621, 1967.

J. W. Goodman. *Introduction to Fourier Optics*, 3rd. ed. Roberts & Company, Englewood, CO, 2005.

E. E. Gossard and R. G. Strauch. *Radar Observations of Clear Air and Clouds*. Elsevier, Amsterdam, 1983.

D. J. Griffiths. *Introduction to Electrodynamics*, 3rd ed. Prentice Hall, NJ, 1999.

B. Gustavsson and T. Grydeland. Orthogonal-polarization alternating codes. *Radio Sci.*, 44(RS6005): doi: 10.1029/2008RS004132, 2009.

Orthogonal-polarization multipulse sequences. *Radio Sci.*, 46(RS1003): doi: 10.1029/2010RS004425, 2009.

J. B. Hagan. *Radio-Frequency Electronics*. Cambridge University Press, Cambridge, UK, 1996.

M. P. M. Hall, L. W. Barclay, and M. T. Hewitt, editors. *Propagation of Radiowaves*. The Institute of Electronic Engineers, London, 1996.

R. C. Hansen. *Phased Array Antennas*. John Wiley & Sons, NJ, 1998.

W. W. Hansen and J. R. Woodyard. A new principle in directional antenna design. *Proc. IRE*, 26(3):333–345, 1938.

R. F. Hanssen. *Radar Interferometry*. Klewer Academic Publishers, Dordrecht, 2001.

R. F. Harrington. *Time-Harmonic Electromagnetic Fields*, 2nd ed. Wiley-IEEE Press, New York, 2001.

F. J. Harris. On the use of windows for harmonic analysis with the discrete Fourier transform. *Proc. IEEE*, 66:51, 1978.

R. L. Haupt. *Antenna Arrays*. John Wiley & Sons, NJ, 2010.

J. A. Högbom. Aperture synthesis with a non-regular distribution of interferometer baselines. *Astron. Astrophys. Supp.*, 15:417–426, 1974.

J. M. Holt, D. A. Rhoda, D. Tetenbaum, and A. P. van Eyken. Optimal analysis of incoherent scatter radar data. *Radio Sci.*, 27:435–447, 1992.

P. E. Howland, H. D. Griffiths, and C. H. Baker. Passive bistatic radar systems. In M. Cherniakov, editor, *Bistatic Radar: Emerging Technology*. John Wiley & Sons, Chichester, UK, 2008.

R. D. Hunsucker. *Radio Techniques for Probing the Terrestrial Ionosphere*. Springer Verlag, New York, 1991.

A. Huuskonen, M. S. Lehtinen, and J. Pirttilä. Fractional lags in alternating codes: Improving incoherent scatter measurements by using lag estimates at noninteger multiples of baud length. *Radio Sci.*, 31:245, 1996.

Chr. Huygens. *Traité de la Lumiere*. Leyden, 1690.

D. L. Hysell and J. L. Chau. Optimal aperture synthesis radar imaging. *Radio Sci.*, doi: 41:10.1029/2005RS003383, RS2003, 2006.

D. L. Hysell, F. S. Rodrigues, J. L. Chau, and J. D. Huba. Full profile incoherent scatter analysis at Jicamarca. *Ann. Geophys.*, 26:59–75, 2008.

J. D. Jackson. *Classical Electrodynamics*, 3rd ed. John Wiley & Sons, New York, 1975.

E. T. Jaynes. Where do we stand of maximum entropy? In R. D. Levine and M. Tribus, editors, *The Maximum Entropy Formalism*, pages 15–118. MIT Press, Cambridge, MA, 1979.

E. T. Jaynes. On the rationale of maximum-entropy methods. *Proc. IEEE*, 70:939, 1982.

R. M. Jones and J. J. Stephenson. A versatile three-dimensional ray tracing computer program for radio waves in the ionosphere. Technical Report 75–76, US Department of Commerce, 1975.

J. B. Keller. Geometrical theory of diffraction. *J. Opt. Soc. of Am.*, 52:116–130, 1962.

A. M. Kerdock, R. Mayer, and D. Bass. Longest binary pulse compression codes with given peak sidelobe levels. *Proc. IEEE*, 74(2):366, 1986.

G. Kirchhoff. Zur Theorie de Lichtstrahlen. *Ann. Phys.*, 254:663–695. doi: 10.1002/andp.18832540409, 1883.

J. A. Kong. *Electromagnetic Wave Theory*. EMW Publishing, Cambridge, MA, 2008.

L. H. Koopmans. *The Spectral Analysis of Time Series*. Academic Press, New York, 1974.

J. D. Kraus. *Radio Astronomy*, 2nd ed. Cygnus-Quasar Books, Powell, OH, 1966.

J. D. Kraus and R. J. Marhefka. *Antennas*, 3rd ed. McGraw-Hill, New York, 2001.

E. Kudeki and S. J. Franke. Statistics of momentum flux estimation. *J. Atmos. Sol. Terr. Phys.*, 60(16):1549–1553, 1998.

L. D. Landau and E. M. Lifshitz. *The Classical Theory of Fields*, 3rd ed. Pergamon Press, New York, 1971.

C. Latham and A. Stobbs. *The Birth of British Radar: The Memoirs of Arnold "Skip" Wilkins*, 2nd ed. Radio Society of Great Britain, 2011.

B. P. Lathi. *Modern Digital and Analog Communication Systems*. Holt, Rinehart and Winston, New York, 1983.

M. S. Lehtinen. Statistical theory of incoherent scatter radar measurements. Technical Report 86/45, Eur. Incoherent Scatter Sci. Assoc., Kiruna, Sweden, 1986.

M. S. Lehtinen and I Häggström. A new modulation principle for incoherent scatter measurements. *Radio Sci.*, 22:625–634, 1987.

M. S. Lehtinen, A. Huuskonen, and M. Markkanen. Randomization of alternating codes: Improving incoherent scatter measurements by reducing correlations of gated ACF estimates, *Radio Sci.*, 1997.

M. S. Lehtinen, A. Huuskonen, and J. Pirttilä. First experiences of full-profile analysis with GUISDAP. *Ann. Geophys.*, 14(12):1487–1495, 1997.

N. L. Levanon. *Radar Principles*. Wiley, New York, 1988.

B. R. Mahafza. *Introduction to Radar Analysis*. CRC Press, New York, 1998.

V. Markel. Introduction to the Maxwell Garnett approximation: Tutorial. *J. Opt. Soc. Am.*, 33(7):1244–1256, 2016.

M. Markkanen and T. Nygrén. A 64-bit strong alternating code discovered. *Radio Sci.*, 31(2):241–243, 1996.

Long alternating codes, 2, Practical search method. *Radio Sci.*, 32(1):9–18, 1997.

M. Markkanen, J. Vierinen, and J. Markkanen. Polyphase alternating codes. *Ann. Geophys.*, 26(9):2237–2243, 2008.

J. S. Marshall and W. Mck. Palmer. The distribution of raindrops with size. *J. Meteor*, 5:165–166, 1948.

J. D. Mathews. A short history of geophysical radar at Arecibo Observatory. *Hist. Geo-Space Sci.*, 4:19–33, 2013.

W. M. Middleton and M. E. Valkenburg. *Reference Data for Radio Engineers*, 9th ed. Butterworth-Heinemann, Woburn, MA, 2002.

R. Mittra, Y. Rahmat-Samii, and W. L. Ko. Spectral theory of diffraction. *Appl. Phys.*, 10:1–13, 1976.

F. E. Nathanson. *Radar Design Principles*. McGraw-Hill, New York, 1969.

N. P. Nguyen. Range cell migration correction for phase error compensation of highly squinted SAR. Proc. 10th Eur. Conf. Synthetic Aperture Radar, Berlin, Germany, 2014.

T. Nygrén. *Introduction to Incoherent Scatter Measurements*. Invers, OY, 1996.

H. Nyquist. Thermal agitation of electric charge in conductors. *Phys. Rev.*, 32:110–113, 1928.

W. K. H. Panofsky and M. Phillips. *Classical Electricity and Magnetism*, 2nd ed. Dover Publications, New York, 2005.

A. Papoulis. *Probability, Random Variables, and Stochastic Processes*. McGraw-Hill, New York, 1984.

P. Z. Peebles, Jr. *Radar Principles*. Wiley, New York, 1998.

C. M. Rader. An improved algorithm for high speed autocorrelation with application to spectral estimation. *IEEE T. Acoust. Speech*, AU-18(4):439–441, 1970.

M. A. Richards. *Fundamentals of Radar Signal Processing*. McGraw-Hill, New York, 2005.

D. M. Riggin, T. Tsuda, and A. Shinbori. Evaluation of momentum flux with radar. *J. Atmos. Sol. Terr. Phys.*, 142:98–107, May 2016.

J. Ruze. The effect of aperture errors on the antenna radiation pattern. *Nuovo Cimento Suppl.*, 9(3):364–380, 1952.

J. D. Sahr and F. D. Lind. The Manashtash Ridge radar: A passive bistatic radar for upper atmospheric radio science. *Radio Sci.*, 32(6):2345–2358, 1997.

F. R. Schwab. Relaxing the isoplanatism assumption in self-calibration; applications to low-frequency radio interferometry. *Astron. J.*, 1984.

A. Segalovitz and B. R. Frieden. A "CLEAN"-type deconvolution algorithm. *Astron. Astrophys.*, 70:335–343, 1978.

C. E. Shannon and W. Weaver. *The Mathematical Theory of Communication*. University of Illinois Press, Urbana, 1949.

I. I. Shapiro. Planetary radar astronomy. *IEEE Spectrum*, March:70–79, 1966.

J. Sheffield. *Plasma Scattering of Electromagnetic Radiation*. Elsevier, New York, 1975.

J. Skilling and R. K. Bryan. Maximum entropy image reconstruction: General algorithm. *Mon. Not. R. Astron. Soc.*, 211:111, 1984.

M. I. Skolnik. *Introduction to Radar Systems*, 3rd ed. McGraw-Hill, New York, 2001.

A. Sommerfeld. Lectures on theoretical physics, v. 4. In *Optics*, pages 198–201, 211–212. Academic Press, New York, 1954.

O. J. Souvers, J. L. Fanselow, and C. S. Jacobs. Astrometry and geodesy with radio interferometry: Experiments models, results. *Rev. Mod. Phys.*, 70:1393–1454, 1998.

R. H. Stolt. Migration by Fourier transform. *Geophysics*, 43:23–48, 1978.

J. A. Stratton. *Electromagnetic Theory*. McGraw-Hill, New York, 1941.

J. Strutt. On the transmission of light through an atmosphere containing small particles in suspension, and on the origin of the blue of the sky. *Philosophical Magazine*, 47(5):375–394, 1899.

W. L. Stutzman and G. A. Thiele. *Antenna Theory and Design*, 3rd ed. John Wiley & Sons, New York, 2012.

M. P. Sulzer. A radar technique for high range resolution incoherent scatter autocorrelation function measurements utilizing the full average power of klystron radars. *Radio Sci.*, 21:1033–1040, 1986.

A new type of alternating code for incoherent scatter measurements. *Radio Sci.*, 28:995, 1993.

E. Talebian and M. Talebian. A general review of the derivation of the Clausius-Mossotti relation. *Optik*, 124:2324–2326, 2013.

V. Tazlukov. *Signal Processing in Radar Systems*. CRC Press, New York, 2013.

A. R. Thompson, J. M. Moran, and G. W. Swenson. *Interferometry and Synthesis in Radio Astronomy*. John Wiley & Sons, New York, 1986.

L. Tsang, J. A. Kong, and R. T. Shin. *Theory of Microwave Remote Sensing*. John Wiley & Sons, Wiley-Interscience, New York, 1985.

M. A. Tuve. Early days of pulse radio at the Carnegie Institution. *J. Atmos. Terr. Phys.*, 36:2079–2083, 1974.

S. V. Uppala and J. D. Sahr. Spectrum estimation moderately overspread radar targets using aperiodic transmitter coding. *Radio Sci.*, 29:611, 1994.

H. C. van de Hulst. *Light Scattering by Small Particles*. Wiley, New York, 1957.

D. K. van Keuren. Science goes to war: The Radiation Laboratory, radar, and their technological consequences. *Rev. American Hist.*, 25:643–647, 1997.

J. Vierinen. Fractional band-length coding. *Ann. Geophys.*, 29(6):1189–1196, 2011.

R. A. Vincent and I. M. Reid. HF Doppler measurements of mesospheric gravity wave momentum fluxes. *J. Atmos. Sci.*, 40(5):1321–1333, 1983.

I. I. Virtanen, M. S. Lehtinen, T. Nygrén, M. Orispää, and J. Vierinen. Lag profile inversion method for EISCAT data analysis. *Ann. Geophys.*, 26:571–581, 2008.

I. I. Virtanen, J. Vierinen, and M. S. Lehtinen. Phase-coded pulse aperiodic transmitter coding. *Ann. Geophys.*, 27:2799–2811, 2009.

J. L. Walsh. A closed set of normal orthogonal functions. *Amer. J. Math*, 45:5–24, 1923.

R. C. Watson. *Radar Origins Worldwide: History of Its Evolution in 13 Nations through World War II*. Trafford Publishing, Victoria, BC, 2009.

R. Wilczek and S. Drapatz. A high accuracy algorithm for maximum entropy image restoration in the case of small data sets. *Astron. Astrophys.*, 142:9, 1985.

R. F. Woodman. Inclination of the geomagnetic field measured by an incoherent scatter technique. *J. Geophys. Res.*, 76:178, 1971.

P. M. Woodward. *Probability and Information Theory, with Applications to Radar*, 2nd ed. Pergamon Press, Oxford, 1964.

H. Yagi and S. Uda. Projector of the sharpest beam of electric waves. *Proc. Imp. Acad. Japan*, 2(2):49–52, 1926.

Index

aliasing
  frequency, 16, 179, 183
  range, 179
ambiguity function
  range-Doppler, 204
  range-lag, 276
Ampere's law, 359, 361
antenna
  aperture, 98
    circular, 105
    efficiency, 107, 112
    rectangular, 103
    reflector, 109, 114
  array, 71
    acoustic, 91
    adaptive, 89
    binomial, 80
    broadside, 75
    endfire, 77
    Hansen–Woodyard, 79
    parasitic, 81
    three-dimensional, 85
    two-dimensional, 83
    uniform linear array, 73
  backscatter gain, 164, 210, 290
  balun, 65
  beam solid angle, 34, 38, 40, 78, 79
  beam synthesis, 88
  current loop, 55
  directivity, 7, 37–39, 43, 47, 75, 77, 79, 85, 108
  effective area, 7, 40, 43
  effective length, 50
  efficiency, 36
  gain, 7, 38–40
  Gaussian beam, 35, 38, 40, 290
  helical, 63
  Hertzian dipole, 29
  impedance, complex, 36, 48
  main beam, 34, 74, 106
  radiation pattern, 33, 35, 47, 57, 74, 79, 84, 103, 106
  beamwidth between first nulls (BWFN), 34
  half-power full beamwidth (HPFB), 34, 77, 106
  radiation resistance, 36, 48
  sidelobe, 34, 74, 106
    grating lobe, 34, 74
  wire, 44
    dipole, 45
    folded dipole, 54
    half-wave dipole, 46
    monopole, 52
    short dipole, 55
Appleton, Edward, 2
Appleton–Hartree equation, 237
Astrom's equation, 236
autocorrelation function, 23, 24, 26, 92, 140, 195, 212, 215, 266, 267, 270, 271, 304–306, 337

Beer–Lambert law, 299
Biot Savart law, 357
blackbody radiation, 121, 125
Booker quartic, 237
Boot, Harry, 5
Born approximation, 163, 277, 290, 318
Bouger's law, 220, 228
Breit and Tuve, theorem of, 226
Breit, Gregory, 3
bright bands, 291
brightness, 129, 326
Bruggeman's mixing formula, 298

central limit theorem, 22, 277
channel capacity, 136
Clausius–Mossotti relation, 296
CLEAN algorithm, 329
clear-air echoes, 314
convolution theorem, 13, 104, 105, 329
Cornu spiral, 252
correlation function, 20
Coulomb's law, 353
current density, 29, 30, 53, 223, 234, 358
  surface, 49, 100, 367

383

Daventry experiment, 4
De Forest, Lee, 2
Debye relaxation model, 295
diffraction
   Fraunhoffer, 102
   Fresnel, 102
   knife-edge, 251
diffraction theory, 98
   geometric theory, 258
   scalar, 99, 251
   spectral, 258
   vector, 99, 255
Doppler radar, 169, 172, 175–179, 183, 185, 188, 202, 204, 265, 304
drop-size distribution, 292, 293
   gamma distribution, 292
   log-normal distribution, 293
   Marshall–Palmer distribution, 292
duality, 57

electromagnetic field, 32, 58, 98, 100
   far field, 37, 102
   near field, 37, 102
   radiative, 32
   reactive, 32
Euler spiral , *see* Cornu spiral

Faraday's law, 355, 360
field equivalence principle, 99, 370
Fourier series, 11, 16, 88
Fourier transform, 12, 104, 304
   discrete, 16, 304
Fresnel equation of wave normals, 235
Fresnel integral, 253
Fresnel zones, 254
Fresnel, Augustin-Jean, 98
Friis transmission formula, 44
Fuller, Leonard, 2

Gauss' law, 354
geometric optics, 98, 112
Green's function, 31, 204, 256, 370, 371

Hülsmeyer, Christian, 1
Habann, Erich, 4
Heaviside, Oliver, 2
Helmholtz equation, 31, 152, 153, 255, 317
Hollmann, Hans Erich, 4
Hull, Albert, 4
Huygens source, 100, 252
Huygens' principle, 98
Huygens, Christian, 98
hydrometeor, 288
   mixed phase, 296
   relative permittivity, 294
Hyland, Lawrence, 3

image theory, 53, 85, 256
index of refraction, 218, 243, 317
interferometry, 4, 86, 186, 323, 327
   InSAR, 343
ionosonde, 3
ionospheric sounding, 224, 244

Kennelly, Arthur, 2
Kirchhoff, Gustav, 98
klystron, 4

lag products, 266, 267, 270, 271, 273, 274, 278
Langevin function, 294
Lippmann–Schwinger equation, 318
Lorentz sphere, 297
Lorentz-Lorenz equation, see Clausius–Mossotti relation 296
Lorenz gauge condition, 29, 32, 68, 362

magnetron, 4
Marconi, Gugleilmo, 2
matched filter, 93, 137, 168, 171, 196, 199, 211, 290, 336, 338
Maximum entropy algorithm, 330
maximum usable frequency (MUF), 227
Maxwell Garnett equation, 298
Maxwell's equations, 29, 53, 57, 100, 235, 255, 257, 353
Maxwell, James Clerk, 98

noise, 8, 116, 123, 307
   antenna, 127
   atmospheric absorption, 129
   bandwidth, 136
   cascadès, 133
   cosmic, 128
   figure, 134
   quantization, 143
   receiver, 134
   self clutter, 283
   system, 127, 131
   temperature, 127
Nyquist noise theorem, 123, 131
   generalized, 124
Nyquist, Harry, 123

Okabe, Kinjiro, 4
overspread targets, 184, 185, 264

Page, Robert, 3
Parseval's theorem, 12, 14, 205, 339
phasor notation, 9
Planck's law, 122, 129
plasma frequency, 224
Poincaré, Henri, 99
polarization, 58, 150, 311
   circular, 59

elliptical, 60
linear, 59
Stokes parameters, 63, 151
Polder–van Santen formula, see Bruggeman's mixing formula 298
Popov, Alexander, 3
potential
  electric, 29, 361
  magnetic vector, 29, 100, 361
power spectral density, 15, 304
Poynting vector, 10, 32, 35, 46, 49, 50, 55, 72, 82, 108, 240
principle of pattern multiplication, 72, 75, 80, 83, 84
probability, 18
  conditional, 21
  normal distribution, 21
propagation, 218
  absorption
    ionospheric, 248
  attenuation
    atmospheric, 298
  Faraday rotation, 249
  reflection
    ionospheric reflection, 224, 244
  refraction, 218
    eikonal analysis, 229
    inhomogeneous media, 229
    ionospheric refraction, 222
    radius of curvature, 232
    stratified media, 231
    troposheric ducting, 220
pulse compression, 141, 195
  frequency hopping and chirping, 201
    Costas codes, 202, 209
    Frank codes, 202
  phase coding, 195
    alternating code, 274
    Barker code, 196
    coded long pulse, 273
    complementary (Golay) codes, 198
    cyclic codes, 199
    Gold codes, 200
    maximum-length codes, 198
    polyphase codes, 201

radar
  aperture synthesis, 327
  Arecibo, 43, 86, 89, 115, 133, 182, 197
  BMEWS, 5
  Chain Home, 4
  continuous wave (CW), 175
  Dew Line, 5
  frequency-modulated continuous wave (FMCW), 177
  Freya, 4
  imaging, 323
  ISAR, 347

  Jicamarca, 180, 182, 333
  moving target indicator (MTI), 188
  passive, 211
  Pave Paws, 5
  planetary, 185
  pulsed, 179
  resolution, 186
  synthetic aperture (SAR), 86, 335
  weather, 288
  Wurzburg, 4
radar equation, 8, 148, 164, 289, 290
radiative balance, 129, 300
rainfall rate, 293
Randall, John Turton, 5
random process, 23, 304
random variable, 18
  even-moment theorem, 26
Rayleigh–Jeans law, 122
reciprocity theorem, 41, 125, 373
reflectivity, radar, 290, 291
refractivity, 218
Ruze formula, 114
Rytov approximation, 319

sampling theorem, 16, 136, 179, 183, 184
scatter
  Bragg, 162
  corner reflectors, 160
  irregular media, 316
  Mie, 152, 288, 300, 301
    phase functions, 157
  Rayleigh, 149, 289
  specular reflection, 158
  Thomson, 161
  volume, 163, 210, 290, 325
scattering, 147
  cross section
    radar, 7, 147, 289
    total, 147
  efficiency, 156
  matrix, 155
  polarization, 150
SETI, 146
Shannon–Hartley theorem, 136
signal processing, 167
  alternating code, 274
  binary phase coding, 270
  coded long pulse, 273
  coherent, 169, 171
  error analysis, 143, 280, 305, 316, 333, 344
  incoherent, 169, 170
  long pulse, 265
  multipulse, 268
  SAR, 336
  wavenumber domain (WD), 339
Snell's law, 112, 220, 227, 231, 246
Sommerfeld, Arnold, 99

spectral moments, 305
spherics, 128
Stephan–Boltzmann law, 122
Stolt mapping, 340
Stuart, Balfour, 2

Taylor, Alfred, 3
time reverse mirror (TRM), 92
Tizard mission, 5
Tuve, Merle A., 3

underspread targets, 184

visibility, 326

Walsh matrix, 275
Watson Watt, Robert, 4
wave normal equation, 236
Wilkins, Arnold, 4
wind field, 308
    least-squares estimation, 310
    momentum flux estimation, 315
    profiling, 315

Young, Leo, 3

Z-R relationships, 302
Zacek, August, 4